A TEXTBOOK OF
MODERN PLANT PATHOLOGY

A TEXTBOOK OF
Modern Plant Pathology

K.S. BILGRAMI
Professor of Botany
University of Bhagalpur

H.C. DUBE
Department of Botany
University of Udaipur

VIKAS PUBLISHING HOUSE PVT LTD
NEW DELHI BOMBAY BANGALORE CALCUTTA KANPUR

VIKAS PUBLISHING HOUSE PVT LTD
5 Ansari Road, New Delhi 110002
Savoy Chambers, 5 Wallace Street, Bombay 400001
10 First Main Road, Gandhi Nagar, Bangalore 560009
8/1-B Chowringhee Lane, Calcutta 700016
80 Canning Road, Kanpur 208004

First Edition : 1976
Reprint : 1976
Reprint : 1979

1V02B3002

ISBN 0 7069 0421 4

Rs.24

[Paper used for the printing of this book was made available by the Government of India at concessional rate.]

Printed at Skylark Printers, New Delhi 110055

PREFACE

DURING the past thirty years, plant pathology has witnessed the emergence of remarkable new knowledge. In 1875 it was a schizophrenic profession interloping in different realms of pure science; it was a cook book. Plant pathologists were squirt gun botanists. Today, happily, no science is more multifarious in scope than this. Spawned in the backyard of mycology, it now embraces bacteriology, virology, nematology, physiology, genetics, climatology, meteorology and many others. Plant pathology is autonomous, standing firmly on its two legs; one planted in pure science and the other in the art of crop culture (agriculture). Understanding the disease process is science. The knowledge, thus gained, becomes power in fields and granaries, when it is used in diagnosing and treating sick plants. Here, plant pathology becomes an art. It is taught, rightly, therefore, both in pure botany and agriculture. It provides exciting opportunities to pure scientists to unravel the mysteries of the deranged physiology and also fulfils its obligation to society by expunging diseases and pathogens from all places except history books. The literature in plant pathology has grown to the status of having advance treatises and an "Annual review".

Plant pathology has a high and proud tradition in India and to continue that we have to do a big job. We have to make our teaching more scientific, which, at present, is obsessed with symptomatology. Our syllabi, even now, in some universities, confuse plant pathology with mycology. This has done considerable damage. Some old departments still stick to the name 'mycology and plant pathology'. Fungi are important but so are the bacteria, viruses, nematodes, mycoplasma and several environmental factors that cause disease. Again, since disease is abnormal physiology, we have to shift the emphasis from the pathogen to the diseased plant, which undergoes suffering and plant pathology is study of suffering plants (Pathos=suffering, Logy=discourse). It is unfortunate that we have no book emphasizing this shift.

More courses in physiology, genetics, bacteriology, virology, nematology, soil science, ecology and climatology have to be given to our students.

Our practical courses should not end up with drawing beautiful camera lucida drawings and 'thin, well stained' sections but should include training in biochemical methods. These technical 'know how' are imperative to understand the malfunctioning diseased plant.

We believe that a new approach to the organization of the subject matter is urgently needed. This book is an attempt in this direction. If we have succeeded the credit should go to our great teachers R.N. Tandon, T.S. Sadasivan and K.S. Bhargava. It will be our pleasure to be told where we have erred. Needless to say, any suggestions to improve the book will be gratefully received and acknowledged.

We are grateful to the Chief Executive Officer, Annual Reviews, Inc. Palo Ato, California, USA for permitting us to use the photographs included in Chapter 1. Thanks are due to Professor R.P. Scheffer, Michigan State University, USA, Dr K.R. Samaddar, University of Kalyani, India and Professor Clayton Person, University of British Columbia, Canada for allowing us to use Figures 9.1 and 12.1. It is our pleasure to acknowledge help received from our colleagues at Bhagalpur and Udaipur.

K.S. BILGRAMI
H.C. DUBE

CONTENTS

PART ONE
THE FUNDAMENTALS

PART ONE

THE FUNDAMENTALS

1

PLANT PATHOLOGY: ITS HISTORY AND PROGRESS

FOSSIL records of parasitic fungi date back to Devonian period, which possibly suggests that plant diseases originated along with the plants; much before man came on earth. When men started growing food they also unwittingly started culturing the parasites. The outbreaks of plant diseases, then, must have occurred much before recorded history. Mention of plant diseases are found in ancient Greek and Indian citations though they knew little about them.

First authentic records of plant diseases are found in the writings of Theophrastus—"The father of Botany". While describing trees, cereals and pulses he also described their diseases. He recorded the harmful effects of wind, weather and location. He and other Greek philosophers of the time believed that the diseases originated from the plants or from the environment. This erroneous belief delayed our understanding of parasitism and disease for two thousand years. Plant pathology along with other fields of intellectual enquiry benefitted considerably from the intellectual climate of the European renaissance of 14th century.

The invention of the microscope in the 17th century gave vision to the hitherto unknown world of microorganisms. Leeuwenhoek discovered bacteria in 1675. An Italian botanist Micheli in 1729 made an extensive study of fungi and their reproductive structures. He discovered the role of spores and experimentally proved that the fungi originated from their spores. This was the first experimental proof that fungi are autonomous organisms which produce seed-like bodies and not capricious creatures of the *spontaneous generation*. This was a big step forward and gave the first blow to the concept of spontaneous generation. Micheli was much ahead of his time and his thinking had little impact on the prevalent beliefs. This was predominantly the age of describing plants and naming them. Linnaeus (1753) gave the bionomial system of nomenclature but his contributions to fungi were insignificant. He too believed in the transcendental origin of the diseases. His students Persoon and Fries applied the bionomial system of nomenclature to fungi and gave a firm footing to the taxonomy of fungi. "These fungi," stated Fries (1821-1832), "depend on a diseased condition of the plant rather than vice-versa. From the study of many examples we learnt that they are hereditary and that they depend on the composition of the atmosphere. Every fear of their propagation by sporidia is superfluous."

After Micheli, Tillet (1755) made the next important contribution. He experimentally proved that wheat seeds dusted with bunt powder gave rise to bunted plants. *This established that bunt disease of wheat was contagious*. Tillet was not aware that the bunt powder consisted of spores of the fungus and he ascribed the disease to some poison present in the dust. He also observed that chemical treatment of seeds prevented the disease. In 1807 Prevost gave first experimental proof that bunt of wheat was caused by a fungus. He developed methods to obtain the bunt fungus in pure culture. He showed that the wheat bunt organism after germination penetrated the wheat plant tissue and caused the disease. This was a revolutionary proposition. He also experimented on the prevention of the disease and found that dipping the seeds in a solution of copper sulphate controlled the disease. His seed treatment was widely practised Prevost did not consider the bunt organism as the sole cause of disease but only as the immediate or direct cause. He recognized the secondary effect of the environment on the host.

Prevost thus laid foundations for several branches of modern plant pathology. Once the cause of one disease was known it should have been easier to find out the causes of other diseases. If it did not happen and required repeated telling to dispel the fog, it was because the autogenetic theory, unfortunately, always had the support of powerful voices. Thus Ugner, Meyen and Leibig fast adhered to the spontaneous generation theory. Meyen (1841) regarded fungi as 'pseudo-organisms' and Leibig likened disease to putrification and fermentation. Tulasne brothers, the great mycologists of France drew the attention of other contemporary mycologists to Prevost's findings.

The devastating epiphytotic of the potato disease in the middle of the 19th century was one of the most tragic events in human history. It swept over the whole of Europe and USA but in Ireland it was catastrophic. Over one million people died and one and a half million migrated. This tragic event was a blessing in disguise, in so far as it hastened the realization of the importance of plant diseases. Many committees and commissions were formed to report on the disease. The causal organism was identified by Speerschneider (1857 *see* Keitt, 1959) and De Bary (1861-63-67). It was *Phytophthora infestans*. The battle against the spontaneous origin of disease was largely won. The science of plant pathology, evolving for a long time, was born. Berkley, the most prominent British mycologist of his time, had, after initial reservations, accepted the causal role of microorganisms. He described (1847) *Oidium tuckeri*, the powdery mildew fungus of vine, as the cause of this disease and this helped in the quick acceptance of De Bary's announcement of *Phytophthora* being the cause of the potato late-blight. It is interesting that the editor of the *Garden Chronicle* which recorded the miseries of the potato epiphytotic in an inimitable style was Dr John Lindley, a firm supporter of the autogenetic origin.

De Bary's other contributions too stand out extremely important. Working with *Sclerotinia* rot diseases of carrots and other vegetables he noted the killing of tissues in advance of the fungus (De Bary, 1886). The juice from the necrotic tissue too macerated healthy tissues, an ability which was lost on boiling. With this demonstration of the role of enzyme in tissue disintegration, De Bary had initiated the physiological era in plant pathology. H.M. Ward, a student of De Bary, emphasized the role of environment in the epidemiology of coffee rust, then devastating the coffee plantation in Sri Lanka (Ceylon). Inspite of these brilliant starts, if the pathogen eclipsed the disease, it was because plant pathologists were struggling hard to get across to the people the role of microorganisms in disease causation. The physiological trend, though firmly founded by De Bary and Ward, was overshadowed by descriptive plant pathology.

The first book on plant pathology was written by Kuhn in 1858, with full knowledge of the role of fungi in disease development. The scattered observations were woven into a coherent mass of theory. Brefeld of Germany (1875) developed the modern techniques of inoculation and growing microorganisms in pure culture, which were later perfected by Koch, Petri and others. It enabled a scientific study of etiology of plant diseases. Parasitic fungi started being grown in sterile synthetic media to study their physiology, nutritional requirements, sporulation and genetics. It was a big leap forward in our science.

In 1878 the downy mildew threatened the vine orchards in France. This saw the emergence of the chemical control of plant diseases. Growers used to spray a "poison" (copper sulphate and lime) on vines to deter the pilferers. P.M.A. Millardet, Professor of Botany at Bordeaux, noticed that the sprayed grapes were free from the disease. He was quick to put two and two together and deduced that the cure for pilferage was cure for the mildew too. This observation led to the invention of the Bordeaux mixture (1885), the fungicide era had begun. Mildew was severe again in 1885 but this time it had lost the sting. Bordeaux mixture saved the vine yards in Europe with spectacular success. This had never happened before. The news spread far and wide and Bordeaux mixture began its triumphant march around the world, controlling diseases wherever they occurred. The notoriety of the potato blight was checked and several diseases fell to the Bordeaux mixture one after the other. Bordeaux mixture changed the whole course of plant pathology. The method of chemical control of plant diseases, initiated by Prevost in 1805, got established in plant pathology.

The other important method of plant disease control viz., use of resistant plants, was not thought of before 1905. Biffen in UK working with stripe-rust of wheat showed that resistance in crosses between resistant and susceptible wheat plants was inherited as a recessive Mendelian factor. Orton in USA bred resistant varieties which could ward off fusarial wilts.

Bacteria as Pathogens

The theory of the spontaneous generation of disease was finally buried by Pasteur and Koch in 1876 when they convinced one and all that the anthrax disease was incited by a bacterium. This had wide implications in related fields of study, specially plant pathology. In 1878 Burrill of Illinois, reported that the fire blight disease of pear and apple was caused by a bacterium. Arthur (1885) proved that the disease could be incited by the bacterium obtained from a pure culture. By 1900, E.F. Smith firmly established the study of bacterial diseases by bringing in the best methods of study available in animal bacteriology. His extensive studies on bacterial diseases stand out as a great contribution to plant pathology. Alfred Fischer of Germany challenged Smith's findings and the period between 1890 to 1900 in plant pathology saw one of the most interesting dialogues, with overtones of political campaigning, regarding bacteria's credentials as pathogens. Smith silenced Fischer.

Erwin Frink Smith

The Virus as Disease Incitant

"Breaking" of tulip flowers was known to the horticulturists for many centuries though the etiology of the disease was not known. Such flowers were in great demand. Adolf Mayer in 1870, working with a serious disease of tobacco, which he named 'mosaic', found that the disease could be transmitted to a healthy host by injecting the juice of the diseased plant. He failed to isolate any bacteria but firmly believed some minute bacteria to be associated as the cause. Bacteria were in vogue. Smith in USA too had the same observations with Peach yellows. He could infect a healthy plant by bud grafting and believed this disease to be similar to tobacco mosaic. Ivanowski (1892) filtered the infective juice through bacteria retaining filters and found that the pathogen passed through. Mayer had earlier claimed to have filtered them out on such filters. But Ivanowski failed to realize the importance of his findings and believed some toxin or very minute bacteria, to be the infecting agent. Beijernick, a Dutch botanist (1896), proved that the disease did not involve a microbial agent or anything corpuscular, but a (*contagium vivum fluidum*) contagious living fluid was responsible for the disease. He used the word "Virus" for the contagious fluid and by 1900 viruses as disease causing agents were established in plant pathology. Since then viruses have made a very important place for them and their pursuers in plant pathology. Fungi which had dominanted in the latter half of nineteeth century as the pathogens of plant diseases lost the glamour to the new entrants—bacteria and the viruses. Bacteriologists and Virologists entered plant pathology and the science of plant pathology started making strides in the twentieth century.

Plant Pathology in Twentieth Century

The major trends initiated in the latter half of the 19th century progressed in several directions. The rediscovery of Mendel's laws of heredity in the beginning of this century had its impact and the genetics of the host and the pathogen attracted attention of plant pathologists. The science of plant pathology grew multifarious in scope. Some of the major aspects of study can be summarised as follows:

(1) **The physiological plant pathology** initiated by De Bary and Ward, was pursued vigorously by L. R. Jones who did pioneering work on pectic enzymes secreted by soft rot bacteria. Professor William Brown's series of papers on physiology of parasitism cut

nany new grounds and gave directions which are still being pursued. After the second world war the biochemical approach was nore intense. Investigations on the role of enzymes and toxins n disease started with exceptional zeal. Earnest Gaumann in Zurich, J.C. Walker at Wisconsin, A.E. Dimond at New Haven, T.S. Sadasivan at Madras, R.K.S. Wood at London, Uritani in Japan and Rubin in USSR, contributed extensively to elucidating the role of enzymes and toxins.

(2) **Genetics of the host plant and the pathogen** was studied extensively in relation to the inheritance of resistance and virulence. When Biffin's rust resistant varieties succumbed to infection in Australia there was a wave of disbelief in genetically bound disease resistance. However, the discovery of the physiological races of the pathogen broke the cobweb. The need for research on the genetics of pathogenic fungi was felt by E.C. Stakman as early as 1914 and he and his associates contributed valuable information on delineating and cataloguing of physiological races. Advances in the knowledge of sexuality of fungi by Blakeslee (1904), Cragie (1927) and Dodge (1927) enabled genetic study of pathogenic fungi. The recent genetic studies with bacteria and viruses pushed up the microbial genetics to the forefront of life sciences. Flor (1955) explained host-parasite interaction in flax rust by assuming gene for gene relationship between rust reaction in the host and pathogenicity in the parasite. He identified 25 such gene pairs in the flax and rust fungus. The gene for gene relationship has been found in some other host-pathogen systems e.g. *Phytophthora infestans* on potato. Several disease resistant varieties have been and continue being produced at several plant breeding stations all over the world. But it is a continuous process as the fungi too breed more virulent races and thus plant breeders always have to be on the alert.

(3) **Fungicide research** developed into an exciting field of endeavour. Several biochemists and organic chemists joined plant pathologists in evolving new fungicides. The team work of plant pathologists and chemists has thrown new light on pathogen toxicant relations. The mechanism of fungicide action was explored by H.D. Sisler, C.E. Cox, R.A. Ludwig, R.J. Luken and some others. J.G. Horsfall's book (1956) entitled ***Principle of Fungicidal Action*** distilled the available knowledge into a compact mass of theory.

The addition of organic mercurials to the arsenal of plant patho-

T. S. Sadasivan

Ernst Albert Gaumann

F. C. Bawden

William Brown

E. C. Stakman

J. C. Walker

logy by Rheim in 1912 was indeed a big contribution. W.H. Tisdale and I. Williams (1934) revolutionized the field of organic fungicides by discovering the alkyldiothiacarbamates, and at present the use of dithiocarbamates outrank all other fungicides. Alkylene bis-dithiacarbamates, the most widely used fungicide, was discovered by Dimond, Heuberger and Horsfall (1943). The discovery of gliotoxin by Weinding and Emerson (1936) from *Trichoderma viridae*, that prevented damping off of citrus seedlings, drew attention to antibiotics in plant pathology. Most of the presentday antibiotics are bactericidal. However, gliotoxin (from *T. viridae*), petulin (from *Penicillium petulium*) and griseofulvin (from *Penicillium janezewskii*) have been used against pathogenic fungi. The most powerful antibiotics came from the actinomycetes e.g. actidione (from *Streptomyces griseus*), nystatin (from *S. noursei*). Dekker(1955)proved that the fungitoxic antibiotic Romocidine (*S. rimosus*) can be used as a chemotherapentant in seed dressing. The antibiotic penetrates the seed skin and kills the internal pathogens such as *Asochyta pisi* in peas. But the initial enthusiasm about antibiotics did not last long; in general antibiotics do not control plant diseases very well. Misato (1969) has reported promising results by blasticidin in controlling blast disease of rice (*Pyricularia oryzae*).

(4) **Environment in relation to disease development** forms the subject of one tenth of the literature in plant pathology. L.R. Jones studied the effect of soil temperature on the cabbage yellows and initiated a strong group of workers at Wisconsin. The "predisposition" of the host is now considered as a major factor in disease development.

(5) **Nature of disease resistance**, challenging as it is, invited great interest of some able biochemists. Link and Walker (1933) reported presence of protocatechuic acid and catechol in the dry pigmented scales of onion bulbs resistant to *Colletotrichum circinans*. The non-pigmented susceptible varieties lacked the chemicals which prevented spore germination. Phenols with antibiotic properties have been persistently associated with disease resistance. Their oxidation products, the quinones, are even more toxic to microorganisms. The infected tissues frequently show increased activity of polyphenol oxidase and peroxidase enzymes and an accumulation of phenols which arrest the growth of fungi and also inactivate the macerating enzymes secreted by the pathogens, e.g. pectic and cellulolytic enzymes. Valuable contributions have come, notably

from G.L. Farkas, Kiraly, Poszar, J. Kuc and I. Uritani. H.M. Ward noticed that immediate death of tissues, surrounding the infection zone in resistant hosts, provided a mechanism to localize the pathogen. Stakman (1912) called it hypersensitivity (cf. Müller, 1959). Now such reactions of the host immediately after the physiological contact between non-congenial host and pathogen are known to be caused by fungi, bacteria, viruses, and insects. The post-infection formation of "Phytoalexins" was invoked by Muller to explain the hypersensitivity.

The report of "Phytoalexins" by K.O. Müller in 1956 raised great hopes that antibody like substances were discovered in plants. Phytoalexins are chemicals produced only in non-hosts by non-pathogens, and not found in healthy tissues. The chemical wards off the fungus and provides protection to further infection by other pathogens. A long list of phytoalexins is now available and except one (Ipomeamarone) all are phenolic in nature.

"Interferon like" substances have been reported in virus infections. Interferon is produced in cells of vertebrate animals following inoculation with active or inactivated particles (*see* Isaacs, 1962). This provides resistance by preventing the replication of the virus. The antibody mechanism confers primary resistance to establishment of infection.

(6) **The biochemistry and physiology of the diseased state of the host.** The various metabolic processes affected have been pried to great depths to unravel the sequence and extent of events in pathogenesis. The general conclusions are that in the diseased state the tissues disintegrate, growth and reproduction are adversely affected, host starves, water becomes deficient and respiration is altered. All these studies have proved that the diseased state has a biochemistry of its own. The techniques and 'know how' from biochemistry and physical sciences have greatly contributed to these studies.

(7) **Tissue culture** enabled plant pathologists to grow organs, tissues and free cells alive *in vitro* for almost indefinitely. It has provided us with better knowledge of abnormal growth in diseased plants. The transformation of a normal cell into a tumor cell has been studied in crown gall disease by P.R. White, A.C. Bräun, and others. A tumor transforming principle, produced by the bacterium, triggers the unchecked growth. Some biosynthetic system, non-functional in normal cells, becomes activated in tumor cells and the

degree of activation determines the rate of tumor growth. The cell membrane has been found to regulate the biosynthetic mechanisms which determine the autonomy of the tumor cell. Braun observed a gradual recovery of tumor cells to normalcy and ruled out mutation as the cause. Virus induced tumors have been grown in tissue culture.

The first artificial culture of a rust fungus and microbe-free culture of viruses and nematodes were done by this method.

(8) **Ecological studies of plant pathogens** received active and conscious study in twentieth century. The root disease investigators realized that they were dealing not with one biological relationship between host and pathogen but with many microbial interactions as well. Taubenhaus and Ezekiel discovered in 1930 that the root disease fungus *Phymatotrichum omnivorum* could not grow in soil although the finest rootlets could be infected by the fungus. This observation had a great future in plant pathology. The success in research depends on asking good questions. Waksman in 1917 asked "is there any fungus flora of soil". This succinctly suggests the state of knowledge about soil fungi at that time. Waksman in 1932 emphasized the important role that the fungi play in the economy of the soil. Till then the fungi were not known to have any role in the decomposition of organic matter in the soil. The fungal spores present in the soil were thought to have drifted from the air. Fungi were later proved to actually precede bacteria in colonizing the dead and dying plant residues. Waksman determined which soil fungi were an integral part of the general soil microflora and which were exotic.

Reinking and Manns in 1933 observed that certain species of *Fusarium* were found in all the soils while others were restricted in distribution to few localities only, dependent upon certain factors like presence of the host plants and soil conditions. The first group of species they named as "soil inhabitants" and the latter as "soil invaders". S.D. Garrett (1950) realised that this pattern of distribution could be made the basis of distinguishing the soil fungi into two groups: (*i*) soil inhabitants, and (*ii*) soil invaders. He renamed the soil invaders as "root inhabiting fungi"—having little or no saprophytic ability to survive in soil. The two groups were distinguished by their "competitive saprophytic ability".

The concept of "Innoculun potential" originated from root disease investigation and was consolidated by Garrett in 1956.

The interesting field of "rhizosphere" studies was initiated in

1904 by Hiltner. Lockhead et al (1940) got evidence that the rhizosphere flora affected the resistance and susceptibility of plants to root infections. Since West's (1939) first study of root exudates from flax seedlings, considerable work has been done on root exudates to show its selective action on rhizosphere microflora. The important discovery of formation of antibiotics in soil *in situ* was of far reaching importance. Fungistatic factors present in secretions of seedlings prevent spore germination in their vicinity.

(9) **Virology** developed into a separate discipline. Stanley dug real gold in 1935 by isolating tobacco mosaic virus in paracrystalline form. F.C. Bawden and N.W. Pirree of Roothamsted Experiment Station, UK, in 1938 proved that viruses were nucleoproteins. During the past 35 years a great deal has been learnt about the nature of viruses and the physiology of virus infected plants. The morphology of viruses, their composition, longêvity *in vitro*, virus-vector relationship and methods of chemical inactivation of their nucleic acid are areas where much headway has been made. Now all viruses (except algal viruses) are known to contain RNA. The discovery of the satellite virus (Kassanis, 1966), the naked virus consisting only of RNA in potato spindle virus (Diener and Raymer, 1967), the double stranded viruses in rice dwarf viruses and potato spindle virus (plant viruses were known to have only single stranded RNA) are landmarks in plant virus research. The mechanism of replication of the genome and other viral constituents within the cell is being worked out. K.K. Reddi (1964) proved that tobacco mosaic virus is synthesized in the nucleus of the host.

(10) **Mycoplasma group of pathogens** have been recently discovered as causal agents of some diseases hitherto described as virus induced. The Japanese workers Doi et al (1967) located in dwarf mulberry, potato witches-broom and Aster yellows, wall-less, pleromorphic mycoplasma like bodies. Such structures have been detected in leaf hoppers too, which serve as vectors.

In twentieth century plant pathology started performing its obligations to society. Disease forecasting services caution farmers against possible incidence of diseases. The shrinking of the world due to modern transport has carried new diseases to disease-free areas. Natural barriers have outlived their efficacy in preventing spread of pathogens through wind. Plant disease control is an international problem and quarantines are in force to check entry of pathogens to new places. Over 200 organizations are engaged in plant protection work. The Food and Agriculture Organization of the United Nations, with its trained bi- and multilingual scientists, study plant diseases wherever they occur and prevent chances of epiphytotics. The tragedy of 1845 cannot repeat now. Plant pathologists, the world over, form a well knit group and maintain contacts through meetings and journals.

In 1855 plant pathology was a 'cook book'. Now it has come of age. It has grown multifarious in scope, embracing all segments of biology. The impact of physical sciences, especially biochemistry, has been tremendous in elucidating the crippled metabolism of the sick plant The use of radio-isotopes and direct radiation is fast increasing particularly in virus research and is worth emulating in other branches too. Tracer techniques have been successfully employed in following the metabolic pathways of the parasites, biosynthesis of toxins, and other aspects of host-parasite interaction. Geneticists are providing new methods towards better understanding of the mechanism of inheritance of resistance and virulence.

REFERENCES

ARTHUR, J.C., 1885, "Proof that bacteria are the direct cause of the disease in trees known as pear blight," *Bot. Gaz.*, **10**: 343-345.

BAWDEN, F.C and PIRREE, 1938, "A plant virus preparation in fully crystalline state," *Nature*, **141**: 513-514.

BEIJERINCK, M.W., 1898, "Concerning a Contagium Vivum Fluidum as cause of the spot disease of tobacco leaves." Translated from the German by J. Johnson in *Phytopathol. Classics* No. 7, 1942. American Phytopathological Society, Cayuga Press, Ithaca, New York, pp 33-52.

BERKLEY, M.J. 1847, *Garden Chronicle. 119.*

BLAKESLEE, A.F. 1904, "Sexual reproduction in the Mucorineae, *Proc. Am. Acad. Arts Sci.*, **40**: 205-319.

BIFFEN, R.H., 1905, "Mendel's laws of inheritance and wheat breeding," *J. Agr. Sci.*, **1**: 4-40.

BREDFELD, O., 1875, "Methoden zur Untersuchung der Pilze. Londwirtsch Jahrb.," **4**: 151-175.

BURRILL, T.J., 1878, "Report on botany and vegetable physiology," Pear blight. Trans. Illinois State Hort. Soc., **11**: 114-116.

CRAIGIE, J.H., 1927, "Experiments on sex in rust fungi," *Nature* (London), **120**: 116-117.

DE BARY, A, 1861, "Die gregenwortig herrschende Kartoffelkrankheit," ihre Ursache and ihre Verhuntung. A. Felix, Leipzig, p. 75.

———, 1863, "Researches sur le development de quelques Champignons parasites," *Ann. Sci. Nat. Botn.*, (4); 20: 1-144,

DE BARY, A., 1876, "Researches into the nature of the potato-fungus," *Phytophthora infestans. J. Bot Brit. and For.*, **14**: 105-126, 149-154.

———, 1886, "Uber einige Sclerotinien und Sclerotien Krankhiten," *Bot. Ztg.*, **44**: 377-387, 393-404, 409-426, 433-441, 449-461, 465-474.

DEKKER, J., 1955, "Internal seed disinfection by an antibiotic from *Streptomyces rimosus*," *Nature* (London), **175**: 689.

DIENER, T.O. and RAYMER, W.B., 1967, "Potato spindle tube virus: A plant virus with properties of a free nucleic acid," *Science*, **158**: 378-381.

DIMOND, A E., HEUBERGER, J.W. and HORSFALL, J.G., 1943, "A water soluble protective fungicide with tenacity," *Phytopathology*. 33: 1095-1097.

DODGE, B.O., 1927, "Nuclear phenomena associated with heterothallism and homothallism in the ascomycete *Neurospora*," *J. Agric. Res*,. **35**: 289-305.

DOI, Y., TERANAKA, M., YORA, K. and ASUYAMA, H., 1967, "*Mycoplasma* or PLT group like microorganisms found in the phloem elements of plants infected with Mulberry Dwarf, Potato Witches Broom, Aster yellows or Paulownia Witches Broom," *Ann. Phytopath. Soc. Japan*, **33**: 259-266.

FLOR, H.H., 1955, "Host parasite interaction in flax-rust, its genetic and other implications," *Phytopathology*, **45**: 680-685.

FRIES, E.M., 1921-1932, "Systeme mycologicum, sisteno fungorium ordines, genera et species, huc usque cognitas," Vol. 1, 3, Sumtibus, E. Mauritil, Gryphiswaldiae. Vol. 2 Ex. officina Benlingiana, Lundae.

GARRETT, S.D., 1950, "Ecology of the root inhabiting fungi," *Biol. Rev.*, **25**: 220-254.

HORSFALL, J.G., 1956, "Principles of fungicidal action," Chronica Botanica, Walthum, Mass.

HILTNER, L., 1904, "Uber neuere Erafhrungen und Probleme auf dem Gebiet der Bodenbakteriologie und unter besonderer Beruck sichtigung der Grundungung und Brache," *Arb. dtsch. Landw. Ges.*, **98**: 59-78.

ISAACS, A., 1962, "The role of Interferon," *Endeavour*, **22**: 96-98.

IVANOWSKI, D., 1892, "Ueber die Mosaikkrankheit der tabakspflanze," *St. Petersbourg. Acad. Imp. Sci. Bull.*, (3), **35**: **67:70**. English translation by J. Johnson in *Phytopathol. Classics* No. 7.

KEITT, G.W., 1959, "History of plant pathology," in *Plant Pathology*, Vol. I, 69-97, Horsfall and Dimond (eds.). Academic Press, London and New York.

KASSANIS, B., 1966, "Properties and behaviour of satellite virus," in *Viruses of Plants*, A.B.R. Beemster and D. Dijkstra (eds.), Norths Holland Publishing Co., Amsterdam, pp. 177-187.

KÜHN, J.G., 1858, "Die Krankheiten der Kulturgewachze, ihre Ursachen und ihre Verhutung," G. Bosselmann, Berlin, pp. 312.

LINK, K.P. and WALKER, J.C., 1933, "The isolation of catechol from pigmented onion scales and its significance in relation to disease resistance in onions, *J. Biol. Chem.*, **100**: 379-383.

LINNAEUS, C., 1753, *Species Plantarum*, Impensis L. Salvii, Hohniae, 2 Vols.

LOCKHEAD, A.G., TIMONIN, M.I. and WEST, P.M., 1940, "The microflora of the rhizosphere in relation to resistance of plants to soil borne pathogens," *Sci. Agr.*, **20**: 414-418.

MEYEN, A., 1886, "Ueber die Mosaikkarankheit des Tabaks," Land Wirtsch. *Vers-Sta.*, **32**: 451-467 (English translation by J. Johnson in *Phytopathol. Classics* No. 7, 1942).

MAYER, F.J.F., 1841, "Pflanzen-Pathologie," *Lehre von dem Kranken Ldbe und der Pflanzen.* Haude und spenersche Buchhamndlung, Berlin, pp. 399.

MICHELI, P.A., 1729, "Nova Plantarum Genera. Tupis B. paperinii," *Florentiae.*, pp. 234.

MILLARDET, P.M.A., 1885, "Traitement due mildiou et due rot," *J Agr. Prat.* 2: 513-516 (English translation by F.J. Schneiderhn in *Phytopathol. Classics* No. 3, 1933).

MISATO, T. 1969, "Mode of action of agricultural antibiotics developed in Japan," *Residul Rev.* 25: 93.

MÜLLER, K.O., 1956, "Eingle einfache versuche zum Nachweis von Phytoalexi-men," *Phytopathol. Z.* 27: 237-254.

———, 1959, "Hypersensitivity," in *Plant Pathology*, Vol. I, 469-519. Horsfall and Dimond (eds.), Academic Press, London and New York.

PREVOST, B., 1807, "Memoire sur la cause immediate de la carie ou charbon des bles, et de plusieurs autres maladies des plantes, et sur less preservatifs de la carie, Bernard, Paris, pp. 80.(English translation by Keitt in *Phytopathol. Classics* No. 6, 1939.

REINKING, O.A. and MANNS, M M. 1933, "Parasitic and other *Fusaria* counted in tropical soils," *Z. Parasitenk,* 6: 23-75.

REDDI, K.K., 1969, "Biochemical approach to the study of host-parasite relationship," in *Host Parasite Relations in Plant Pathology*, eds. Kiraly and Ubrizsy, p. 51.

STAKMAN, E.C., 1914, "A study in cereal rusts, physiological races," *Univ. Minn Agr. Expt. Sta. Bull.*, 138.

STANLEY, W.M., 1935, "Isolation of a crystalline protein possessing the properties of tobacco mosaic virus," *Science* , 81: 644-645.

TAUBENHAUS J.J. and EZEKIEL, W.N. 1930, "Recent studies on *Phymatotrichum* root rot," *Am. J. Bot.*, 17: 554-571.

TILLET, M., 1755, "Dissertation on the cause of the corruption and smutting of the kernels of wheat in the head," translated from the French by H.B. Humphrey in *Phytopathol Classics* No. 5, pp. 189.

TISDALE, W.H. and WILLIAMS, I., 1934, *Disinfectant,* US Patent No. 1972, pp. 961.

TULASNE, L.R. and TULASNE 1861-1865, *Selecta fung rum carpologia,* 3 vols. Imperial typograph; Parisics. (English translation by W.P. Grove, 1931), Oxford Univ. Press, London and New York.

WAKSMAN, S.A. 1917, "Is there any fungus flora of the soil?" *Soil Sci* , 3: 565-589.

———, 1932, *Principles of Soil Microbiology*, ed. 2. Baillière, Tindall and Cox, London.

WARD, H.M,, 1882, "Researches on the life history of *Hemileia vastatrix,* the fungus of the coffee leaf disease." *J. Linn., Soc.*, 19: 299.

WEINDING, R. and EMERSON, O.H., 1936, "The isolation of a toxic substance from the culture filtrate *Trichoderma,*" *Phytopathology*, 26: 1068.

WEST, P.M., 1939, "Excretion of thiamine and biotin by the roots of higher plants," *Nature*, 144: 1050-1051,

2

SOME CONCEPTS AND DEFINITIONS

PLANT pathology is the study of the suffering plant. It aims at knowing the cause of the suffering with a view to alleviate them. Since plant as an individual is less important than as a community, plant pathology is usually considered as a "community aspect or plant public health service". As Horsfall has rightly pointed out, this may be true for the 'art' of plant pathology—the diagnosis and treatment of the diseased plant but not for the science of plant pathology, which is concerned with the nature of disease in an individual, as much as with a population of individuals.

The diseased plant is the central theme: but what is disease? Though difficult to define, it is very easy to recognize. A normal plant lives in a balanced dynamic equilibrium with the environment. Any interruption of this balanced equilibrium disturbs the normal metabolism which appears in the form of external visible symptoms. So disease is a malfunctioning process. When this disturbance is a temporary phase, brought about by a factor passing by, it is injury. Cutting the hedges causes injury not disease. It is the same injury as we get when we cut our finger.

Disease results from continuous irritation. So disease is a malfunctioning process caused by continuous irritation. Disease should not be confused with the symptoms; symptoms are only manifestations of the invisible improper 'functioning' of the plants. Disease, again, is not the pathogen: pathogen causes disease. Plants never catch 'disease'; they catch the pathogens and develop disease.

Some Definitions

Pathogen. The agent that by its persistent association causes the disease is a *pathogen*. It stokes the fire and keeps it burning. It may be living, including viruses (infectious) or non-living (non-infectious).

Pathogenesis. The chain of metabolic events that bring about the disease is pathogenesis. It involves the action of the pathogen, susceptibility of the plant and impact of the environment.

Pathogenicity is the capability of the pathogen to cause disease It is different from virulence. Successful expression of pathogenicity is virulence. A pathogen is avirulent when it fails to cause disease.

Predisposition. The environmental effects that make plants more amenable to infection are called predisposition. There is no opposite word for the effects that make them resistant.

Host. The plant that is diseased is the host. The other name suscept, could not be popular.

Disease Syndrome. The sum total of all symptoms and signs is syndrome. Symptoms are visible expressions of the host's response to infection, while sign is a manifestation of the pathogen—some structural part of it.

Approach to Plant Pathology

There are different avenues to look at the diseased plant and the use of these avenues depends upon .the viewer's training and approach. The three main approaches are:

(1) Division of the diseases on the basis of the causal organism; the etiological or classical approach. The pathogen, rather than the diseased plant, becomes the central theme.

(2) Division of diseases based on symptoms (symptomatological approach).

(3) Physiological approach, in which the malfunctioning process is emphasized.

Emphasis on the pathogen amounts to putting the cart before the horse. The symptomatological approach becomes unwieldy because of bizarre symptoms produced in any disease. This is no less true for the third approach, in which too single process alone is seldom affected. Thus a compromise has to be made. Some diseases e.g. downy mildews, powdery mildews, rusts and smuts can be easily grouped on the basis of the pathogen. In some other one major effect, symptom or the most affected physiological process, can be chosen as the basis of naming the disease. The focus of attention, nevertheless, must be the diseased plant.

Causes of Disease

Plant diseases are caused by any of the several factors of the environment viz., physical, chemical or biological. Those caused by biotic agents viz., viruses, bacteria, fungi, nematodes and parasitic flowering plants are called infectious diseases. Such diseases show a sigmoid curve; first increase as the biotic agent reproduces and later diminish on the non-availability of hosts. Diseases caused by physical or chemical factors e.g. air pollutants, water, frost, nutrition etc. are called non-infectious diseases. Such maladies do not exhibit the sigmoid curve. Because the majority of our important crop diseases are caused by living agents, the infectious diseases have received greater attention and are better understood than those caused by abiotic factors.

Many plant diseases result from multiple factors—some biotic and others abiotic. The role of each factor is determined by the environment. Such diseases are called disease complex. The doctrine of specific etiology—the causation of disease by one specific virulent microorganism (given by Louis Pasteur and Robert Koch) and along with this the celebrated 'Koch's Postulates', though apply well to literally hundreds of plant diseases, fail with several others.

Important plant disease complexes, involving certain fungi and plant parasitic nematodes, have been reported. In the 'tomato root-knot' nematode wilt fungus complex, the nematode predisposes the host which is resistant to the fungus.

3

AGENTS OF INFECTIOUS DISEASES
(DISEASES CAUSED BY BIOTIC AGENTS)

FUNGI, bacteria, viruses, nematodes or parasitic flowering plants cause infectious diseases. An understanding of plant diseases involves several disciplines like mycology, bacteriology, virology, nematology, soil science etc. A plant pathologist has to be, then, an initiated mycologist, a knowledgeable microbiologist and a keen physiologist. Since disease control measures are dependent upon the knowledge of pathogen, life cycle and their behaviour in nature, a description of these is relevant.

Fungi as Pathogens

Fungi exhibit greater diversity in form, function and life history than other pathogens. They have also been known as plant pathogens for a longer time than other organisms. All the major groups of fungi have important plant pathogens. Thus *Olpidium*, *Synchytrium*, *Plasmodiophora*, *Phytophthora*, *Pythium*, *Aphanomyces* and the downy mildew fungi (e g., *Peronospora*, *Plasmopara*, *Bremia* etc.) are the best known pathogens from phycomycetes. In Ascomycetes the powdery mildews, *Penicillium*, *Claviceps*, *Monilia*, *Ceratocystis* cause serious plant diseases. Basidiomycetes include the elite among pathogens—the rusts, and those that transform the 'all important' cereal grains into nothing less than a mass of black powder—the smuts. In addition, the wood-rotting fungi *Armillaria*, *Fomes*, *Merulius*, *Poria* and so many have led to total collapse of wooden buildings. Last, but not the least, are the Deuteromycetes which include the *Fusarium*, the *Verticillium*, the *Rhizoctonia*, the *Piricularia* and several others, whose ravages will be evident in pages to come.

Fungi have made history in all parts of the world by causing catastrophic diseases. The late blight disease of potato in Ireland, the powdery mildew of grapes in France, the chestnut disease in USA and the coffee rust in Sri Lanka, record the temerity of only some fungi. The coffee loving people of Western Europe had to reconcile with tea. *Hemileia vastatrix* did it

Fungi are ubiquitous and versatile. The enormous types of spores utilize all possible means of transport to reach long distances. Their heterothallism ensures an expanding range of genotypes, and in its trail-variability. Their great mutability and heterokaryosis, in addition to genetic recombination, ensure the formation of new and frequently more virulent individuals. Many physiologic races of *Puccinia graminis tritici* are the examples.

The Bacteria as Pathogens

The first bacterial disease, fire blight of pear, was described in 1878 by Burrill, only two years after Koch's demonstration of the role of *Anthrax bacillus* in animal diseases. Smith by 1900 established the role of bacteria as plant pathogens.

Species in all major families of higher plants are known to be attacked by one or more bacterial pathogens.

The Bacterial Cell

Bacteria are prokaryotes, characterized by lack of a definite membrane-bound nucleus containing chromosomes. They have a nuclear body (nucleoid) which comprises a single linear DNA molecule that folds back upon itself to form a circle. It is concentrated in the cytoplasm in a gel-like matrix. Cairns (1963) suggested that the chromosome is attached at one point to the cytoplasmic membrane. There is a basic difference at the fundamental level of DNA organization. In addition, there are several other differences in the cellular organization, which are:

(1) The cell wall is made up of mucopeptide rather than cellulose.
(2) Lack of metabolically functional organelles like mitochondria, chloroplast, golgi body and lysosomes.
(3) Ribosomes are of 70S type and not 80S.
(4) Mesosomes are the site of respiration and membrane lipid synthesis
(5) Photosynthetic vesicles and lamellae are the site of photosynthesis.

The Wall

The mucopeptide consists of chains of alternating molecules of β 1-4-linked N-acetyl glucosamine and N-acetyl muramic acid (Fig. 3.1). These chains are cross linked by short chain amino acids

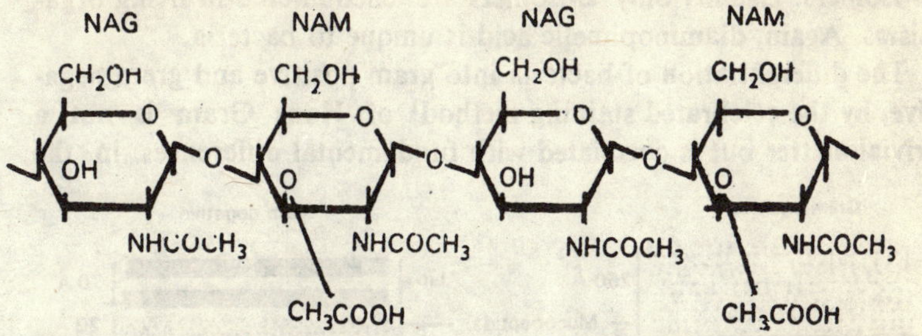

Fig. 3.1. Part of mucopeptide backbone polymer showing alternating units of N-acetyl glucosamine (NAG) and N-acetyl muramic acid (NAM).

joined to the carboxyl group of the muramic acid (Fig 3.2). Only a few amino acids are used for this; these are alanine, glutamic acid, glycine and a diaminoacid such as lysine, diaminopimelic acid, diaminobutyric acid, ornithine or diaminohydroxypimelic acid. It

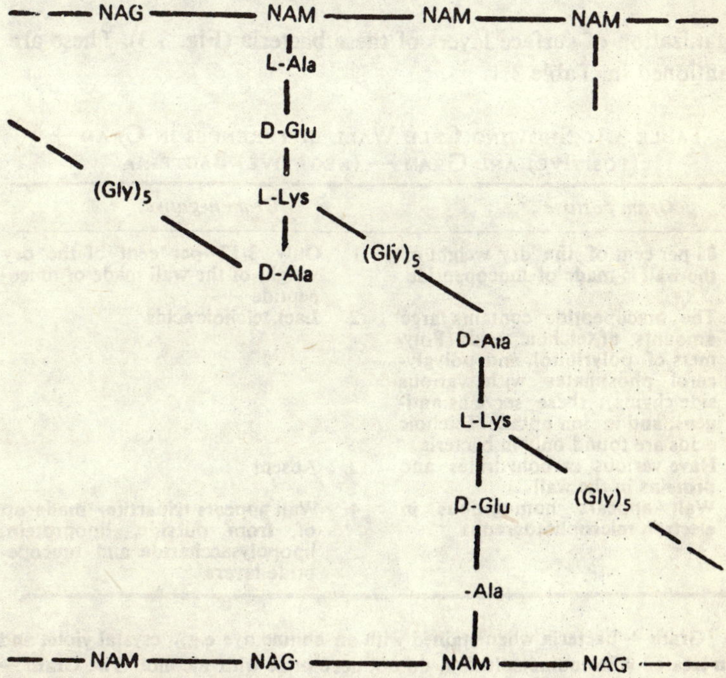

Fig 3.2. Cross-linking of backbone polymers by short chains of amino acids (Alanine, glutamine, lysine and glycine).

is interesting that some of the amino acids (glutamic acid, some of the alanine and occasionally other amino acids) are present as D-isomers. Usually only L-isomers are encountered in living organisms. Again, diaminopimelic acid is unique to bacteria.

The differentiation of bacteria into gram positive and gram negative, by the celebrated staining method[1] of Hans Gram is not a trivial matter but is correlated with fundamental differences in the

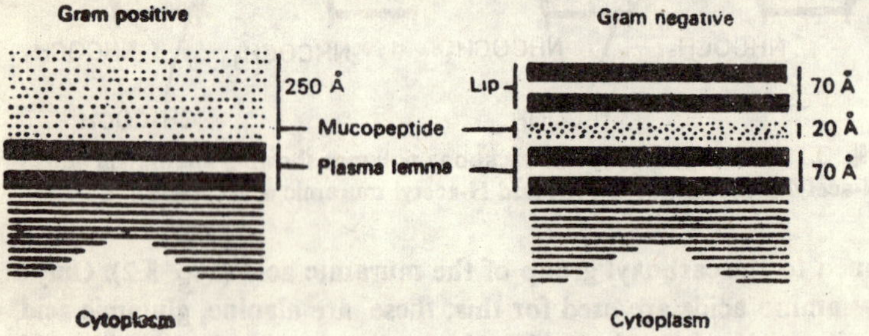

Fig. 3.3. Differences in the outer layer of cell walls in Gram positive and Gram negative bacteria Lip=lipoprotein-lipopolysaccharide complex.

organization of surface layers of these bacteria (Fig. 3.3). These are mentioned in Table 3.1.

TABLE 3.1. SHOWING CELL WALL DIFFERENCES IN GRAM + (POSITIVE) AND GRAM — (NEGATIVE) BACTERIA

Gram positive	*Gram negative*
1. 85 per cent of the dry weight of the wall is made of mucopeptide	1. Only 3-12 per cent of the dry weight of the wall made of mucopeptide
2. The mucopeptide contains large amounts of teichoic acids (Polymers of polyribitol and polyglycerol phosphates with various side chains); these serve as antigens, and in ion uptake. Teichoic acids are found only in bacteria	2. Lack teichoic acids
3. Have various carbohydrates and proteins in the wall	3. Absent
4. Wall appears homogenous in electron microphotographs	4. Wall appears tripartite, made up of, from outside, lipoprotein, lipopolysaccharide and mucopeptide layers

[1]Gram + bacteria when stained with an aniline dye e.g., crystal violet and then treated with iodine solution, do not decolorize with alcohol. The Gram — bacteria lose the stain The colourless cells can be counterstained with carbol fuchsin (red).

Symptoms

The symptoms produced by bacterial pathogens are not markedly different from those incited by fungi. The main examples are:

Wilting — *Pseudomonas solanacearum* is the most important which causes wilt of several crop plants.

Blight — Usually caused by species of *Xanthomonas* and *Pseudomonas*.
X. oryzae causes leaf blight of rice.
P. tabaci—wildfire of tobacco.

Leaf spot—Angular leaf-spot of cotton *Xanthomonas malvacearum.*

Canker —Soft-rot of vegetables (*Erwinia carotovora*). Soft-rot is a special kind of breakdown that does not desiccate/or mummify tissues rather they make it wet and slimy.

Gall—'Grown gall', on over 142 angiospermic genera, is caused by *Agrobacterium tumefaciens*. Bacteria are found in the peripheral zone of the galls. Since growth patterns resemble those of animal and human cancer tissue, bacterial galls are most extensively studied.

Scab—Potato scab caused by *Streptomyces scabies.*

Although most bacterial pathogens induce one major symptom, some diseases show more symptoms. In the bacterial blight of paddy (*X.oryzae*), the disease syndrome comprises blight and wilt symptoms.

Bacterial lesions are characterized by an overlying waxy layer. In wet weather bacterial masses ooze out from the lesions and on drying, get deposited as waxy scales. The bacteria cannot be dislodged unless softened by dew or rain.

THE IMPORTANT GENERA

The important plant pathogenic genera and their important characteristics are given in Table 3.2.

Characteristics of Bacterial Pathogens

Some important features of bacterial pathogens are:

1. All are non-spore forming rods.
2. All except *Corynebacterium* are gram negative.
3. Only *Pseudomonas* and *Xanthomonas* are pigment producers. About half of the pathogenic bacteria belong to the genus *Pseudomonas.*

TABLE 3.2. CHARACTERISTICS OF IMPORTANT PLANT-PATHOGENIC BACTERIA

Genus	*Symptoms*	*No. of species*	*Gram stain*	*Colony colour*	*Motility*
1. *Agrobacterium*	Galls	5	—	White	Motile
2. *Corynebacterium*	Wilt, blight, canker	10	+	,,	Non-motile
3. *Erwinia*	Blight, soft rots	6	—	,,	Motile/non-motile
4. *Pectobacterium*	Soft-rot	1	—	,,	,, ,,
5. *Pseudomonas*	Wilt, spots	85	—	Characteristic pigments	,, ,,
6 *Xanthomonas*	Wilt, spots, blights	47	—	Yellow	Motile

4. Prevalence of water is the key factor of the environment that influences disease development.
5. Bacterial cells have a passive entry into host through stomata, lenticels, wounds, insect bites or through implements. Such direct entry enables them to escape the toxic effects of chemicals.
6. They colonise the intercellular spaces, vascular bundles and spaces ('bacterial pockets') formed by dead cells.

Dissemination and Survival

Wind dispersal of bacterial cells is not of any significance, because they lose their viability when carried by winds. Wind-driven rain is the major source of dispersal of foliage pathogens. The bacterial cells that come out in oozings through stomata, lenticels or breaks in host tissue, are spattered to adjacent plants. Irrigation water is major source of dispersal of soil bacteria Spread through implements is another important method of dispersal. Insects cause spread and also make wounds on the host through which infection occurs. Pathogens of wilt of Cucurbits and Stewart's wilt of corn are examples of insect dissemination.

Bacterial phytopathogens do not form spores and thus compared to fungi, their survival is inefficient. The bacterial cells are easily killed by the unfavourable environmental factors. Those that cause foliage diseases usually survive under layers of slime on leaves or seeds. Nevertheless, there are examples of their survival upto 24 years under a laboratory jar. Bacterial pathogens usually do not survive long in soil because of lack of the saprophytic ability.

Mycoplasma Group of Orgauisms (PPLO Pleuropneumonia Like Organisms)

Japanese plant pathologists Doi et al (1967) reported *Mycoplasma* to be the causal agents of some diseases hitherto known as viral e.g. Mulberry Dwarf, Potato Witches broom and Aster Yellows. They observed the presence of *Mycoplasma* in the phloem—80-800 mμ in diameter with 8 mμ thick limiting membrane, instead of a cell wall. These bodies disappeared on treatment with tetracycline antibiotics. The isolation and culturing of these bodies and proving the Koch's postulates, which are the next logical steps, have been achieved Also, these bodies have been detected in leaf hoppers which act as vectors of Aster Yellows.

Many diseases, which were confirmed as virus diseases, are now revealed to be caused by *Mycoplasma* (Fig. 3.4). Sandal spike disease, Withches broom of Opuntia; Corn stunt and several others (about 30 diseases) are now confirmed *Mycoplasma*-incited diseases.

Characteristics and Classification

Principal characteristics of *Mycoplasma* according to Davis and Whitecomb (1971), are as follows:

(1) They can be grown on cell-free media.
(2) Are small.
(3) Are highly pleromorphic and are bounded by a triple layered unit membrane without a rigid cell wall.
(4) Usually require sterol for growth.
(5) Are highly resistant or insensitive to penicillin but are inhibited by tetracyclines.
(6) Are inhibited by specific antibody, and
(7) Have no history of reversion or derivation from a bacterial parent.

Relationships

The genus *Mycoplasma* was included in ***Bergey's Manual of Determinative Bacteriology*** by Freundt in 1957.

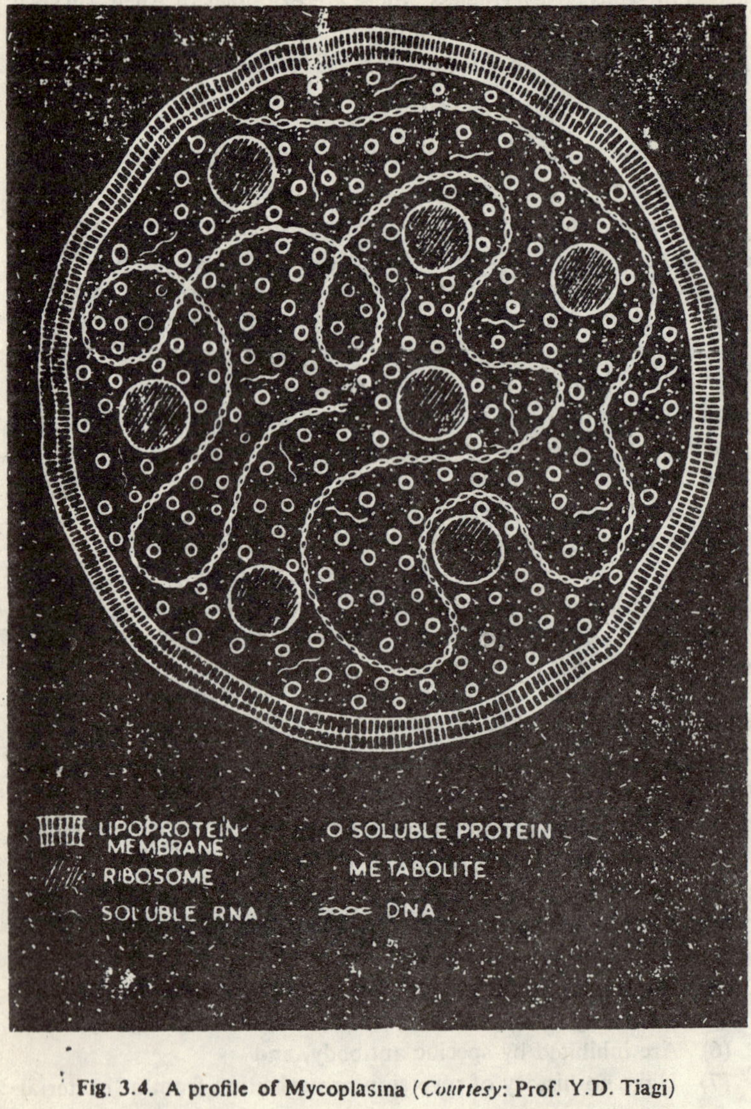

Fig. 3.4. A profile of Mycoplasma (*Courtesy*: Prof. Y.D. Tiagi)

In 1966, Mycoplasmas were recognized by the International Committee of nomenclature of bacteria, to be sufficiently different from bacteria, to form a separate class—*Mollicutes* (*see* Edward and Freundt, 1967). The class Mollicutes consists of two genera

Mycoplasma and *Acholeplasma*, placed in families Mycoplasmataceae and Achloesplasmataceae (Edward and Freundt, 1970). The latter include those mycoplasmas that have no sterol requirement. The species and subspecies are identified on the basis of serological reactions.

It is now generally agreed that *Mycoplasma* are unrelated to L-forms (L for Lister institute) with whom they resemble in the lack of a rigid wall. No L-form is reported to be pathogenic.

Transmission

Mycoplasmas are disseminated through: (*a*) direct contact, (*b*) aerosol, and (*c*) through vectors.

Corn Stunt, a yellows-type disease affecting corn (*Zea mays*) is transmitted by several species of leaf-hoppers (*Daubulus maidis*).

Viruses as Plant Pathogens

'Broken' tulips were favourite of painters 300 years before horticulturists grew them for their beauty and people under the spell of 'Tulipomania' little realized that their obsession was sick. Now viruses are known to be next only to fungi in importance as plant pathogens.

Viruses are defined as sub mircoscopic particles, which are obligate intracellular parasites. They lack enzymes and are formed only of protein and nucleic acids. The protein has a structural and protective function while the nucleic acid is infectious. Viruses are smaller than the limit to physical size of an independent cell, which has been calculated to be about 5000Å in diameter. Thus viruses are unable to carry out most of the living processes of cells. In order to multiply, viruses invade living cells. The infecting virus is undressed before it enters the host cell; the naked nucleic acid is exposed to an embarrassing and hazardous period before it 'takes over' the biological 'assembly lines' and diverts them from making the normal host products to produce its own materials. The virus proteins and nucleic acids, thus synthesized, later combine to produce virus particles. The development of infection is largely due to repetition of this cycle several times and in increasing number of host cells. Since they cannot grow except in living cells, viruses are not organisms. These are neither living 'organism' nor chemicals, but something between and betwixt. As Andre Lwoff, the eminent French microbiologist would have it, '*a virus is a virus*'.

Characteristics of Plant Viruses

(1) They are nucleoproteins.

(2) They contain RNA except the algal viruses and cauliflower mosaic virus which contain DNA

(3) Multiplication is patterned on their genetic material, RNA—which is the infectious portion of the virus particle.

(4) The RNA and protein subunits are formed separately in the host cell and combine to form the full particles.

(5) Lack enzyme systems, and

(6) Show wide differences in shape and size.

The RNA viruses so far analyzed have nucleic acid, with a molecular weight of over two million. It has been proposed that only about one-tenth of the nucleotides are needed to code for the structural protein and, therefore, many nucleotides are available for coding for other activities. The discovery of the satellite virus has shown that ability to code only for the structural protein is not enough for a virus to be pathogenic. The satellite virus (the smallest plant virus) has only few nucleotides left after coding for the protein. This virus fails to induce symptoms in host tissues unless aided by the tobacco necrosis virus, which once existed in nature as a 'contaminant'. The tobacco necrosis virus is, hence, called 'activator' and the association—activation. The activator alone multiplies in plants without inciting satellite symptoms.

Symptoms of Virus Infections

Viruses produce a wide array of symptoms; however, three general categories can be made out—mosaics, yellows and ring spots.

(1) **Mosaic diseases**. Chlorotic mottling appears on leaves and the virus is transmitted by mechanical means or through aphids.

(2) **Yellows diseases**. General yellowing, leaf curling, vein thickening, vein banding, enantions, dwarfing and excessive branching are the symptoms that are included in yellows. These diseases are not mechanically transmissible and are transmitted through insects, leaf-hoppers.

(3) **The ring spot diseases.** Localized necrotic or chlorotic spots, which tend to recover, are produced. These are not insect transmitted. Mechanical or seed transmission, especially the latter, is more common.

Interaction of Viruses in the Host

Plants can be infected by two or more viruses simultaneously. In such cases the effect may be independent of each other, synergistic or inhibitory. In some mixed systemic infections, the symptoms are additive. The symptoms of one, more frequently, dominate the other. The dual infection may cause more serious damage, as happens in double virus streak of tomato infected by tobacco mosaic virus and potato virus X (PVX). Potato virus Y (PVY) stimulates PVX infection in potato.

The previous viral infection may protect the host against infection from a second closely related strain. This phenomenon called **cross protection** was first described in 1929 by Mc Kinny. Now this phenomenon is known with several fungal pathogens also. Potato virus X (PVX) inhibits symptoms produced by Potato-leaf-roll virus; tobacco ring spot virus reduces symptoms of TMV. Bawden and Kassanis (1941, 1945) reported the suppression of one plant virus by another. They found that tobacco severe etch virus prevents the multiplication of potato virus Y (PVY) or of henbane mosaic virus. Cross protection has been utilized as a test in identifying virus strains and in demonstrating relationships of viruses from different sources.

Morphology of Viruses

The dramatic power of the electron microscope made many 'new things familiar and familiar things new'. The methods of ultracentrifugation enabled the sedimentation of so small particles as viruses, and the improved technique of shadowing of particles (Fig. 3.5) revealed structural details down to the literal molecules. The 'spherical' viruses turned out in the electron photomicrographs to be polyhedrons. Two morphological types are common:

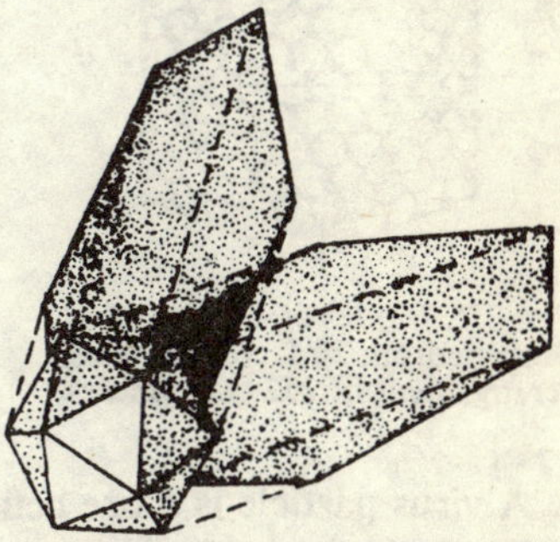

Fig. 3.5. Technique of 'shadowing' of particle in the electron microscope.

(1) **Rod shaped.** The rod shaped viruses e.g., tobacco mosaic virus (TMV) are now known to be 'helical' (Fig. 3.6.) That is, they are superficially cylindrical hollow tube of spiralling protein

sub-units embedding the spiral ribonucleic acid. The spiral has 49 protein sub-units in every three turns and a total of 2,130 units in each virus particle. Every one turn in the spiral has 49 nucleotides.

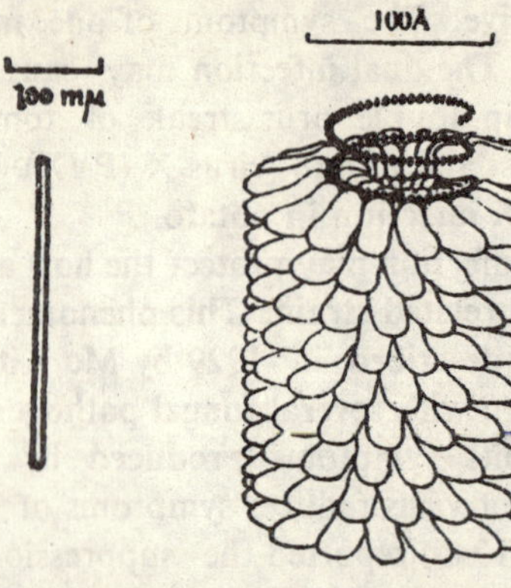

Fig 3.6. Model of Tobacco mosaic virus (TMV) showing protein sub-units (capsomeres) tagged on the embedded RNA helix.

Each protein sub-unit is a polypeptide chain made up of 158 amino acids. It is also known that the nucleic acid thread consists of more than 5,000 nucleotides.

(2) **Polyhedral**. These polyhedral particles are mostly 20 faced (icosahedrons). The surface of the icosahedrons (Fig. 3.7) is made up of regularly spaced protein sub-units. From the number of spheres on one face, the total number of spheres in the protein coat can be calculated by the following formula:

$$10\,(n-1)^2+2$$

Where n = number of spheres present on one edge.

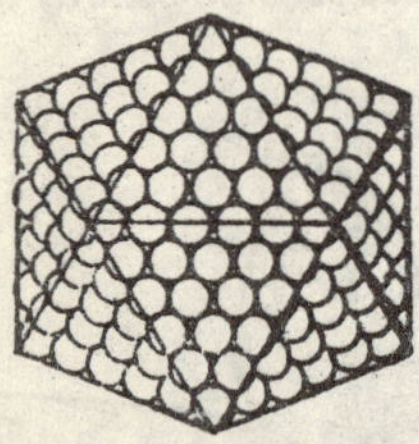

Fig. 3.7. Model of a polyhedral (icosahedron, 20 faces) showing arrangement of 252 capsomeres.

A virus particle is more befittingly called a virion; the nucleic acid as nucleoid; and the protein coat as capsid. The capsid (protein coat) further consists of capsomeres (the protein sub-units).

Transmission

A knowledge of the way in which a plant virus is transmitted determines how it spreads in crops and the kind of control measures that may be effective against it. The transmission is determined by the innate character of viruses. While some viruses are spread on fomites (any particle in close contact with infection), others require vectors, to carry them from one host to another. In plants different types of insect vectors are known.

Transmission by Vegetative means and Grafting

All plant viruses which are systemic in host are invariably transmitted through vegetative organs like tubers, bulbs, rhizomes, cuttings etc. This results in widespread occurrence of virus diseases of vegetatively propagated plants. Almost all fruits, several ornamentals and important crops like potato, hops and strawberry, which are vegetatively propagated, get virus infections by this method.

Grafting is another frequent method of transmission. Natural grafting of roots in adjacent plants is sometimes the only method of virus transmission. *Cuscuta* causes natural grafts between plants and sometimes transmits viruses to incompatible plants.

This characteristic transmission by grafting serves in some diseases, in the absence of other causes, as the only evidence of virus infection.

Transmission through Seeds

An appreciable number of viruses pass from one host generation to the other through seeds. Virus particles may be carried on the surface of the seeds or in parts outside the embryo. Viruses that occur in the phloem, reach the endosperm of seeds during food storage in seed. Tobacco mosaic virus has been observed in the seed coat and endosperm of tomato seeds. The manner in which the surface contaminants, or those present in parts other than the embryo, are transmitted to seedlings, is not clear. Transmission of viruses through infected embryo of seeds is quite understandable in systemically infected virus diseases.

Mechanical Transmission through Sap

Many of the mosaic type of diseases are transmitted mechanically by rubbing juice from diseased plants onto healthy leaves. Prior injury, which in artificial inoculations is made by adding celite or carborundum in the inoculum, increases the amount of infection. But only few viruses are transmitted by this method. The tobacco mosaic virus and the potato virus X, which have no insect vectors, nevertheless ubiquitous, are easily transmitted in fields by frequent rubbing of adjacent plants or through hand implements, clothes, and by grazing animals.

Transmission through Soil

Interest in soil transmission arose with the report of Hewitt et al (1958) that the ectoparasite nematode *Xiphinema index* transmitted the virus causing fan-leaf virus of grape vine. The soil borne viruses have been divided into two groups (Cadman, 1963).

(*i*) Those that are nematode transmitted, and (*ii*) those that are transmitted by fungi.

The two groups differ in several important respects. Some of them are shown in Table 3.3.

TABLE 3.3. VIRUSES TRANSMITTED BY NEMATODES AND FUNGI

Group 1 (Nematode-transmitted viruses)	—	Arabis mosaic
	—	Raspberry ring spot
	—	Strawberry latent ring spot
	—	Tobacco ring spot
	—	Tomato black ring
	—	Tomato ring spot
	—	Cherry leaf roll
	—	Tobacco rattle
Group 2 (Fungi-transmitted viruses)	—	Tobacco stunt
	--	Lettuce big vein
	—	Tobacco necrosis
	—	Wheat mosaic
	—	Oat mosaic
	—	Wheat yellow mosaic
	—	Barley yellow mosaic
	—	Sugarcane chlorotic streak
	—	Tobacco mosaic

Transmission through Vectors

This is the most natural spread of plant viruses. Insects, mites and fungi act as common vectors of plant viruses. Aphids and leafhoppers belong to the order Hemiptera of insects, which have sucking mouth parts and do not ingest any plant tissue. No other group of diseases is so much dependent on vectors for transmission as are the viruses. They insert needle-like structures, mandibles and the maxillae in the host tissue, which have two fine canals.

Through one canal the insect 'receives' the sap and through the other injects its saliva, and alongwith it the viruses. Turgidity of plant cells forces the sap up the canal and hence, if the head of a feeding insect is cut the sap continues dripping out through the stylets left *in situ.* It has become a good method of collecting phloem sap.

Aphids, leaf-hoppers, white flies, spittle insects, mealy bugs, thrips and beetles are the different insect vectors.

Aphids

Majority of plant viruses are transmitted by Aphids. Of the 250 species known as vectors, *Myzus persicae* is the most common. The aphid-transmitted viruses may be *non-persistent* (lost usually after one feed) or persistent (retained for longer periods—frequently for the rest of the life). Since the same aphid transmits some viruses in a persistent manner and some others in a non-persistent manner, the relationship is obviously determined by the virus. The virus does not behave differently. The non-persistent viruses are mechanically transmitted. Starved insects (e.g. Beet mosaic virus) are more efficient vectors—perhaps because they move to more plants and puncture only the epidermis. Non-persistent viruses remain in epidermis only. These viruses are so rapidly transmitted that they cannot do more than adhering to the stylets, at the tips, in the grooves or wedged between scales and bristles.

That it is not so simple a transmission is proved by the fact that: (*i*) they are specifically transmitted by different species, and (*ii*) contagious viruses like TMV and PVX are not transmitted similarly. So transmission is not a straight 'flying pin'.

In the persistent viruses a biological relationship is established between the viruses and the vector. There is frequently an incubation period between the acquisition of the virus by the aphid and its becoming infective. The aphid must feed for many hours. Aphids that feed on infected plants for a day are more infective than those having fed for a few hours. Freitag (1936) and Rochow (1959) observed that when fed for one hour on infected plants, each aphid infected only one or two plants out of twenty fed on (leaf-hoppers infected four) but when fed for twentyfour hours the aphids infected fourteen plants (the hoppers infected twenty). Usually non-persistent viruses are aphid-borne and cause mosaic disease.

Leaf hoppers

Leaf-hoppers usually show greater biological relationship with the viruses and, therefore, are usually vectors of persistent viruses. The leaf-hopper-transmitted viruses have certain general characteristics: (*i*) are never mechanically transmitted, (*ii*) are never transmitted by any other vector, and (*iii*) produce the same general symptoms, mostly malformations of flowers and leaves—never leaf mottling (the characteristics of aphid-borne virus).

The leaf-hopper *Cicudulina*, vector of maize streak virus, has two races which are morphologically alike but show difference in transmission of the virus. One can, the other cannot. However, interesting as it is, if a puncture is made in the gut, the non-vector behaves as a vector. Thus, apparently, the gut prevents the passage of the virus. Now it is known that this capability of virus transmission is due to a sex-linked gene.

Hereditary transmission of viruses in leaf-hoppers is fascinating. The virus is passed on to offspring *via* the egg upto several generations. Fukushi in 1935 (Cf. Smith, 1962) first demonstrated that the Rice dwarf disease virus, transmitted by *Nephotettix apicalis*, was passed on (*via* the female parent) upto 6 generations. Similar observations were made by Black (Cf. Smith, 1962) with the transmission of clover club-leaf virus upto thirtyone generations. Such viruses that multiply in the host are known as "propagative viruses". Direct evidence of multiplication inside vector was obtained by Dutch workers with Aster yellow virus. They ground up some infected leaf-hoppers and inoculated this into virus-free vectors. This was repeated several times ("serial transmission technique"). It was shown that the virus persisted through several transmissions, which was not possible without multiplication. Dilutions involved would be too great for the survival of the virus.

The non-persistent and persistent are 'much abused terms'. These are extremes between which there are several intermediates. Thus some authors use the term stylet-borne, circulatory, propagative (transmission) to designate the virus-vector relationship.

Do viruses harm the vectors?

Evidence is at hand that shows that the insects carry the viruses at great cost. Jensen in USA found that leaf-hoppers transmitting a virus of peach trees have a shorter life. Watson and Sinha, in England, found that the infective females of leaf-hopper *Delphacodes pellucida* carrying the virus of wheat striate mosaic, produced 40 per cent lesser progeny, and also many of the embryos were affected and died at a late stage of their development.

Thrips. Virus causing tomato spotted wilt is transmitted by thrips. The insect feeds with a kind of pick-axe movement of the single mandible. It becomes infective only when it has fed at the larva stage. It cannot pickup viruses at adult stage.

Beetles. Several viruses are spread by beetles. Beetles (and Caterpillars) have biting mouth parts and ingest the leaf tissue. Regurgitation, while feeding, inoculates the viruses, obtained from infected plants, into healthy tissues. Grass-hoppers and earwings too transmit viruses by a similar mechanism.

Mites. Mites have piercing and sucking type of mouth parts and transmit at least four plant viruses viz., black currant reversion, fig mosaic, peach mosaic and wheat streak mosaic. The relationship of the virus with the vector is not much studied. The wheat streak mosaic virus is picked up by the mite only at the nymph stage and not at the adult stage.

Vectors other than Insects

Nematodes and fungi are vectors that usually transmit the viruses in the soil. This has been seen earlier in the chapter.

Three ectoparasitic plant nematodes—*Longidorus. Xiphinema* and *Trichodorous* transmit about a dozen viruses. Grape fan leaf virus was the first report of nematode (*Xiphinema*) transmission of viruses.

The exact biological relationship between the nematodes and the viruses is not clear. Since most of the nematodes retain infectivity throughout life, it is believed that the viruses multiply in their body. The nematodes acquire infectivity from the epidermal cells of diseased leaves. Due to their slow movement nematodes are poor spreaders.

The chytrids *Olpidium brassicae*, *Synchytrium endobioticum* and *Polymyxa graminis* (members of Plasmodiophoromycetes) are known to serve as vectors to several (about a dozen) viruses Some are as follows:

O. brassicae —transmits tobacco necrosis virus (TNV), tobacco stunt virus (TSV) and lettuce big-vein virus (LBV).

S. endobioticum—transmits potato virus X, (PVX) and potato mop-top virus.

P. graminis — transmits wheat mosaic virus (WMV), barley yellow mosaic virus (BYMV), oat mosaic virus (OMV).

Tobacco mosaic virus (TMV), tomato bushy stunt virus (TBSV), chlorotic streak of sugarcane are the other viruses which have fungal vectors: but these are not confirmed.

Nematodes as Plant Pathogens

Though diseases of plants caused by nematodes have been known for a long time, it was not until the mid 1940's that their importance was realized. Fumigation of heavily nematode-infested soils by two nematocides D-D (mixture of dichloropropane and dichloropropene) and EDB (ethylenedibromide) showed startling improvement in crop yield, by killing the nematodes. This resulted in immediate attention towards understanding the role of nematodes in causation of plant diseases. In the comparatively short period much has been known about their role as independent pathogens and also in causing diseases in association with other pathogens. Their role as virus vectors is another factor which has given an important niche to nematodes in plant pathology. Now phytonematology is as fascinating and rewarding as are studies with fungi, bacteria or viruses.

General Characters of Nematodes

Nematodes belong to class Nematoda and are ubiquitous in nature; mostly infest soil. Almost all nematodes feed on living organisms like higher plants, fungi, bacteria, insects, and animals. No plant nematode has yet been cultivated *in vitro*. Plant parasitic nematodes, of which 111 are known, are generally obligate parasites. These are transparent or have surface striations, minute, (.01-0.5X0.5-3.0mm), elongated or eel-shaped (Fig. 3.8) and move with snake-like motion. They possess a stylet or spear with which they puncture intact plant cells and inject the saliva. Cell contents, partially dissolved by the saliva, are ingested. Though morphologically similar, sexes are usually separate. In some species the females become swollen and saccate e.g. *Meloidogyne* (Fig. 3.9).

Life Cycle

It consists of the egg, four larval stages (L_1, L_2, L_3 and L_4) and the adult form. Usually the larvae resemble the adult, but for the sex organs. The female can produce fertile eggs either after or

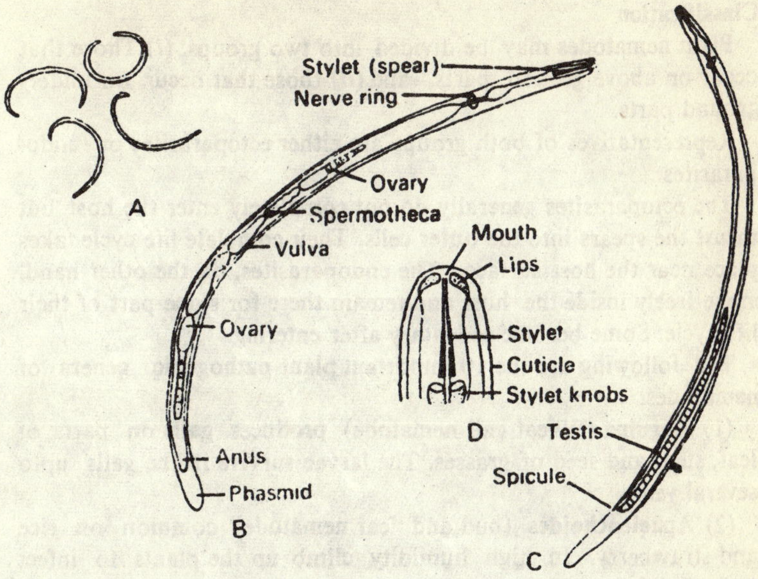

Fig 3.8. Morphology and anatomical characteristics of nematodes A. Appearance under a dissecting microscope B. Female. C. Male. D. Head with spear.

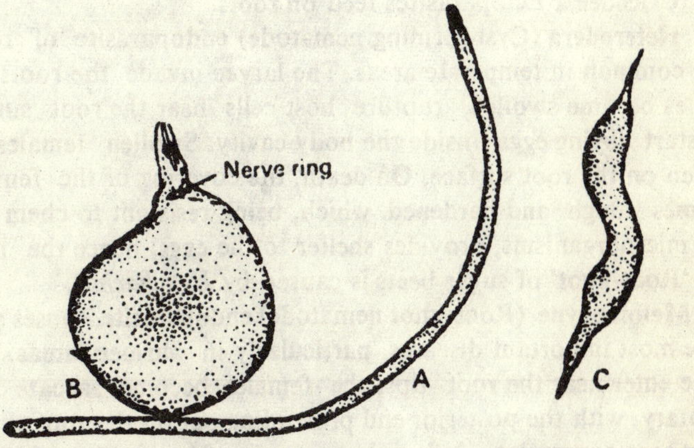

Fig. 3.9. Meloidogyne, a 'root knot' nematode. Note the morphological differences between A. mature male, B. mature sedentary female, and C. sausage shaped larva.

without mating. The females produce sperms themselves; two months are needed for completion of the life cycle. Larvae infect the host plants. Sometimes *L.* and *L.* larvae are unable to infect.

Classification

Plant nematodes may be divided into two groups: (*i*) Those that occur on above ground parts, and (*ii*) those that occur on underground parts.

Representatives of both groups are either ectoparasites or endoparasites.

The ectoparasites generally do not completely enter the host but thrust the spears into the outer cells. Their complete life cycle takes place near the host surface. The endoparasites, on the other hand, move freely inside the host and remain there for some part of their life cycle. Some become sedentary after entering.

The following are the 10 important plant pathogenic genera of nematodes:

(1) **Anguina** (Wheat gall nematode) produces galls on parts of leaf, stem and seed of grasses. The larvae survive in the galls upto several years.

(2) **Aphelenchoides** (bud and leaf nematode) common on rice and strawberry. In high humidity climb up the plants to infect buds and foliage.

(3) **Criconemoides** (Ring nematodes) both the genera are marked by cuticular striations.

(4) **Criconema** Ectoparasites feed on roots.

(5) **Heterodera** (Cyst forming nematode) endoparasite of roots, more common in temperate areas. The larvae invade the roots; the females become swollen, rupture host cells near the root surface and start laying eggs inside the body cavity. Swollen females can be seen on the root surface. On death, the covering of the females, becomes tough and hardened, which, being resistant to chemicals and microorganisms, provides shelter to the eggs; hence the name cyst. 'Root-knot' of sugar beets is caused by *H. schachtii.*

(6) **Meloidogyne** (Root-knot nematode) endoparasite, causes some of the most important diseases, particularly in warmer areas. The larvae enter near the root tips; the females become saccate and sedentary, with the posterior end protruding out on the root surface where eggs accumulate; induce hyperplasia of host cells and 'Root-knots' appear.

(7) **Pratylenchus** (Lesion nematode) endoparasite, moves throughout the cortex and lays eggs in several scattered pockets of damaged cells caused by feeding. Characteristic necrotic root lesions appear.

(8) **Trichodorus** (Stubby root nematode) ectoparasite, feeds on

root tips, making root system 'stubby'. Also a vector of plant viruses.

(9) **Tylenchulus** (Citrus nematode) Saccate, sedentary females are root parasites. The larvae which survive in soil infect the roots. The females swell and deposit eggs in a gelatinous secretion. Swellings are not formed.

(10) **Xiphinema** (Dagger nematode) ecto-root parasites of many plants; important also as virus vectors.

Symptoms

The most common and important damage occurs to roots. The above ground symptoms are those that are caused by root damage.

Symptoms on Roots

Root-knots (Fig. 3.10), root lesions, excessive branching, ceased or retarded root growth, branching and swelling of root-tips are the various effects of feeding on roots by nematodes.

Symptoms on Above Ground Parts

Galls, necrotic lesions and rots, twisting of stems, leaves, ear stalks, and abnormal development of floral parts are caused by migratory nematodes.

Host-parasite Relations

The fact that no plant pathogenic nematode has yet been cultivated *in vitro*, suggests that a delicate balance exists between the nematodes and their living food. It seems to be more true with endoparasites causing galls. Their wide host range suggests that their nutritional requirement is not very specific and is present in almost all living organisms. Nonetheless, it is not known. Maybe they are deficient in some enzyme cofactors, vitamins etc. This can be more confidently said only when they are cultivated in the laboratory.

Root feeding species impair the water uptake capacity of roots that manifests in mineral deficiency symptoms in the leaves. As Christie (1961) has remarked, mechanical damage caused by feeding on roots or withdrawal of a little host tissue is insignificant, as compared to more involved biochemical interactions between the host and the nematode. The injury caused by biting, nonetheless, makes the host susceptible to many pathogens.

The chemicals, secreted or excreted, impair the host physiology.

Nematode enzymes, particularly those present in the saliva, are intimately related with the parasitic relation. Several enzymes, e.g., amylase, invertase, protease, polygalacturonase, β-glucosidase etc., have been detected in the saliva of several nematodes. Many other, yet undetected, enzymes may be active in cell penetration and hydrolysis of protoplasm.

Relationships with Other Pathogens in Disease Causation

The nematodes in association with fungi and bacteria, cause several diseases, which because of involvement of two pathogens, are called disease complex. Several nematode fungus and nematode bacterium relationships offer a very challenging and important area of research. Some ordinarily weak fungal and bacterial pathogens, once they gain entry into the root because of the avenues provided by nematode feeding, can cause serious diseases. Thus, nematodes increase the chances of infection by wilt pathogens e.g. wilt of cotton and *Albizzia*. In fact nematode influence on *Fusarium* wilt expression is so profound that wilt control in a number of crops is based upon concomitant root-knot nematode (*Meloidogyne* sp.) control. Similar effects have been noted for *Verticillium* wilt, *Pythium* damping off, *Rhizoctonia* and *Phytophthora* root rots. The role of nematodes in transmission of plant viruses is already discussed.

Control of Nematode Diseases

The control measures involve cultural practices (rotation of crops and sanitary practices), use of nematocides, use of resistant varieties and biological control. Some nematodes (species of the genera

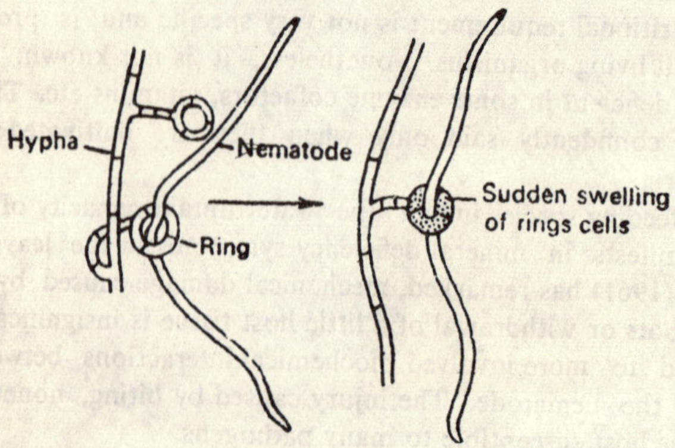

Fig 3.11. Shows trapping of nematode by a predatory fungus.

Fig. 3.10. 'Root knot' symptom caused by a nematode.

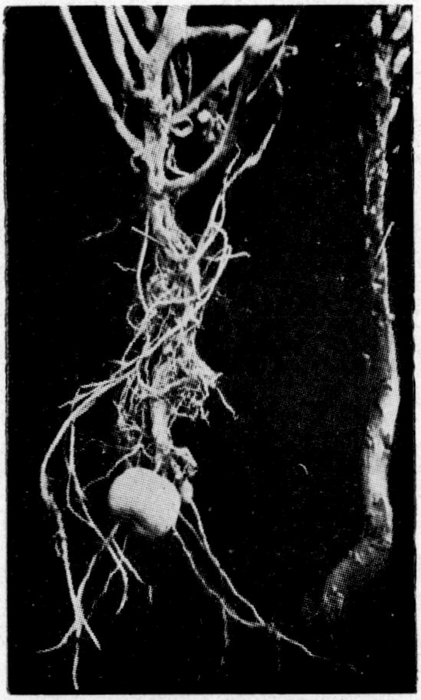

Fig. 3.12. *Orobanche* in association with its host.

Fig. 3.13. *Striga* parasitizing on roots of maize. (*Courtesy* P. P. Mehta)

Monochus, Dorylaimus and *Aphelenchoides*) and predatory fungi (the *Arthobotrys oligospora, Dactylella cionopaga* etc.) can be used to control the number of nematodes. The mycelium forms sticky or mechanical traps (loops and branches) to which nematodes adhere. Other hyphae produce enzymes which digest the prey. *D. bembicodes* forms 'rings' (Fig. 3.11) whose cells swell instantly when a nematode gets into it. The animal is held tightly in the ring; other hyphae invade and digest it. In Hawaii, the root-knot nematode of pine apple was controlled by use of this fungus. The nematode population fell from 611 to 2 nematodes per 2500 g of soil.

Parasitic Green Algae as Pathogens

Parasitic green algae occur on cultivated plants but, because of their green colour, are usually overlooked. All belong to chlorophyta (green algae) and may be purely epiphytic (*Cephaleuros*), endophytic (*Cholrochytrium*) or parasitic *Rhodochytrium*. Species of the genera *Chlorochytrium* and *Phyllobium* cause little damage. Only *Cephaleuros* causes damage of economic importance. *Cephaleuros*, which causes serious damage to tea, guava and citrus, is the best studied algal pathogen. Other parasitic algae are not properly studied.

The disease caused by *Cephaleuros* is called 'red rust', a misnomer indeed. The red rust of tea, caused by *C. parasiticus* and *C. mycoidea*, is found in nearly all the tea gardens in India but is more common in North East.

Parasitic Flowering Plants as Plant Pathogens

These may be broadly placed in four groups: (*i*) *Mistletoes*, (*ii*) the *dodders*, (*iii*) the broom rapes, and (*iv*) the witchweeds.

Mistletoes (Loranthaceae)—May be leafy or leafless, and are green-yellowish or brownish plants, growing on stems by establishing vascular contacts through their roots. Mostly (about 900 species) they attack trees and are usually more prevalent in tropical countries. However, the 'dwarf mistletoes' of the genus *Arceuthobium* cause serious damage to coniferous trees in Western North America. Species of *Viscum* and *Phoradendron* are the other important seed plants that sap the vitality of several trees.

The Dodders (*Cuscuta*—Convolvulaceae) have slender twining orange-yellow, leafless stems; form conspicuous tangled nets on several hosts. It causes considerable damage to flax, sugar beets, onions and other crops. They also transmit viruses to different hosts.

The broomrapes (*Orobanche*—Orobanchaceae). Over 150 species are known; appear as clumps of whitish yellowish or brownish stems. which arise from the roots of infected plants (Fig. 3.12). ***O. cernuo, O. ramosa*** and ***O. minor*** cause considerable damage to tobacco crop.

Witch Weeds (*Stiga*, Scropulariaceae). Of the 50 species of *striga*, only five are important as parasites of crops. *Striga asiatica*—on maize (Fig. 3.13.) *S. hermonthica* (Sorghum), *S. euphrasioides* (Sugarcane), *S. gesnerioides* (Tobacco) and *S. lutea* (Corn).

Pathogenesis in these parasites is established by means of haustoria or roots which make vascular connections with the host. Primary spread is by minute seeds which are produced in large numbers. Wind is the primary source of spread; contamination of crop seeds is another method of dissemination.

Control

Most control measures aim at the prevention of formation and distribution of the seeds. Use of 'trap crops', herbicides, heavy dose of fertilizers and resistant varieties are the other methods of control.

REFERENCES

BAWDEN, F.C. and KASSANIS, B. 1941, "Some properties of tobacco etch viruses," *Ann. Appl. Biol.*, **28**: 107-118.

———, 1945, "The suppression of one plant viruses by another," *Ann. Appl. Biol.*, 32: 52-57.

CADMAN, C.H. 1963, "Biology of soil borne viruses," *Ann. Rev. Phytopathol.* 1: 143-172.

CAIRNS, J. 1963, "The bacterial chromosome and its manner of replication as seen by autoradiography," *J. Mole. Biol.*, 6: 208-213.

CHRISTIE, J.R. 1961, "Some interrelationships between nematodes and other soil borne pathogens," in *Nematology, Fundamentals and Recent Advances*, 432-436. (Sasser, J.N and Jenkins, R.W., eds.), Univ. North Carolina.

DAVIS, R.H. and WHITECOMB, R.F. 1971, "Mycoplasmas, Rickettsiae and Chlamydiae: Possible relation to yellows diseases and other disorders of plants and insects," *Ann. Rev. Phytopathol*, 9: 119-154.

DOI, Y., TERANAKA, M., YORA, K. and ASUYAMA, H. 1967, "Mycoplasma or PLT group-like microorganisms found in the phloem elements of plants infected with mulberry dwarf, potato witchesbroom, aster yellows, or Paulownia witchesbroom," *Ann. Phytopathol, Soc. Japan*, 33: 259-266.

EDWARD, D.G. and FREUNDT, E.A. 1967, "Proposal for Mollicutes as name of the class established for the order Mycoplasmatales," *Int J. System. Bacteriol.*, 17: 267-268.

———, 1970, "Amended nomenclature for classification of strains related to *Mycoplasma laidlawii*," *J. gen. Microbiol.* 62: 1-2.

FREITAG, J.H. 1936, "Negative evidence on multiplication of curly-top virus in the beet leaf hopper *Eutetti tenellus*," *Hilgardia*, 10: 305.

HEWITT, W.B., RASKI, D.J. and GOHEEN, A.C. 1958, "Nematode vector of soilborne fanleaf virus of grape veins," *Phytopathology*, 48: 586-595.

ROCHOW, W.F 1959, "Transmission of strains of barley yellow dwarf virus of two aphid species," *Phytopathology*, 49: 844.

4

AGENTS OF NON-INFECTIOUS PLANT DISEASES

FOR many years plant pathologists have been obsessed by living disease agents. Problems of mineral excesses and deficiencies, herbicide damage, salinity, light effects, climatic effects, and of soil physics and chemistry were thought to be domain of the plant physiologists, agronomists and soil chemists. Research on abiotic diseases, however, has gained recognition and is now an important part of our science.

The abiotic factors sometimes cause injury which is different from disease.

A plant lives in a state of dynamic equilibrium with the environment. Both chemical and physical factors must remain within certain limits. Materials required for growth from soil must be available in quantity that is neither more nor less and at right times. Similarly, the chemical composition of the air must be free from the "volatile corruption." Any kind of disturbance to balanced state would result in damage to plants.

We shall study the various abiotic factors under the following heads.

Air Pollution

Stack gases, emitted from industrial processes and automobile exhausts, are toxic to plants. The term "smog" is used for mixtures of smoke (or gases) and moisture in the air. The smog may be reducing or oxidizing. Conversion of sulphur dioxide gas to its acid aerosol, is reducing. Automobile exhaust gases, by the action of sunlight, form phytotoxic oxidants which are toxic to plants. This is oxidizing "smog". Ozone and peroxyacetylnitrate (PAN) are the two most important photochemical phytooxidants. Nitrogen dioxide and hydrocarbons of the automobile exhausts are energized by the ultraviolet light to react with oxygen resulting in production of ozone at a rate faster than its decomposition to oxygen. Similar photochemical reactions between oxides of nitrogen and hydrocarbons result in the formation of PAN. The occurrence of ozone damage to crops in areas away from the population centres is explained to be due to the production of oxygen by electrical discharges such as lightning flashes and by the intense action of ultraviolet light on the volatile hydrocarbons produced by large masses of vegetation such as forests. Table 4.1 shows the various toxicants and their action.

TABLE 4.1. AIR POLLUTANTS AND THEIR EFFECTS ON PLANTS

Compound	*Formation*	*Symptoms/examples*
Ozone	Photochemical reaction between oxygen and hydrocarbons of automobile exhausts.	Bleached areas on the upper leaf surface: "weather fleck" on tobacco, Smog damage to citrus and other crops.
Peroxyacetylnitrate (PAN)	Photochemical reaction between hydrocarbons and oxides of nitrogen both present in automobile exhausts.	Silver glazing of the lower leaf-surface; spinach and *Petunia* affected.
Sulphur dioxide	Burning of coal, oil etc. ore reduction.	Tissues in marginal areas become necrotic, e.g. SO_2 damage to alfa-alfa, cotton and forest trees.
Ethylene	Burning of natural gas	Adverse effect on plant growth,-epinasty, leaf drop and irregular flower opening; damage orchids.
Fluorine/ Fluorides	Waste products in ceramic, and fertilizer industry.	Marginal necrosis: fluorine damage to gladioli and pines.

Toxic Effects of Chemicals Improperly Used

The incorrect use of fertilizers, fungicides and pesticides can cause serious damage. Thus before release, the phytotoxic property of chemicals for different plant parts, under different environmental conditions, must be determined. The herbicide 2,4-D, when drifted to broad leaf crops, does much damage.

Mineral Elements

Deficiency of Mineral Elements. Plants have a characteristic "mileu interior"—need a proper 'nutrient balance'. The ratio of one element to another is very important. Any deviation from this causes metabolic disorders which manifest as disease symptoms. Soil pH, moisture, temperature, microbial activities, are the factors that affect the availability of certain elements.

If any of the elements is deficient in soil or unavailable due to factors mentioned above, the plant expresses symptoms characteristic of the particular substance. The symptoms of various mineral deficiencies are summarized in Table 4.2.

TABLE 4.2. THE SYMPTOMS AND DISEASES DUE TO MINERAL DEFICIENCY

Deficient mineral element	*Deficiency symptoms/examples of disease*
Nitrogen	Stunting, loss of green colour (however, nitrogen deficiency imparts resistance).
Boron	Poor growth of tops and roots, internal tissue breakdown, e.g. heart and dry rot of sugar beets, internal cork of apples and internal browning of cauliflower.
Calcium	Distortion and death of young leaves. Tip burn of leaves, meristematic tissues lack normal development. Tip burn of cabbage, black heart of celery, blosom end rot and stem and leaf die-back of tomato.
Copper	Die-back symptoms. Fruit trees like apple,citrus, pear show die-back, called 'exanthema'. Cereals deficient in copper show stunting and leaf wilting
Iron	Chlorosis, varying from light mottling to almost complete absence of chlorophyll.
Magnesium	Chlorosis of older leaves. Necrotic flecks and spots appear on foliage of cotton, tobacco, corn etc.
Phosphorus	Growth is retarted but foliage remain green.
Potassium	Marked reduction in growth, foliage remains green. Chlorosis, necrosis and streaking appear first in older leaves.
Sulphur	Stimulates nitrogen deficiency. Tea yellows disease.
Zinc	Little-leaf, rosetting, e.g. Pecan rosette, citrus mottling, and little-leaf of stone fruits and grapes; Khaira disease of paddy.

Mineral Excesses

Sometimes salts accumulate in such excess that crop production becomes uncertain. Such soils have been classified into three groups: Saline, Saline-alkali and non-saline alkali. Reclamation is resorted to by addition of chemicals, which by 'ion-exchange' remove the excess mineral. Thus reclamation of saline alkali or non-saline alkali soils is done by adding chemicals, which cause exchange of sodium from the soil complex. Salinity in soil is rarely eliminated. Growing salt-tolerant crops is the only solution.

Temperature

Sudden rise and fall in temperature cause injury. Harmful effects of chilling, freezing and sun-burn are well known. The plants get adapted to their climate and show chill or frost hardiness.

(1) **Chilling injury** causes surface pitting, necrosis, and failure in ripening of fruits. The seedlings become susceptible to infections. This injury is important in case of some warm season fruits.

(2) **Freezing injury** is caused by formation of ice. Since water is more pure in intercellular spaces, ice crystals are first formed there than in the cells. The crystals damage various cell organelles.

(3) **Injury from high temperature.** Sun-scald injury occurs when shade loving plants are suddenly exposed to direct sun. Sunken brown areas on apples, water-soaked areas on tomatoes, are examples of high temperature injury.

(4) **Soil moisture.**

Low soil moisture causes accumulation of toxic ions such as manganese and boron, which damage tissues and cause stomatal closure. This adversely affects plant

High soil moisture results in diminished oxygen in the soil water, which affects root growth. Plants become succulent, which results in their increased susceptibility to certain pathogens.

(5) **Unfavourable light intensity.**

Insufficient light —causes etiolation, stunting, reduction in flowering.

Improper photoperiod —results in abnormal shape, erratic flowering etc.

High light intensity —leads to scorching and rolling of leaves, drying of flowers.

Light governs so many plant responses that its role in abnormalities must be taken into consideration when growing plants in a new geographical area or in artificial laboratory culture.

(6) **Miscellaneous factors.** Wind action can cause root injury. Lightning, hail and ice damage the crops and cause wounds.

5

INOCULUM AND INOCULUM POTENTIAL

THE INOCULUM

THE part of the pathogen, which on contact can cause infection in the host, is called inoculum. The various types of propagules produced by fungi e.g., spores, conidia, sclerotia, rhizomorphs serve as inoculum. Bacteria and viruses, as such, serve as inoculum. Fungi are more versatile in this respect. The inoculum, disseminated by water, wind, insects etc., spread the disease to far off areas. Their number is a great factor in disease occurrence. In the absence of the host, the pathogens form structures—chlamydospores, sclerotia, thick resistant hyphae, etc., which lie dormant (perennate) until the host is available. On the other hand several pathogens have the capacity to live saprophytically in the absence of the host. The strong pathogens have less of what is termed by Garrett the "competitive saprophytic ability". This determines their survival in soil. The pathogens that lack saprophytic ability usually produce thick walled, resistant propagules which perpetuate the fungus. In still other cases, some pathogens have a wide host range and inoculum is available from collateral hosts.

Table 5.1 summarizes the inoculum of various kinds of pathogens.

INOCULUM POTENTIAL

F.D. Heald (1929) first drew attention to the fact that an increase in the amount (quantity) of the inoculum increased the intensity and frequency of disease. He demonstrated that a considerable number of bunt spores on seed were necessary to cause disease This was the first evidence showing a relationship between amount of inoculum and disease incidence. He used the term "spore load" and emphasized the synergistic action among the spores to be the reason for the successful infection. Horsfall (1932) coined the term 'inoculum potential' to explain the chemical mass effect and defined it as the number of infecting particles present in the environment of the infected host. This concept carries with it the idea of mass action—greater the mass of the inoculum more severe will be the disease. It can thus be said that the inoculum potential of *Phytophthora infestans* increases during the development of an epiphytotic of potato blight. Horsfall's (1932) inoculum potental thus was the "spore load" of Heald, and it is in this sense that the phrase has been most widely used in phytopathological literature. The concept however, has undergone great change eversince, and the confusion is well evident in the literature dealing with the subject.

TABLE 5.1.

Pathogens	*Kind of inoculum*
Viruses	Virus particles
Bacteria	Bacterial cells; in actinomycetes e.g. *Streptomyces*, as conida
Fungi	1. Active mycelium—*Rhizoctonia*
	2. Dormant mycelium—Loose smut of wheat
	3. Specialized mycelium
	Sclerotia—*Clavicep*
	Microsclerotia—*Verticillium dahliae*
	Rhizomorphs—*Armillaria mellea*
	4. Spores
	Asexual spores—
	Motile zoospores—Downy mildews
	Aplanospores—Mucorales
	Conidia—Powdery mildews
	Sexual spores—
	Zygospore—Mucorales
	Oospore—Downy mildews
	Basidiospores—Rusts
	—Smuts.

Zentmyer, Wallace and Horsfall (1944) expanded Horsfall's original (1932) definition and considered inoculum potential as disease producing power of the host environment. Thus environment entered the concept of inoculum potential. The earlier concept of Horsfall was concerned only with the virulence of the pathogen—while the later modification included the effect of the various factors of the environment on infective capacity of the pathogen.

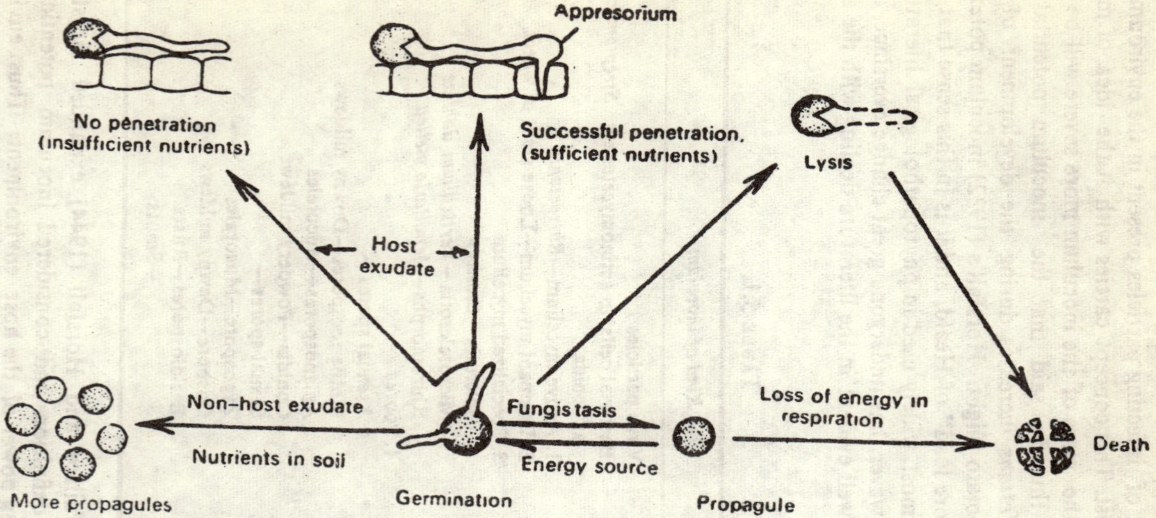

Fig. 5.1. Diagram showing major factors that influence the energy level of inoculum.

S.D. Garrett (1956 a,b) felt that the term inoculum potential describes the degree of infectivity. He thus redefined inoculum potential "as the energy of a pathogen available for infection of a host at the surface of the host organ to be infected". So, according to Garrett, inoculum potential expresses a form of potential energy. Degree of infectivity can be viewed as a concept involving biological energy available to the pathogen for colonization of the host. This energy of the pathogen would be a function of inoculum density (number of units of inoculum), available nutrient (both internal and external) to the propagule, environmental factors and genetic capacity of the organism. Inoculum potential thus can be increased by either one or both of the following factors: (*i*) Increase in the number of propagules of the pathogen per unit area of the host surface, and (*ii*) by increase in the nutritional status of such units. Some of the major factors that influence the energy level of inoculum have been summarized in Fig 5.1.

For germination of soil-borne plant pathogens, the soil provides the carbon and nitrogen to the propagules. If fungistasis prevents germination the energy is conserved over long periods. Additional energy sources needed to overcome fungistasis and initiate spore-germination come from the plant exudates and residues. Normally the non-host exudates help in production of more resting bodies. Once the spore germination occurs, the chances of infection definitely increase. The juxtaposition of the pathogen and the host at the host-surface is one of the most critical moments in the development of disease, and the energy available to the pathogen is decisive in the establishment of infection. The energy involves the genetic makeup of the propagules, their density and nutritorial status. Under a given set of environmental conditions there is a critical level of inoculum potential for the establishment of any disease. This is proved by the following observations:

(1) Inoculation experiments indicate that a decrease in the number of infective propagules lowers the disease frequency till critical level comes, below which no infection occurs.

(2) In some fungal diseases the nutritional status and the volume of the mycelial inoculum determines the success or failure of infection.

(3) In some root-infecting fungi, infection is achieved by mycelial strands and rhizomorphs. The aggregation of hyphae into compact structures increases the invasive power or the inoculum potential.

Dimond and Horsfall (1960) introduced one more factor, the susceptibility of the host and redefined the inoculum potential as the resultant of the action of the environment, the vigour of the pathogen, the susceptibility of the host and the amount of the inoculum present.

The term inoculum potential would then include four separate but independent concepts.

(1) The original concept of inoculum potential (=inoculum density)—(Horsfall 1932).

(2) The concept of infection potential involving biological energy available at the infection court. (Garrett's concept, 1956).

(3) The concept of the susceptibility of the host.

(4) The concept of disease proneness brought about by the environment.

A large number of plant pathologists are not in favour of including so many factors in a single term. Wilhelm (1966) suggested that the phrase inoculum potential should be used only to indicate inoculum density. Host resistance, distance between hosts and the influence of the environment are factors that are beyond the concept of inoculum. Since they come in the disease phase a new term "disease potential" could be useful. Wilhelm (1966) maintains that potential can not be equated with the expression of potential. Further, the energy can not be measured while the number of propagules can be easily counted, and the relation between the amount of propagules and disease can be shown. This is a practical approach and it is in this way that inoculum potential has generally been referred in literature.

Disease Severity

Disease severity depends on inoculum potential and disease potential. Baker et al (1967) have expressed the relationship by the following formula:

Disease severity = Inoculum potential × Disease potential.

Inoculum potential being the energy available for colonization and disease potential—the ability of the host to contract disease.

REFERENCES

BAKER, R. 1968, "Mechanisms of biological control of soil borne plant pathogens," *Ann, Rev. Phytopathol.* **6**: 263-294.

DIMOND, A.E. and HORSFALL, J.G., 1960, "Prologue. Inoculum and the diseased population. I.," *Plant Pathology—An Advanced Treatise,* **3**: 1-22, Academic Press.

GARRETT, S.D. 1956a, *Biology of Root Infecting Fungi,* Cambridge University Press, London and New York.

———, 1956b, "Rhizomorph behaviour in *Armillaria mellea* (Vahl) Quel II. Logistics of infection," *Ann. Botany* (London), **20**: 193-209.

HEALD, F.D. 1921, "The relation of the spore load to the percentage of smut appearing in the crop," *Phytopathology,* **11**: 269-278.

HORSFALL, J.G 1932, "Dusting tomato seed with cooper sulfate monohydrate for combating damping off," *Exp. Sta. Tech, Bull ,* **198**: 1-34.

WILHELM. 1966, "Chemical treatments and inoculum potential of soil," *Ann. Rev. Phytopathol,* **6**: 53-78.

ZENTMYER, G.A , WALLACE, P.P. and HORSFALL, J.G 1944, "Distance as a dosage factor in the spread of Dutch elm disease," *Phytopathology,* **34**: 1025-1033.

6

EFFECT OF ENVIRONMENT ON PLANT DISEASES

THREE things are essential for occurrence of any disease: a susceptible host; a virulent pathogen and favourable environment. In the presence of the first two the environment plays the decisive rôle. It is due to the environment that some diseases are prevalent in certain areas and their activity is restricted to certain part of the year only. In fact, before the acceptance of the germ theory of plant disease, environment was thought to be the sole cause of all animal and plant diseases. It was Prevost who in 1807 for the first time proved that the bunt disease was caused by a fungus. But he did not consider the fungus as the sole cause of disease and called it only the "direct" or "immediate" cause. He was cognizant of the secondary or conditioning effect of the environment. Thus, the environment was relegated from primary to secondary position. H.M. Ward (1880) emphasized the role of environment in the epidemiology of coffee rust. When the pathogen was riding high in the minds of plant pathologists, L.R. Jones at Wisconsin initiated a new field of research on the effects of soil temperature, moisture and other environmental factors upon the epidemiology of disease. He affirmed that "the role of environment to the predisposition of the host as well as to the virulence of the parasite, can not be overemphasized".

The Environment

The plants with their roots sunk deep into the soil, and the shoot system above ground, encounter two kinds of environment viz the soil and the aerial environment. The aerial environment affects mostly diseases that initiate in the leaves or stem, while those that originate in the roots are vastly affected by the soil environment.

The environment is an everchanging factor. The complex of factors which affect disease development are temperature, humidity, light, atmospheric pressure, wind, rain, dew and soil. These factors can not be taken in isolation and it is the interaction of all the variables that operates in nature. The various environmental factors affect disease development by their effect on the pathogen, on the host or on the host-pathogen interaction. The survival of the pathogen from one crop season to other, building up of inoculum, dispersal of inoculum and germination of propagules, all are dependent on the favourable environmental factors. On environment depends the growth and vigour of the host plants. The predisposition of the host is brought about by the environment. The built-in resistance of resistant cultivars succumbs before the mighty forces of the environment. Once the infection is established, everything is not over, and sudden changes in the environment may retard or accelerate the disease process. The environment can outwit the pathogen at any stage of disease development and plant pathologists make use of such mechanisms to get rid of some diseases.

Effect of Temperature

Pathogens differ in their optimum temperature for growth, sporulation and infection. The temperature thus determines the distribution of pathogenic fungi and the diseases caused by them. Of the two most important wilt pathogens, *Fusarium* is found in warmer regions while *Verticillium* is confined to cooler temperate areas. This restricts the fusarial and verticillial wilts to warmer and cooler zones respectively.

In most cases the same range of temperature suits the growth of the pathogen and also disease incidence. Disease can be avoided if the crops can be grown at temperatures unsuitable to the pathogen. Black-stem rust of wheat (*Puccinia graminis tritici*) spreads rapidly if a suitable temperature (24°C) is available for uredospore germination. At low temperatures uredial production is retarded and the frequency of fresh infections greatly diminishes due to lack of sufficient inoculum. Temperature induced "predisposition" is the most important factor in disease incidence.

Host susceptibility in paddy to blast disease (*Piricularia oryzae*) is to a very great extent determined by temperature. Infection in susceptible varieties occurs only at a low night temperature (20°C). At 26°C night temperature even susceptible varieties do not take infection and at lower temperatures (15°C) even the resistant varieties are rendered susceptible. Thus, a low night temperature is critical for infection. This important contribution came from the

Madras school of Sadasivan. They have been able to pindown the biochemical aspects of this predisposition, and have found the accelerated nitrogen metabolism responsible for the susceptibility.

In several virus diseases a high temperature treatment, before inoculation, increases susceptibility of host. Kassanis (1952) found that post-inoculation heat treatment reduced the lesion number. However, Yarwood's fundings (1955) are different; dipping of inoculated leaves into hot water at (50°C) for 20 seconds increased the number of lesions by bringing out the latent infections to surface.

Soil Environment

Soil is the natural medium of plant growth, which has an environment of its own: more complex and variable than the aerial environment. Majority of plant diseases have their root in the soil. L.R. Jones at Wisconsin, around 1920, initiated research into the effects of soil temperature, moisture content and other environmental factors upon the epidemiology of soil borne diseases. The famed Wisconsin soil temperature tank devised by Jones (which could regulate soil temperature) proved very successful in such studies.

Soil water. The impact of soil moisture in disease development is not well known. The soil water can be classified into four categories: (*i*) gravitational water; (*ii*) capillary water; (*iii*) hygroscopic water; and (*iv*) water vapour. Only capillary water is taken up by the plants. Different types of soils retain varying amount of water. Fusarial wilts, which are so much influenced by soil temperature, are little affected by soil moisture. Zoospores of *Phytophthora* and *Pythium* are more severe in presence of abundant water, because of their rapid dispersal. The root infections become severe with increasing amount of water and become most severe at the saturation point. On the other hand the common scab of potato, caused by *Streptomyces scabies*, is more prevalent in dry soils. In some diseases e.g. club-root of crucifers, high moisture, needed in the early stages of infection, is not essential at the later stages.

Moisture

Moisture is the precondition for fungal infections, though the degree of dependence on free water for spore germination varies greatly. Bacterial pathogens enter host tissue only through a film of water. Virus infections are adversely affected by free water.

Most of the fungal pathogens require free water for spore germination. Once the infection is established, the water requirement is met by the host tissues. However, *Phytophthora infestans* and other downy mildews require high atmospheric humidity throughout the disease period. Dry weather, following infection, retards spread of the pathogen. This is the reason why late blight of potato usually does not rise to epiphytotic proportions, though it occurs frequently in the hilly regions of India. Other group of important fungal pathogens, though germinate well in free water can also grow in the absence of liquid water. *Venturia inaeaualis*, *Ustilago hordei*, *Puccinia graminis* and *Alternaria solani* are examples of this group. In still others e.g. powdery mildews, spore germination is adversely affected by free water. Thus, powdery mildew diseases are generally prevalent in dry areas.

For bacterial pathogens a continuous film of water is essential for infection. Stomata are the main portals of entry and the stomatal chambers, with high humidity, provide place for proliferation. Rain and dew form a film of water on the leaf surface and bacterial pathogens riding on this, find their way into the stomatal chamber. A rainy day with gusty winds is the day of bacterial infections.

Virus infections are inhibited by free water. Yarwood (1953) found that dry inoculum or quick drying of moist inoculum, increased the number of lesions several folds. In his opinion the inhibitory effect of water was due to removal of ions, mainly phosphate ions, which are necessary for binding of virus particles to their specific sites.

Light

The effect of light on the virulence of the pathogens is not significant. Insufficient light adversely affects the vigour of the plants and thus increases the susceptibility to diseases. In several fungal diseases, (e.g. root rots, cankers and leaf spots), reduced light predisposes plants to infection. Powdery mildew pathogens germinate much better in light but ecologically they are more prevalent in shady localities. The germination of certain rusts is much greater in light than in darkness.

In virus infections, the number of lesions increase if the plants are kept in dark before inoculation. A common practise followed in laboratories is to keep the plants in darkness for one or two days before inoculating viruses. Bawden and Roberts (1947) found that shading makes the leaves fragile and the chances of wounding and therefore for infection are increased. They also suggested that in light some product (s) of photosynthesis may be interfering with infection This was confirmed by Wiltshire (1956), who found light treatments in the absence of carbon dioxide (photosynthesis checked) had similar effects as darkness on susceptibility. Reduced carbohydrate level and changes in nitrate nitrogen were noticed in dark treated plants.

Atmospheric pressure. There are few reports about the role of atmospheric pressure on disease.

The germination of zoosporangia of *Phytophthora infestans* is adversely affected when the atmospheric pressure falls, as happens on a windy day (Fisher and Gaumann, 1929). Bortels (1947) reported that a fall in barometric pressure increased susceptibility of beans, tobacco and potato to the bacterial pathogens *Pseudomonas medicaginis*, *P. tabaci* and *Erwinia carotovora* respectively. The reasons for such changes are not elucidated. Bortels also reported better growth and increased nitrogen fixation for saprophytic bacterium *Azotobacter chroococum* with a rise in the atmospheric pressure. A fall in the pressure retards both growth and the nitrogen fixation.

Rainfall and Dew

Infection by air borne parasites occurs optimally in drops of water. Rain and dew are the main sources of such water. Dew plays more significant role than is generally recognised. The normal duration of five to eight hours of dew deposition is enough to cause infection. Rain, accompanied with wind, is best for spread of bacterial pathogens

Wind

Wind is the main source of dispersal of the inoculum. It also affects the moisture on the leaf surface by its drying action. The wind can carry the inoculum to great distances and control measures against wind-disseminated pathogens in the absence of natural barriers like mountains and oceans will be futile. A good example of spread of disease by wind is the recurrence of wheat black rust (*Puccinia graminis tritici*) in the plains of India The uredospores blown downwards from the hills initiate infection.

Winds bring about dissemination of insect vectors, and during rains also of the bacterial pathogens.

Soil Reaction (pH)

The hydrogen ion concentration of the soil affects disease development through its effect on the pathogen or the host. The pathogens differ widely in their pH requirement and acidophilic and basophilic species are known in the same genus. Naked cells such as myxamoebae or zoospores are more sensitive to pH changes than are the walled spores. Diseases which are sensitive to pH changes can easily be controlled by shifting the pH of the soil to unfavourable range. The incidence of club root of cabbage (*Plasmodiophora brassicae*) is 100 per cent at pH 5.7 and falls to 20 per cent at pH 6.2. Thus the disease can easily be controlled by adding lime to the soil. The influence of pH in this case is on the swarming myxamoebae. On the other hand several diseases are favoured by alkaline soil reaction. Potato scab (*Streptomyces scabies*) becomes severe between pH 5.7—8.2. It does little damage below 5.4 and can be controlled by growing potatoes in acid soils having natural pH around 5.0 and the disease can be easily avoided by adjusting the pH of the scab-infested soil with sulphur or sulphuric acid. Pathogens like *Fusarium*, which can tolerate wide range of soil pH, can not be overcome by such methods.

Soil reaction affects the plants by conditioning the availability of the nutritional elements. The pH greatly influences the uptake of catonic plant nutrients; in acid soils potassium uptake is reduced; alkaline soils adversely affect the availability of calcium, manganese and boron. The ratio of calcium to potassium in plant tissues is important and the pH can modify it to the detriment of the plant.

Soil reaction may also affect the biological balance between plant parasites and saprophytic soil inhabiting fungi and bacteria.

Soil fertility

Numerous investigations on the influence of soil nutrients upon infection suggest that excessive nitrogen decreases resistance of plants. Potassium in general increases resistance while the effect of phosphorus is variable.

Excessive nitrogen favours infection of cereals to rests, rice to blast disease and of several plants to *Verticillium*. In pine and cotton, excess of nitrogen is known to aggravate the damping off of their seedlings. The root damage due to damping off of the seedlings caused by species of *Pythium*, can be minimised by adding phosphorus to the soil. The roots continue to be susceptible but rapid development of new roots makes up for the loss. Nitrogen prolongs the vegetative phase (which benefits the pathogen) and phosphorus shortens it. Thus a balanced nitrogen : phosphorus ratio can do considerable good to the crop.

Potassium confers resistance. Potassium deficiency increases the severity of many diseases. Formation of galls and outgrowths are favoured by increase in potassium level. *Fusarium* wilt of cotton is severe in potassium deficient soils; the disease can be overcome by increasing the potash level. However, cotton wilt caused by *Verticillium albo-atrum* increases with addition of potassium and thus, behaves differently from *Fusarium* wilt. Cereal rusts, powdery mildews and several other diseases are checked or retarded by additon of potash fertilizers to the soil. Potassium induces mechanical resistance in cereals by forming layers of potassium silicate on the host epidermis. Potassium fertilizers provide resistance by stimulating cell wall thickness, heavier cuticle, strong mechanical tissues. A deficiency in potassium results in excess of *N* and *P*, thin walled cells, decreased amino acid synthesis (due to reduced nitrate reduction) accumulation of carbohydrates (due to their non-utilization in amino acid synthesis) and retarded growth of meristematic tissues which could replace diseased tissues. Potassium regulates the permeability of membranes which is now supposed to be an initial decisive factor in determining susceptibility or resistance of the host. In general, an imbalance of fertilizer that results in excessive growth and succulence, favour the highly specialized parasites like rusts, mildews and bacteria.

Organic Matter

Living and dead plant materials constitute a major component of the soil environment and are the primary source of organic matter.

The organic matter in the soil is an another important regulant of soil fertility, and like pH, this also affects the availability of the nutrient elements. It contributes nutrients like nitrogen and potassium after its decomposition, regulates the soil reaction and provides the humus. It affects the balance of soil micro-organisms by favouring the growth of saprophytic soil inhabitants and, therefore, pathogens with poor 'competitive saprophytic ability' are excluded. The decomposition of the organic matter some time liberates certain toxic substances which affect the growth of plants.

Soil Type and Texture

In India wilt of cotton caused by *Fusarium* occurs only in black heavy soils. Root rot of cotton is unknown in such soils, and it is prevalent in light soil. Both the soil types have alkaline pH but differ in texture. Similar restriction of diseases to heavy and light soils impresses upon the role of soil type and texture in the occurrence of diseases.

REFERENCES

BAWDEN, F.C. and ROBERTS, F.M. 1947, "The influence of light intensity on the susceptibility of plants to certain viruses," *Ann. Appl, Biol*, **34**: 286-296.

BORTELS, H. 1940, *Cb. I. Bact.* II. Abst. 1C2; 129-153.

FISCHER, E.D. and GAUMANN, E. 1929., "Biologie der Pflanzenbenbewohnenden parasitischen Pilze." G. Fischer, Jena.

KASSANIS, B. 1952, "Some effects of high temperature on the susceptibility of plants to infection with virus," *Ann. Appl. Biol.*, 39: 358-369.

PREVOST, B. 1807, *Phytopathol Classics*, No. 6. 1939.

SADASIVAN, T. S., SURYANARAYANAN, S. and RAMAKRISHNAN, L. 1965, "Influence of temperature on rice blast disease." in *The Rice Blast Disease*, 163-171, Johns Hopkins Press, Baltimore, Maryland.

SURYANARAYANAN, S. 1958, "Role of nitrogen in host susceptibility to *Pyricularla oryzae*," *Curr. Sci.*, 27: 447-448.

WILTSHIRE G.H. 1956, "The effect of darkening on the susceptibility of plants to infection with viruses. I. Relation to changes in some organic acids in the French bean," *Ann. Appl. Biol.*, **44**: 233-248. "II Relation to changes in ascorbic acid content of French bean and tobacco," *Ibid.*, **44**: 249-255.

YARWOOD, C.E. 1953, "Pressure effects in fungus and virus infections," *Phytopathology*, **43**: 70-72.

———. 1955, "Deleterious effects of water in plant virus inoculations," *Virology*, **1**: 268-285.

7

NUTRITION OF PLANT PATHOGENS

NUTRITION refers to the food requirement of any living system. So far as phytopathogenic organisms are concerned, although they express a similarity in broader behaviour for their basic nutritional needs, yet, they maintain their individuality for the choice of specific substances. We could best sum up this behaviour as that of fundamental unity with maximum diversity. Microorganisms possess a remarkable capacity to utilize a wide range of organic and inorganic substances. The entire host-parasite relationship actually depends on the capacity of the plant pathogens to derive nutrition from their hosts. This feat is accomplished with the help of various enzymes. As is well known, the green plants form a number of food substances including carbohydrates, amino acids and organic acids etc., which are partly consumed to produce energy and are partly condensed to build up pectic and cellulosic material for the cell walls, fat and starch reserves as well as proteins and nucleoproteins for formation of protoplasm and nucleus, which form the living part of the cell. Like other biological forms, the phytopathogens are also continuously struggling for their survival and they procure their nutritional quota out of the reserve food of the host. The relationship between the two is, therefore, based on nutrition in which the pathogens are the gainers while the green plants, besides being always on the defensive, are the practical losers. Infection impairs the synthesis of carbohydrates and other metabolites, results in increased respiration, which inhibits the synthesis of carbohydrates, cell permeability is disturbed due to which distribution pattern of metabolites is affected, transport of metabolites is impeded on account of diseased conducting vessels. In case of infection to roots even mineral nutrition of the plants is disturbed. The mere initiation of infection reflects that pathogen has got a foot-hold which obviously is not possible till it starts getting its nutrition at the expense of the host.

Most of the nutrients usually exist in nature in crude, complex and mixed state, the efficiency of the phytopathogens is, therefore, judged on basis of their capacity to derive their nutrition from the host, which like any other living system always tries to resist to part with its food reserves. The role of some of the important nutrients is discussed here.

CARBON SOURCES

Living organisms require about 40 elements for their successful and normal growth, among which carbon plays the key role. In its functional role carbon is of very great significance, because fungi and most of the bacteria being chemotrophs obtain all their energy requirements from catabolic degradation of one or the other carbon containing ingredients of cell. Phytopathogenic organisms exhibit carbon heterotrophy and meet their carbon-requirement from various vegetable sources. The basic carbon components of the host are, therefore, important for pathogenic microorganisms.

Utilization of Carbon Sources

Carbohydrates being the major constituents of the host are of great significance. Glucose, fructose, sucrose, maltose and starch are some of the common carbohydrates present in the green plants. Out of these, glucose is the most preferred source for majority of the microbes. This hexose sugar sometimes interferes with the utilization of other carbohydrates. The phenomenon is referred to as "glucose effect". It is believed that repression by glucose is due to its effect upon certain degradative enzymes. Recently the phenomenon has been renamed as catabolic-repression. In this phenomenon, glucose (or any other readily metabolized C-source) causes a reduction in the rate of synthesis of certain degradative enzymes. Such a role of glucose is of special significance for pathogenic fungi and bacteria which usually encounter complex carbohydrates in nature. Two more mechanisms in the control of carbohydrate utilization by glucose have been reported: (*i*) transient repression, and (*ii*) catabolite inhibition, the latter involves the inhibition of enzyme action.

Value of fructose is, however, slightly inferior to that of glucose for a large number of phytopathogens (Agarwal and Agnihotri, 1970; Hasija, 1965, 68, 70; Herrick, 1940; Sarbhoy, 1965; Singh and Tandon, 1966). Mannose and galactose may be slightly or significantly inferior to glucose, while L-sorbose, an isomer of D-fructose with respect to carbon 5, is poor and toxic for majority of fungi. In *Ustilago maydis* genetical explanation has been put forward to explain this behaviour. The sexually compatible monosporidial lines and their solopathogenic diploid derivatives, exhibit varying ability to metabolise sorbose. Pentose sugars (D-xylose, D- and L-arabinose, L-rhamnose and D-ribose) have found only limited application in nutrition of pathogenic organisms. Their values differ considerably. Some fungi distinguish between D- and L-enantiomorphs of arabinose and generally it is the L-isomer which is more conducive to fungal growth. Sugar alcohols, which are obtained by reduction of monosaccharides, also vary in their nutritive value but they are invariably inferior to their corresponding sugars. Some of the sugar alcohols have been obtained from plants or plant parts. Among the sugar alcohols, mannitol is the most acceptable source to fungi (Sarbhoy, 1965; Singh and Tandon, 1967; Tandon and Chandra, 1962; Tandon and Chaturvedi, 1962). Reports also suggest that the organisms, which thrive well on galactose, are able to use dulcitol.

Certain non-fatty organic acids like citrate, succinate and malate, which are the usual products of glucose metabolism are acceptable to pathogenic fungi only to a limited extent. In nature, fungi and bacteria come across a variety of complex oligosaccharides and polysaccharides. Some of the polysaccharides like cellulose, pectins and chitin etc., are the constituents of host cell walls and act as complex physical and chemical barrier for the invading pathogen. These structural polysaccharides are generally complex, insoluble linear polymers quite suitable for their protective role. Enzymes capable of degrading the cell wall materials, are widely distributed both among the pathogenic and nonpathogenic fungi. Role of such enzymes during initiation of infection has been established in several cases. Starch, which is the most common nutritional polysaccharide has comparatively better solubility. The general pattern of utilization of complex carbohydrates involves their initial breakdown to their monosaccharide units through the agency of various specific enzymes, generally of extracellular nature. A large number of fungal pathogens are known to utilize starch, sucrose and maltose, which are commonly present in the green plants. The initial degradation of these substances involves the rupture of glycosidic linkages, which bind the various monosaccharide units. It is well established now that the so-called hydrolytic enzymes act like group transferases and bring about trans-glycosidation. A sugar-enzyme complex is supposed to be formed, which is subsequently transferred to a suitable acceptor. In case of oligosaccharides the basal complex carbohydrate serves as the acceptor which results in formation of higher oligosaccharides of varying chain lengths. Synthesis of transient oligosaccharides is also known in many fruits under pathogenesis (Ghosh et al. 1964, 65). In nature transglycosidation is of special significance, as it enables the pathogen to use only that much of carbohydrate which is required while the surplus is reserved in a "self locked system"

Although any cell constituent containing breakable carbon bond may serve as substrate for respiration, the carbohydrates generally serve as the principal source of energy. Even when the basic substrate for respiration is a complex carbohydrate, it is degraded to monomeric forms to serve as the source of energy.

It is now generally accepted that Embden-Meyerhof (EM) pathway of glycolysis is operative in many fungi, and that over all respiratory mechanism is generally similar to that operative in other organisms involving the tricarboxylic acid (TCA) cycle as well as the terminal oxidation with the help of electron transport or respiratory chain. As many as ten enzymes associated with EM pathway occur widely in fungal mycelia and spores which indicates that fungal cells have potential to use this pathway. The only enzyme specific to EM pathway is phosphofructokinase and its presence in a cell may be a definite indication of this pathway. However, in a few exceptional cases e.g. spores of *Tilletia caries*, this pathway is completely inoperative. It is the Entner-Doudoroff (ED) pathway, which brings about glycolysis. In addition to the above two pathways, the pentose phosphate (PP) pathway is also operative in many fungi. The overall significance of these glycolytic pathways is to furnish the cell with: (*a*) energy, (*b*) precursors for biosynthetic reactions, and (*c*) a system of oxidation-reduction reactions through which these precursors are incorporated. However, there are immense possibilities that a number of new pathways might be operative among fungi and bacteria.

The EM pathway and TCA cycle combindingly produce maximum numbers i.e. 38 ATP molecules through terminal electron transfer and oxidative phosphorylation. The presence of a functional TCA cycle in fungi has long been a subject of controversy and in fact the matter has only recently been settled after some comprehensive studies on different aspects of the problem. On the basis of intensive work carried out during the last fifteen years it is now established that the TCA cycle is functional in a large number of pathogenic fungi, including *Puccinia graminis* and *Fusarium oxysporum*. Much of the nutritional and physiological aspects of fungi have been covered with the molds, there is, however, an urgent need to carry on these studies with pathogenic fungi and bacteria, which encounter a much more hostile environment in search of their food.

NITROGEN SOURCES

Nitrogen nutrition of pathogenic fungi has been well illustrated (Lilly and Barnett, 1951; Cochrane, 1958). Like carbon, nitrogen is also employed both for structural and functional purposes by fungi. Degree of virulence in some fungi has been correlated with the nitrogen contents of the mycelium. Studies dealing with nitrogen metabolism of pathogenic microorganisms are full of gaps and exact mechanisms of nitrogen assimilation into amino acids and proteins and synthesis of specific peptides as well as the metabolism of purines and pyrimidines leave many questions unanswered. Plant pathogenic fungi are known to use nitrates very well (Misra and Mahmood, 1960; Suryanarayanan, 1958; Thind and Randhawa, 1957). The capacity to use these salts is sometimes lost by mutation. Capacity to use nitrate depends on nitrate reductase activity of fungi. Form of nitrate and corresponding change in the pH of the medium during the incubation to a great extent determine the value of nitrogen source. pH of the culture solution is also a decisive factor in determining the nutritional utility of nitrite, which is toxic when the pH is acidic. Since nitrite is an intermediate product in utilization of nitrate, ordinarily it is expected that the organisms, which utilize nitrates well should be able to use nitrites equally well if not better. Value of amino acids, which exist both in free and bound state in the host, differs substantially for pathogenic micro-organisms. This is partly because some of the microbes meet their carbon requirement also from the amino acids. Other factors, like permeability as well as their contamination with vitamins, are also attributed as the possible reasons for such behaviour. Common amino acids present in host in free state generally supported good growth of pathogenic species of *Phyllosticta* (Bilgrami, 1964). Several pathogens (Srivastava and Tandon, 1969; Tandon and Chaturvedi, 1962) respond better towards mixture of amino acids than an individual source. Their availability in mixed state in nature seems to be the obvious reason ror such a behaviour. *Xanthomonas* sp. is incapable of growing in a synthetic medium that contains asparagin as the only source of both carbon and nitrogen.

As mentioned earlier, nitrate reduction in fungi is carried out by nitrate reductase. This enzyme was formerly considered to be only assimilatory in nature but subsequently its dissimilatory nature was also established (Walker and Nicholas, 1961 a, b and c). Most of the experimental work on nitrate reduction in fungi has been carried out with *Neurospora crassa* (Evans and Nason, 1952; Nason and Evans, 1953) or *Aspergillus niger* (Nicholas and Fielding, 1951; Steinberg, 1937). This aspect has been left totally untouched so far as pathogenic fungi are concerned. Nitrate reductase activity in fungi is intimately linked with cytochrome system, particularly C-cytochrome. Reduction of nitrate to ammonia involves a series of steps and each stage is controlled by specific enzymes, which are generally metalo-flavo-proteinaceous in nature.

Ammonia, which is the key intermediate in nitrate metabolism combines with carbon compounds to form organic nitrogenous substances. The details of the mechanism by which this key process proceeds is not known in most of the fungi. Studies on biosynthetic processes in germinating spores of some fungi lead to conclusion that skeletal material for amino acid synthesis is obtained as a result of catabolism of carbohydrate. Uredospores of *Uromyces phaseoli* and *Puccinia helianthi* furnished with ^{14}C-2 acetate produced labelled glutamate, asparatate and alanine within few seconds. There is overwhelming evidence to show that glutamic acid is a central product in amino acid metabolism and many other amino acids are derived from it by known pathways in fungi. The interconnection of glutamic acid with proline, hydroxyproline and ornithine is well established in *Neurospora crassa* and *Torulopsis utilis*. Some recent investigations, however, go to suggest that ^{14}C-fructose supplied to *Candida utilis* resulted in formation of a group of amino acids from which other amino acids were later derived. Such results led to the concept of amino acid families (viz., glutamic acid, aspartic acid, alanine and serine), which exhibited intimate bio-synthetic relationship among them. This concept is, however, at variance with the other alternative because it does not attach primary importance to glutamic acid. Intensive investigations on amino acid metabolism of pathogenic fungi might be both rewarding and revealing because the nutritional environment available in the host is fully saturated with various types of amino acids.

Phosphorous is essential for the growth and development. Its quality and concentration in the medium determines the utilization of nitrogen and synthesis of certain vitamins. Sulphur is also needed both for functional and structural purposes. Activity of many enzymes is dependent on SH group. Some amino acids like cystine and methionine as well as vitamins like biotin and thiamine contain sulphur. Potassium is known to activate enzymes in many fungi.

Among the metallic elements iron, zinc, manganese, molybdenum, magnesium and copper etc. are needed in minute doses for the growth of fungi and bacteria. They play specific or general roles in vitamin synthesis, enzymic activation or other functional and structural purposes.

Phytopathogenic fungi and bacteria may be self sufficient with respect to vitamins or they may be partly or totally deficient for vitamins. Capacity of microorganisms to meet the vitamin requirements is also determined by the nature of the carbon and nitrogen source available to the growing organism. Certain pathogenic bacteria like *Xanthomonas oryzae* do not appear to be vitamin dependent for their multiplication.

The mode of obtaining food from the hosts by nematodes is quite different from those of fungi and bacteria. Usually the nematodes pierce the hard parts of the host by the stylet and at least a part of cell contents is directly sucked in, several glands secrete saliva which flows in the oesophagus and is ejected through the stylet. A number of hydrolysing enzymes like amylase, cellulase, protease and chitinase have been obtained from the saliva of several phytopathogenic nematodes.

During the last three decades substantial contributions have been made on the nutritional aspects of phytopathogenic fungi. Such studies are important from the point of view of determining the host parasite-relationship, which is extremely variable and is to a great extent determined by the nutrition, because the parasites maintain a close and intimate connection with their hosts in order to obtain food. In majority of obligate parasites the concentration of easily assimilable simple sugars and other metabolites increases in the infected cells. Regions around haustoria of many obligate fungi are specially rich in food substances. As a result of infection the physiology of the diseased cells is affected in a manner which goes to the advantage of the parasite and to the disadvantage of the host. Studies on nutrition of phytopathogenic organisms may lead to discovery of new metabolic pathways which are not yet properly understood.

REFERENCES

AGARWAL, G.P. and AGNIHOTRI, R.C., 1970, "Assimilation of sugars by two follicolus fungi, *Pestalotia sapotae* and *P. versicolor*." *Proc. Nat. Acad. Sci. India*, 40(B), IV: 239-249.

BILGRAMI, K.S., 1964, "Studies on growth response of some pathogenic species of *Phyllosticta* in different nitrogen media." Flora, Bd. 154: 1-11.

BROWN, W., 1965, "Toxins and cell-wall dissolving enzymes in relation to plant diseases," *Ann. Rev. Phytopathology*, 3: 1-18.

COCHRANE, V.W., 1958., *Physiology of Fungi*. New York. John Wiley and Sons, pp. 524.

DURAIRAJ, V., 1956, "Growth of *Colletotrichum capsici* in pure culture," *J. Indian Botan. Soc.*, 35: 409-413.

EVANS, H.J. and NASON, A., 1952, "The effect of reduced triphosphopyridine nucleotide on nitrate reduction by purified nitrate reductase," *Arch. Biochem. Biophys.*, 39: 234.

FOSTER, J.W., 1949, *Chemical Activities of Fungi*, New York. Academic Press, pp. 648.

GHOSH, A.K., TANDON, R.N., BILGRAMI K.S. and SRIVASTAVA M.P., 1964, "Studies on fungal diseases of some tropical fruits, II. Post-infection, changes in sugar contents of some fruits," *Phytopathol, Z.* 50: 283-298.

—— –, BHARGAVA S.N. and SRIVASTAVA M.P., 1965, "Formation of a new oligosaccharide in Mango fruits under pathogenesis," *Curr. Sci.*, 34: 465.

HASIJA, S.K., 1965, "Carbon-requirements of *Colletotrichum inamdarii*," *Indian Phytopath.* 18: 21-25.

———, 1968, Assimilation of sugars by *Curvularia pallescens*, *Alternaria citri* and *A. tenuis*," *Indian Phytopath*, 21: 49-61.

———. 1970., "Physiological studies of *Alternaria citri* and *A. tenuis*," *Mycologia*, 62: 289-295.

HERRICK, J.A., 1940, "The carbon and nitrogen metabolism of *Stereum gausapatum*," *Ohio. Jour. Sci.*, 40: 123-129.

KAPOOR, I.J. and TANDON, R.N., 1969, "Formation of a new oligosaccharide in Guava fruits under pathogenesis," *Curr. Sci.*, 38: 444.

LILLY, V.G. and BARNETT, H.L., 1951, *Physiology of the Fungi*, New York, McGraw-Hill Book Co., pp. 464.

MISRA, A.P. and MAHMOOD, M., 1960, "Effect of carbon and nitrogen nutrition on growth and sporulation of *Colletotrichum capsici*," *J. Indian Botan. Soc.*, 39: 314-321.

NASON, A. and EVANS, H.J, 1953, "Triphosphopyridine nucleotide nitrate reductase in *Neurospora*," *J. Biol. Chem.*, 202: 655-73.

NEWBURGH, R.W. and V.H. CHELDELIN, 1958, "Glucose oxidation in mycelia and spores of the wheat smut fungus *Tilletia caries*," *J. Bacteriol*, **76**: 308-311.

NICHOLAS, D.J.D. and FIELDING, A.H., 1951, "The use of *Aspergillus niger* for determining magnesium, zinc, copper and molybdenum available in soils to crop plants," *J. Hort. Sci.*, **26**: 125-147.

PERLMAN, D., 1948, "On the nutrition of *Sclerotium delphinii*," *Am. Jour. Bot.*, **35**: 360-363.

SARBHOY, A.K., 1965, "Some nutritional studies of the genus *Zygorhynchus*," *Canad. Jour. Microbiol*, **11**: 297-308.

SINGH, B.P. and TANDON, R.N., 1966, "Carbon requirements of certain isolates of *Alternaria tenuis* Auct," *Proc. Nat. Acad. Sci India*, **36(B)**: 199-204.

———, 1967, "Carbon and nitrogen requirements of *Alternaria tenuis* Auct," *Proc. Nat. Acad. Sci. India*, **37(B)**: 227-232.

SRIVASTAVA, M.P. and TANDON, R.N., 1969, "Utilization of amino acids in mixture by four isolates of *Botryodiplodia theobromae*," *Proc. Nat. Acad Sci. India*, **39(B)**; 65-72.

STEINBERG, R.A., 1937, "Role of molybdenum in the utilization of ammonium and nitrate nitrogen by *Aspergillus niger*," *J. Agr. Res.*, **55**: 891-902.

SURYANARAYANAN, S., 1958, "Growth factor requirements of *Pyricularia* spp. and *Sclerotium oryzae*." *Proc., Indian Acad. Sci.*, **48**: 154-188.

TANDON, R.N. and CHANDRA, S., 1962, "The utilization of oligosaccharides by some fungi causing leaf spot diseases," *Flora Bd.*, **152**: 241-253.

TANDON, R.N. and CHATURVEDI, C., 1962, "The utilization of amino acids by three imperfect fungi," *Phytopath. Z.*, **45**: 237-242.

TANDON, R.N. and GREWAL, J.S., 1954, "Studies on carbon selectivity, from some organic compounds, and its effect on growth and sporulation of *Alternaria tenuis* strain B causing Core Rot of apples," *Proc. Nat. Acad. Sci.* **24(B)** 149-158.

THIND, K.S and RANDHAWA, H.S., 1957, "Studies on the nutrition of fungi I," *Proc. Nat. Acad. Sci.*, (India) **27**: 39-46.

WALKER, G.C. and NICHOLAS, D. J. D., 1961a, "Hydroxylamine reductase from *Pseudomonas aeruginosa*," *Biochim. Biophys. Acta.*, **49**: 361-368.

———, 1961b, "An iron requirement for a dissimilatory nitrate reductase in *Neurospora crassa*," *Nature*, **189**: 141-142.

———, 1961c, "Nitrite reductase from *Pseudomonas aeruginosa*," *Biochim. Biophys. Acta*, **49**: 350-360.

WOOD, R.K.S., 1967, *Physiological Plant Pathology*. Blackwell Scientific Publications, Oxford, p. 570.

8

ENZYMES AND PLANT DISEASES

THE establishment of the pathogen in the host and the subsequent development of disease involves a chemical interaction between host and the pathogen. The pathogen employs offensive chemical weapons to breach host barriers which are resisted by the host with all its might. Parasitism and disease resistance run parallel and are inseparable. A kind of bio-chemical 'tug of war' ensues between the host and the pathogen in which one tries to outwit the other. Enzymes, toxins, growth regulators, polysaccharides, antibiotics are the important chemical weapons of the pathogen. Viruses, however, on account of their basic nature can not produce any of these substances.

Disintegration of tissues is invariably associated with disease syndromes of various types, though in varying degree. Physiological plant pathology started with De Bary's work on extracellular enzymes of *Sclerotinia.* Studying the mode of invasion by *S. libertiana* De Bary (1886) recorded two types of actions: dissolution of middle lamella and the killing of the cells. The two factors are known as macerating and killing factors. He attributed cell killing to oxalic acid secreted by the pathogen *in vitro.* De Bary is rightly called the father of experimental plant pathology. Marshal Harry Ward of Cambridge (1888) working with a disease of lily caused by *Botrytis cinerea,* obtained cellulase in the culture filtrate.

The role of oxalic acid as a killing factor was again proposed by Smith (1902) for *B. cinerea* which was later disproved by Professor William Brown of London (1915). L.R. Jones (1905) reported the production of pectic enzymes by soft-rot bacteria. Brown conclusively demonstrated (1915, 17) that a pectic enzyme—Protopectinase, was involved in maceration of tissue by *B. cinerea.* He could not separate the killing factor from the macerating factor. Pectic enzymes have received great attention after 1950 and much of the important information about these enzymes is of comparatively recent origin.

COMPOSITION OF THE CELL WALL

Epidermal cells have an array of lamellae made up of different chemical substances. Many species of plants have a layer of wax which is followed by cuticle, impregnated with wax. The cutine gradually decreases with the depth of the epidermis and is replaced by pectin—which occurs as a homogeneous layer in some plant species. Subsequently in the secondary wall of the cell pectin is replaced by cellulose. Cellulose layers have been reported to contain appreciable amounts of protein which increases in quantity as the phospholipid membrane is reached.

Plant waxes. The wax projections present on plant surfaces provide a strong barrier to infection as till date no pathogen is known to secrete enzymes that would degrade wax. Only mechanical entry can enable a pathogen to breach this layer.

Cutin. It is a polymer of 1-4 βN-acetyl glucosamine units (Fig.8.1) A layer of cutin lies beneath the wax layer, the upper part is

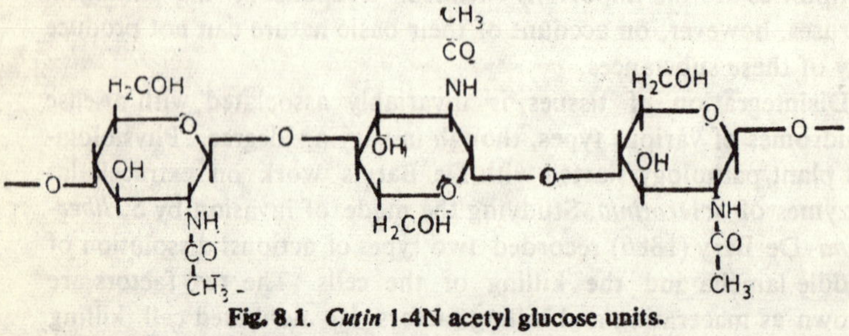

Fig. 8.1. *Cutin* **1-4N acetyl glucose units.**

impregnated with wax: in the lower part it is admixed with pectin and cellulose, until it finally merges with the outer wall of the epidermal cells.

There is close chemical similarity between cutin and suberin. Pathogens lack in the chemical 'know how' to degrade them. Heinen (1960, 63) and Linskens et al. (1965) have reported two cutin degrading enzymes, produced by *Penicillium spinulosum*: (*a*) Cutin esterase, and (*b*) Carboxy cutin esterase (=Carboxy cutin peroxidase). The extent of distribution of cutinases among pathogens and their role in pathogensis is yet to be elucidated. At present cutin is regarded only as mechanically penetrable. A suberin-degrading enzyme has been reported to be produced by *Armillarea mellea* growing on bark of *Brachystegia spiciformis* (Swift, 1965).

Lignin. Lignin forms 15-38 per cent content of woody plants and is next only to cellulose in abundance. It occurs mainly in epidermal cell walls and xylem vessels.

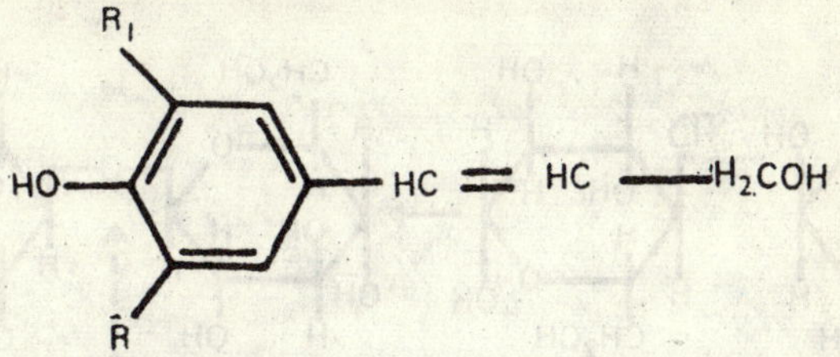

Fig. 8.2. Phenylpropane molecule; R and R_1 either OCH_3 or H.

Lignin is an aromatic substance with molecular weight ranging between 800 to 12,000. It is a polymer of phenyl-propane units (Fig. 8.2), formed from shikmic acid; but lignin-chemists are not yet certain about its chemical structure and composition.

Lignin-decomposing enzymes which although have not been described for pathogenic fungi are known to be produced by wood-rotting fungi. Wood-rotting, 'white rot fungi', numbering about 500 (all belonging to Basidiomycetes) produce an enzyme trans-methylase, which modifies lignin for further degradation by cellulolytic enzymes. Thus these fungi bring about lignin decomposition indirectly by removing methoxyl groups and exposing cellulose microfibrils to enzymatic attack by cellulases.

Cellulose is the insoluble polysaccharide that forms the structural framework of cell walls. It is a polymer of β glucose units joined by β-1-4 glycosidic linkages (Fig. 8.3). About 280 to 800 chains of cellulose molecules are bound laterally by hydrogen bond and other physical forces to give rise to micro-fibrils which in turn aggregate to form the fibrils. The polymer chains at some places in the fibrils show maximum orientation, being very tightly packed, while at other places they are relatively loosely arranged. Places of compact orientation are called crystalline areas while regions with loose chains are known as amorphous areas. A single cellulose chain passes through several crystalline and amorphous areas in the fibril. Crystalline areas are resistant to enzyme attack as high orientation prevents entry of big enzyme molecules.

The cellulose fibrils by intertwining make a porus lattice-like structure which forms the structural framework of the cell wall.

The submicroscopic spaces between the fibrils are filled with pectic materials, lignin or hemicellulose in varying amounts. These spaces form an inter-connecting system extending throughout the cell wall.

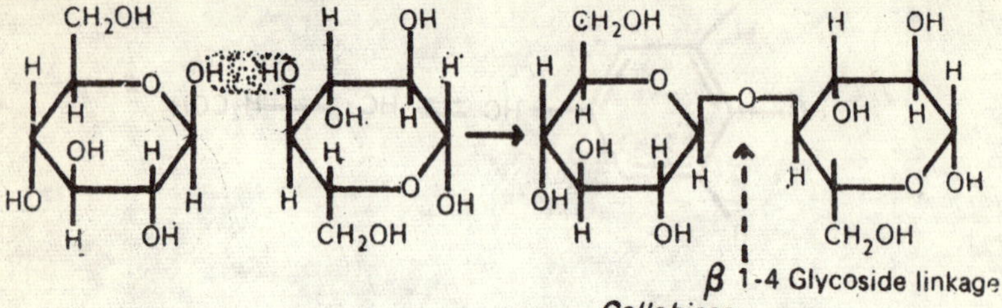

Fig. 8.3 (*a*) Formation of β, 1-4 glycosidic linkage between two β glucose molecules.

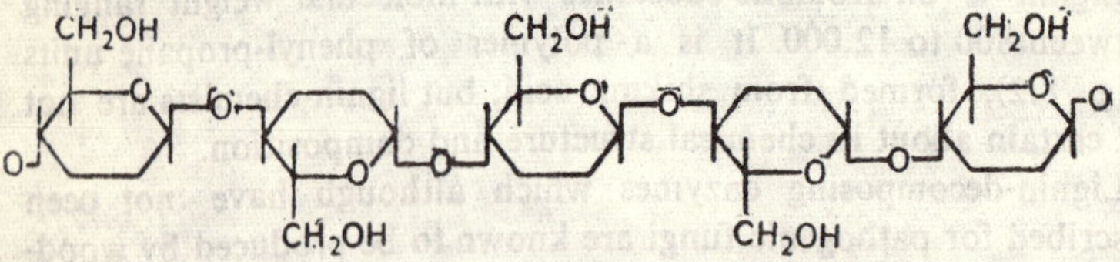

Fig. 8.3 (*b*) Cellulose. β, 1-4 linked glucose units

Cellulolytic Enzymes

Information about cellulolytic enzymes has come mostly from studies on wood decay and fungal deterioration of manufactured cellulose. A number of fungi and bacteria are capable of producing cellulolytic enzymes. Two groups of enzymes, designated as C_1 and C_x are involved in cellulose degradation.

C_1. This enzyme loosens the cellulose fibrils of the crystalline area. C_1 enzyme is an enigma and its *modus operandi* is not known. It is a large molecule: retained in dialysis sacks; precipitated by solvents and excluded by sephadex G_{75}. It is an inducible enzyme. An important feature of this enzyme is that it is not recovered from the culture medium. It may be located on the surface of the mycelium, or adsorbed on the solid culture media. How it loosens the crystalline areas and whether the solubilization of native cellulose is by hydrolytic mechanism or some other process is not yet clear. It is suggested to be a small C_x molecule able to penetrate the cellulose lattice but the data of Mandels and Reese (1964) indicate that they do not differ in size from C_x. That it could be a hydrogen bondage is suggested by Siu (1954).

The best substrate for assaying C_1 activity is cotton fiber, which is unaffected by C_x enzymes.

C_x. The C_x group of enzymes hydrolyse the cleavage of β, 1-4 glycoside linkages but a prior modifying action of C_1 enzyme is essential for their activity. They can not attack native cellulose. Several enzymes have been characterized in the C_x group of enzymes, which attack the glycoside bonds randomly or terminally and accordingly called C_x endo or C_x exoenzymes respectively. The endo-C_x-enzymes fragment the long cellulose chains into soluble, low molecular weight smaller pieces of 6 or lesser glucose units. The C_x-exoenzymes by acting on terminal bonds further break these smaller chains into cellobiose which may further be hydrolyzed by another enzyme cellobiase to liberate glucose units

C_x activity is assayed by measuring—loss of viscosity or the production of reducing sugars from carboxymethyl cellulose. C enzyme has no effect on it.

There are two theories to explain the mechanism of cellulose degradation: the unienzyme theory (Whittaker 1953, 1957; Aitkin et al, 1950) and the multienzyme theory (Reese, 1959).

The unienzyme theory envisages complete degradation of cellulose into glucose units by a single enzyme. Whittaker (1957) obtained an electrophoretically pure enzyme from *Myrothecium verrucaria*, which degraded native cellulose to glucose. Aitken et al, (1956) while agreeing with the above theory, implicate the involvement of cellobiase enzyme to release glucose from the cellobiose molecules produced by a single cellulase enzyme. This occurs in certain wood rotting fungi.

Thus the cellulose degradation can be represented as given below:

$$\text{Native cellulose} \xrightarrow{\text{Cellulase}} \text{Cellobiose} \xrightarrow{\text{Cellobiase}} \text{Glucose.}$$

The multienzyme theory explains cellulose degradation in a series of steps—as represented below:

$$\text{Native cellulose} \xrightarrow{C_1} \text{Reactive cellulose} \xrightarrow[\text{(hydrolytic) cleavage}]{C_x \text{ Endoenzymes}} \text{Shorter cellulose chains.}$$

$$\xrightarrow[C_x \text{ exoenzymes (hydrolytic)}]{} \text{Cellobiose} \xrightarrow[\text{(hydrolytic)}]{\text{Cellobiase}} \text{Glucose}$$

Both mechanisms may be operative under different situations with different organisms. Cellulolytic enzymes are adaptive enzymes with a trace always available to the organisms to act as the starter.

Hemicelluloses. These are present in the spaces between the cellulose microfibrils of the cell wall. Chemically they are complex and heterogeneous polysaccharides formed by linear union of smaller homogeneous sugar chains like, xylans, arbans, mannans, and galactans (Fig. 8.4). These names are based on the composition of the shorter component chains. Xylans are made of xylose units; Arabans by arabinose; mannans by mannose; galactan by galactose and so on. Branched chains, though rare, do exist.

Arabinose Arabinose Arabinose Arabinose	Arban α, 1-3 bond
Xylose Xylose Xylose Xylose	Xylan α, 1-3 bond
Galactose Galactose Galactose Galactose	Galactan β, 1-4 bond
Mannose Mannose Mannose Mannose	Mannan β, 1-4 bond

Fig. 8.4. A portion of hemicellulose molecules.

Several bacterial and fungal pathogens, as well as saprophytic and wood rotting fungi, produce hemicellulase enzymes which degrade the polymer by hydrolytic cleavage of the bonds. Since the polymer has several kinds of bonds, several enzymes are needed to cope with the variety of bonds.

Pectic Substances and Pectolytic Enzymes

Of all the enzymes associated with plant diseases, pectic enzymes are the best known and are implicated almost in a routine manner in the elucidation of host-parasite interactions.

Pectic substances form the middle lamella exclusively. They form a layer below the cuticle in the epidermal cells and fill the intermicellar spaces of the cellulose fibrils in the primary cell wall. The different pectic materials and the enzymes concerned with their degradation are detailed below.

(1) **Pectic acid** is a linear polymer of D-galacturonic acid residues joined laterally by α 1-4 glycosidic linkages (Fig. 8.5). It is water soluble and easily precipitated by calcium and magnesium ions to form insoluble salts. It occurs mainly as calcium and magnesium pectates, chiefly in the middle lamella.

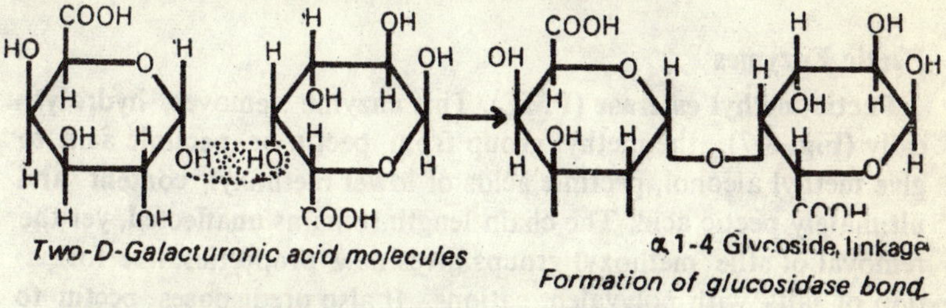

Fig. 8.5 (*a*) **Formation of α 1-4 glucoside bond.**

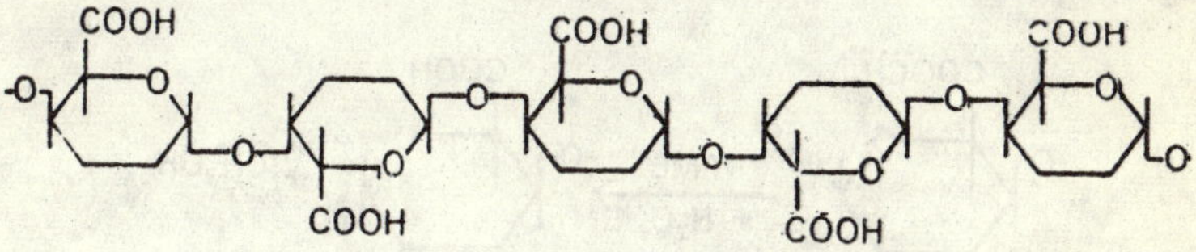

Fig. 8.5 (*b*) **A portion of pectic acid chain.**

(2) **Pectin** is the methyl ester of pectic acid (Fig. 8.6). The esterification may be to different degree. If 75 per cent or more carboxyl groups are esterified the polymer is referred to as pectin; chains with lesser methoxyl content are designated as pectinic acid.

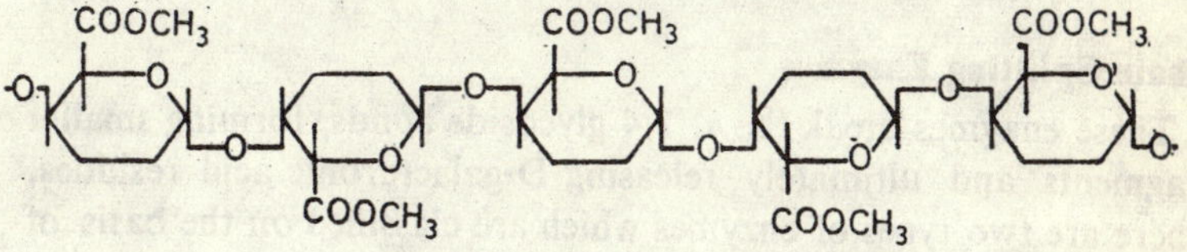

Fig. 8.6. A portion of pectin chain.

Pectin is water soluble and neutral, because most acid groups are esterified and can not be precipitated by polyvalent cations.

(3) **Protopectin.** The insoluble native pectic material as occurs in the primary cell wall is called protopectin. It consists of very long pectin chains—consisting of over 1,000-2,000 units. Thus it differs

from pectin only in chain size which consists of only 200 units in pectin.

The various types of pectic chains may be cross linked among themselves or with those of other types or with cellulose chains. The association with other sugars is also demonstrated.

Pectic Enzymes

Pectin methyl esterase (PME). This enzyme removes, hydrolytically (Fig. 8.7), the methyl group from pectin or pectinic acid to give methyl alcohol, pectinic acids of lower methoxyl content and ultimately pectic acid. The chain length remains unaffected, yet the removal of the methoxyl groups gives new properties, like formation of salts with polyvalent cations. It also predisposes pectin to cleavage by chain-splitting enzymes which otherwise would not degrade pectin.

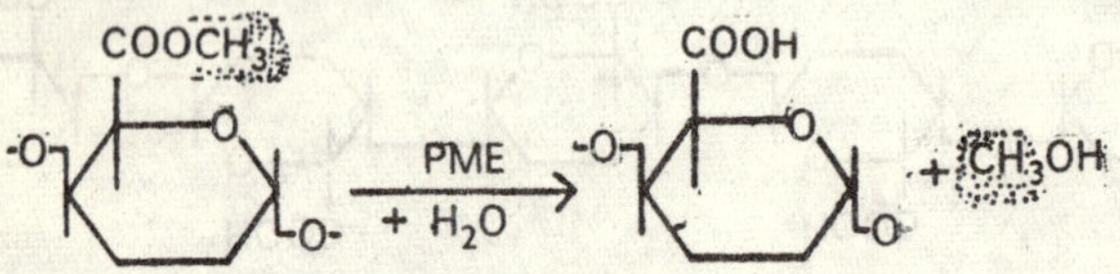

Fig. 8.7. PME action—removal of methyl groups from pectin.

PME is widely distributed among higher plants, but to a much lesser extent in the micro-organisms. The enzyme of microbial origin has lower pH optima as compared to those of higher plants.

PME. PME is assayed by measuring the decreasein pH of the enzyme-substrate reaction mixture, by titration methods.

Chain Splitting Enzymes

These enzymes break the α, 1-4 glycoside bonds, forming smaller fragments and ultimately releasing D-galacturonic acid residues. There are two types of enzymes which are classified on the basis of hydrolytic or trans-eliminative split of the bond.

Glycosidases [=Polygalacturonase and Polymethyl galacturonase (PG, PMG)]. These enzymes break the glycoside bonds hydrolytically. Their nomenclature is based on the preference for the substrate (pectin or pectic acid) and site of splitting (random or terminal) If the enzyme shows greater activity on pectic acid, the enzyme is called Polygalacturonase (PG); it is designated as polymethyl galacturonase (PMG), if the preferred substrate is pectin. The two nzymes are further characterized as endo or exoenzymes; those attacking terminal bonds are called exo-PG (or Exo-PMG) and those attacking linkages at random are designated as endo-PG (or endo-PMG). The endo-PG or PMG enzymes are more efficient enzymes as they break the integrity and chain length. The endo-PG or PMG enzymes thus bring about a loss in viscosity of pectic substrates and are easily assayed viscometrically. Viscosity loss after different periods of enzyme action are depicted in Fig. 8.8. Enzyme activity is calculated by the formula:

$$\frac{1}{\text{time for 50 per cent loss in viscocity}} \times 1000.$$

The exoenzymes are assayed by measuring the release of reducing sugar by enzyme action.

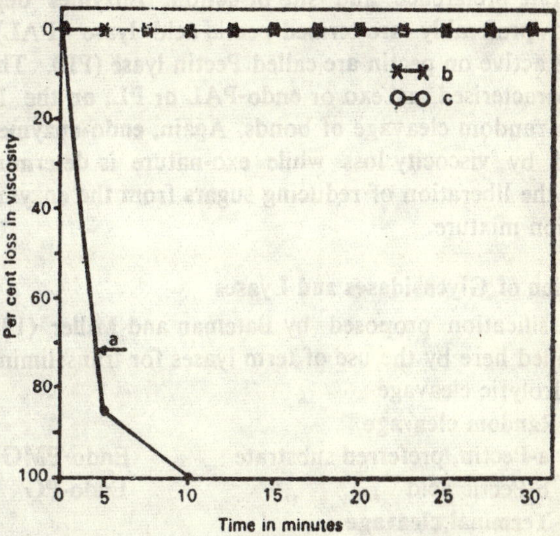

Fig. 8.8. Showing rate of loss in viscosity of NaPP caused by (*a*) Culture filtrate of *Curvularia lunata* grown on Richard's pectin medium. (*b*) Culture filtrate containing sucrose. (*c*) Boiled culture filtrate.

Transeliminases or lyases. This mode of cleavage was first reported by Albersheim et al. in 1960 and has since then been reported with several pathogens.

Transeliminative cleavage of the bond involves removal of the proton of C_5 of one residue—resulting in the formation of an unsaturated bond between C_4 and C_5 (Fig. 8.9). The unsaturated product absorbs maximally at 230/235 nm and at 547/550 nm on treatment with thiobarbutaric acid reagent.

The nomenclature of these enzymes has been changed from transeliminases to lyases, on the suggestions of the commission on enzymes. However, both names are in use.

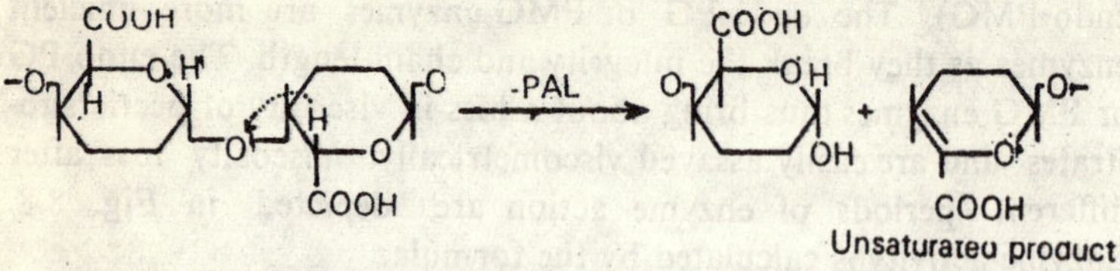

Fig. 8.9. *Trans*-eliminative cleavage of glycoside bond of a portion of pectic acid by a pectic lyase.

The lyase enzymes are also classified further, on the basis of their substrate preference and site of action. Enzymes degrading pectic acid preferably are termed pectic acid lyase (PAL), while those more active on pectin are called Pectin lyase (PL). These are further characterised as exo or endo-PAL or PL, on the basis of terminal or random cleavage of bonds. Again, endo-enzyme action is detected by viscocity loss while exo-nature is determined by measuring the liberation of reducing sugars from the enzyme-substrate reaction mixture.

Classification of Glycosidases and Lyases

The classification proposed by Bateman and Miller (1966) has been modified here by the use of term lyases for transeliminases.

(A) Hydrolytic cleavage

1. **Random cleavage**

a-Pectin, preferred substrate	Endo-PMG
b-Pectic acid ,, ,,	Endo-PG

2. **Terminal cleavage**

a-Pectin preferred substrate	Exo-PMG
b-Pectic acid ,, ,,	Exo-PG

(B) Transeliminative bond cleavage

1. **Random cleavage**

a-Pectin preferred substrate	Endo-PL
b-Pectic acid ,, ,,	Endo-PAL.

2. **Terminal cleavage**

a-Pectin preferred substrate	Exo-PL
b-Pectic acid ,, ,,	Exo-PAL

Detection of the Mode of Cleavage

The hydrolytic or trans-eliminative cleavage can be known spectrophotometrically with the help of a colour reaction between the products of enzyme action and thiobarbutaric acid. A peak in the absorption spectrum of the enzyme-substrate reaction mixture, at 510 nm indicates hydrolytic cleavage while a peak at 547 or 550 nm suggests trans-eliminative cleavage (Fig. 8.10).

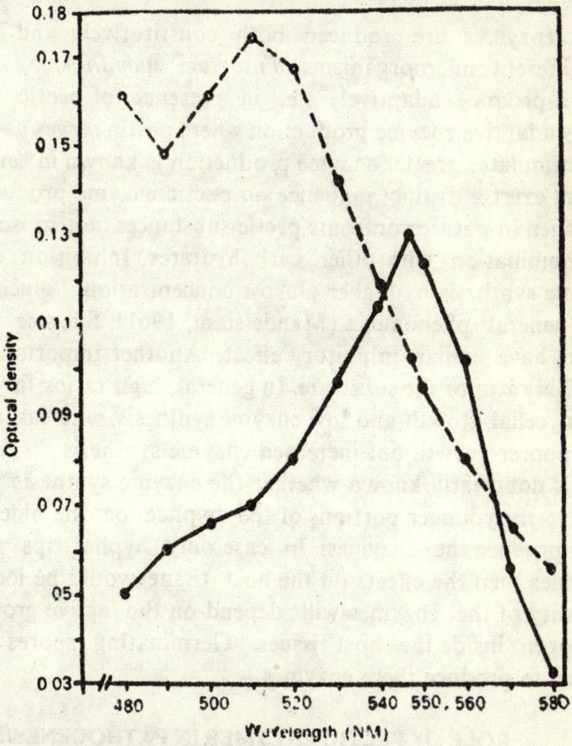

Fig 8 10 Showing typical absorption spectra of (*a* hydrolytic, and (*b*) *trans*-eliminative cleavage [*u* ----- *b* ——]

Wall Modifying Enzyme (WME)

Karr and Albersheim in 1970 reported a new pectic enzyme from pectionol R-10, a commercial preparation from *Aspergillus niger* and called it wall modifying enzyme (WME). As is evident from the name itself, the enzyme modifies the native pectic material for subsequent degradation by other enzymes in the mixture. Removal of this factor results in failure of other enzyme action.

Protopectinase. The term is used for crude enzyme preparations that attack parent pectic material, the protopectin. It yields pectin chains of shorter length. It can be a type of PG or PMG acting in concert with PME. The concept of this enzyme is hypothetical and its use is discouraged.

DYNAMICS OF PECTIC ENZYME PRODUCTION

Pectic enzymes are produced both constitutively and adaptively by different microorganisms. *Fusarium moniliforme*, *Sclerotium rolfsii* produce adaptively i.e. in presence of pectic substrates. Partly adaptive enzyme production where pectin serves as an inducer and stimulates greater enzyme produetion is known in several cases. Sugars exert a distinct influence on pectic enzyme production. The pathogen in nature confronts pectic substances not in isolation but in combination with other carbohydrates. Inhibition of induced enzyme synthesis by higher glucose concentration "glucose effect" is a general phenomena (Mandelstam, 1961). Sucrose and other sugars have similar inhibitory effect. Another important factor is the C/N ratio of the substrate. In general, high ratios favour increased mycelial growth and low enzyme synthesis, whereas lower ratios give poorer growth but increased enzyme synthesis.

It is not clearly known whether the enzyme synthesis is confined only to the younger portions of the hyphae or the older portions also produce the enzymes. In case only hyphal tips secrete the enzymes then the effects on the host tissue would be localized and quantity of the enzymes will depend on the active growth of the pathogen inside the host tissues. Germinating spores have been shown to produce these enzymes.

ROLE OF PECTIC ENZYMES IN PATHOGENESIS

It is difficult to assess the role of these enzymes in disease. There is no correlation between the abuity of micro-organisms to produce pectic enzymes and their parasitic vigour. Many non-pathogenic organisms also secrete pectic enzymes. Therefore, capability of enzyme production, though an important attribute of the pathogen, is not all that is required for pathogenecity. It is only one of the several properties that make an organism pathogenic. What makes an organism pathogenic is a difficult question to answer on basis of the present knowledge. Biochemists though agree on the chemical basis of pathogenicity, are unable to pitch pathogenicity to a single factor. A few cases are known where pathogenesis is determined by a single factor but such cases are rather very scant. Nevertheless, they are important sign posts in our quest of chemical basis of pathogenicity. A more elaborate picture about th' aspect will be presented when we take up toxins.

In certain diseases pectic enzymes have a decisive role. In diseases involving rotting of tissues, these enzymes play a key role. They alone bring about maceration. Pathogens of higher order that do not injure tissues e.g. rusts, rarely employ pectic enzymes in their mechanism of pathogenesis.

In order to evaluate the role of pectic enzymes at least the following two conditions must be met, i.e. production of the enzymes *in vivo* and a conducive environment in the tissue for their activity.

The *in vivo* production is known by three lines of evidences.

(1) Higher activity of these enzymes in diseased tissues.

(2) Histological and histochemical examination of infected tissues for their activity.

(3) Chemical changes in pectic substances or detection of their break-down products.

Even the *in vivo* occurrence of the enzymes does not suggest that they are active too. It is of course a significant pointer to probe further. On the contrary, failure in extraction of an enzyme from diseased tissue does not necessarily mean its absence: it could be in non-extractable form as can happen if it is bound to some insoluble component or inactivated by host tissues.

Several factors inside the host tissues stimulate or inhibit these enzymes and thus influence the extent of damage. Phenolic compounds and their oxidation products are potent inhibitors of enzyme production. The pathogen itself unwittingly triggers these deactivating mechanisms. Indole-acetic-acid regulates the PME activity in plant cell wall. Calcium also affects pectic enzyme activity. A high calcium content in plant tissues prevents maceration by PG but is stimulatory to lyase enzymes. A high calcium content in bean tissues prevents maceration by *Rhizoctonia solani*, which produces PG dominantly and shows weak transeliminase activity. *Sclerotium rolfsii* which produces oxalic acid in addition to PG removes calcium from calcium pectate as calcium oxalate and then hydrolyses the pectic acid chain easily by PG activity,

Bateman (1968). *R. Solani* lacks this mechanism to overcome calcium-resistance.

Hydration of tissues is another important factor. Gregg (1952) noted that raising the water content of potato tubers enhanced the susceptibility of the tissues. The amount of sugars and more so the C/N ratio, as already mentioned, contributes greatly in the regulation of the enzyme production.

Thus the ability to produce pectic enzymes though important is not enough, it must be superimposed by the ability of the pathogen to grow and metabolize in the host tissues.

MACERATION

The separation of cells along the line of the middle lamella is called maceration. The term "cell separation" is more appropriate. The tissue loses coherence but the individual cells retain their identity for some time.

Middle lamella is the zone between primary walls of adjacent cells half of which belongs to the wall of one cell and half to the next. It is made up of calcium and magnesium pectates; which is a highly hydrated firm gel and shows no structure in the electron microscope (see Wood, 1972).

The enzymes involved in maceration are exclusively the chain splitting pectic enzymes. These enzymes can bring about cell separation *in toto*. Cellulolytic or proteolytic enzymes have conclusively been eliminated from any role. Preparations with no cellulolytic activity cause maceration. Maceration by chain splitting pectic enzymes is well documented in the following cases.

Enzyme	*Produced by*
Polygalacturonase (PG)	*Rhizoctonia solani*
	Corticium patricola
	Rhizopus tritici
	R. stolonifer
Pectic acid lyase (PAL)	*Erwinia aroideae*
	E. atroseptica
Pectin lyase PL)	*Sclerotinia fructigina*

Other Enzymes

Proteolytic enzymes. Pathogens degrade proteins hydrolytically.

The production of proteolytic enzymes by plant pathogens and their role in disease is little known. Such enzymes would of course help the pathogens in penetration and degradation of the plasma membrane. Pectic enzymes have no target in the protoplast and thus can not do much in killing the cells.

Phosphatidase. Tseng and Bateman (1968) showed that *Sclerotium rolfsii* produces a phosphatidase which attacks the lecithin molecules of cell membranes and brings about disruption of the membrane system.

REFERENCES

AITKIN, R. A., EDDY, B. P., INGRAM, M., and WEURMAN, C., 1956, "The action of culture filtrates of the fungus *Myrothecium verrucaria* on β-glucosans," *Biochem. J*, 64: 63-70.

ALBERSHEIM, P., NEUKOM, and DEUEL, H., 1960, "iiber die Bilding von imgesattingten Abbauprodukten durch ein pekt inabbauendes Enzym," *Helvet. Chim. Acta*, 43: 1422-1426.

BATEMAN, D.F. and MILLER, R.L, 1966, "Pectic enzymes in tissue degradation," *Ann. Rev. Phytopathol*, 4: 119-146.

BATEMAN, D F., 1968, "Alteration of cell wall components during pathogenesis by *Rhizoctonia solani*," in *Dynamic Role of molecular constituents in Plant-parasite Interaction*, pp 58-79, C.J. Mirocha and I, Uritani (eds.) St Paul Minnesota. The Bruce Publishing Co.

BROWN, W., 1915, "Studies in the physiology of parasitism I. The action of *Botrytis cinerea*," *Ann. Botany* (London), 29: 318-348.

———, 1917, "IV—on the distribution of cytase in cultures of *Botrytis cinereo*," *Ann. Bot.*, 31: 489-498.

DE BARY, A., 1886, "iiber einige sclerokenien und Sclerotienkrankheiten.' *Botan. Z.*, 44: 377-474.

GREGG, M., 1952, "Studies in the physiology of parasitism XVII. Enzyme secretion by strains of *Bacterium caratovorum* and other pathogens in relation to parasitic vigour, *Ann. Bot.*: London N.S. 16: 235-50.

HEINEN, W., 1960, "iiber den enzymatischen Cutinabbau: I. Nachweiss eines Cutinase systems," *Acta Bot.*, Neerl, 9: 167-190.

———, 1963, "iiber den enzymatichen cutinabbau IV. Irennung der cutinase non oxidative Begleitfermenten," *Enzymologia*, 25: 281.

JONES, L.R., 1905, "The cytolytic enzymes produced by *Bacillus caratovorus* and certain other soft rot bacteria *Zentr. Bakteriol., Parasitenk. Abt.* II 14: 257-272.

KARR, A.L. and ALBERSHEIM, P., 1970, *Pl. Physiol*, 46: 69-80.

LINSKENS, H. F., HEINEN, W. and STOFFERS, A L., 1965, "Cuticula of leaves and the residue problem in F.A. Guneter (ed.). *Residuc Reviews*, 8: 136-178.

MANDELSTAM, J. 1961, "Induction and repression β-galactosidase, in non-growing *Escherichia coli*" *Biochem. J.*, 79: 489-496.

MANDELS, M. and REESE, E.T. 1964, "Fungal cellulases and the microbial decomposition of cellulosic fabrics, in *Developments in Industrial Microbiology Soc. of industrial microbiol.* Publ.

REESE, E T., 1959, "Cellulosic decomposition: fungi," in *Marine Boring and Fouling Organisms*. pp. 265-291, D.L. Ray, Friday Harbor (eds.), Symposia Univ. of Washington Press Seattle.

SIU, R G.H., 1954, "Problems and speculations on the decomposition of cellulose by fungi," Trans. N.Y. *Acad. Sci.* (2) 17: 37-44.

SMITH, R.E., 1902, "The parasitism of *Botrytis cinerea*," *Botan. Gaz*, 33 421-436.

SWIFT, M.J., 1965, "Loss of suberin from bark tissue rotted by *Armillarea mellea*," *Nature*, 207: 437-437.

WARD, H.M. 1888, "A lily disease," *Ann. Bot.* 2: 319-382.

WHITTAKER, D R. 1953, "Purification of *Myrothecium verrucaria* cellulase," *Arch. Biochem. Biophys*, 43: 253-268

———, 1957, "The mechanism of degradation of cellulose by *Myrothecium verrucaria*, cellulase," *Can. J. Biochem. and Physiol*, 35: 733-742.

WOOD, R.K.S., 1972, "The killing of plant cells by soft rot parasites, in *Phytotoxin in Plant Disease*" Wood, et al. (eds.) Academic Press, London, New York.

9

TOXINS AND PLANT DISEASES

THE idea that a toxin plays a role in death of cells was first enunciated by De Bary (1886) who claimed that oxalic acid secretedby *Sclerotinia* sp., was responsible for killing of cells. Smith (1902) supported this by implicating oxalic acid secreted by *Botrytis cinerea*, in causing death of living cells. However, Brown (1915), working with the same fungus, subsequently disproved any such role to oxalic acid. He could not separate the killing factor from the macerating factor. Since then the elucidation of the role of toxins in plant disease has been a challenging subject of study and speculation. Gäumann (1954) gave a general conclusion that "microorganisms are pathogenic only if they are toxigenic".

Information about the role of toxins in diseases caused by insects, mites, nematodes and parasitic phanerogams is negligible. It is unlikely that viruses, because of their very nature, would produce toxins in the strict sense of the word. Toxins are readily produced by fungal and bacterial pathogens in a variety of environments.

WHAT IS A TOXIN?

Toxin in human pathology is a protein of microbial origin, has antigenic properties and has an antitoxin. In addition, administration of small doses of the toxin results in immunity in the host. Real poisons lack the above attributes.

In plant pathology, the concept of toxin is not limited to one group of compounds. All kinds of substances, produced by the pathogen, which are capable of reproducing symptoms similar to that found in natural infections, are toxins. These are simple molecules with low molecular weight, in contrast to the high molecular weight toxins of human pathogens. Because of the small size, they are extremely mobile and easily reach the subcellular levels of the host.

Toxins are different from enzymes in the fact that they do not attack the structural integrity of the tissues but affect the metabolism in a subtle manner. It is the subtility of action that differentiates toxins from enzymes (Buddenhagen and Kelman, 1964).

Toxins act directly on the protoplasts of the cells and substances having same end effect but an indirect method of action, are not toxins. Thus high molecular weight polysaccharides, secreted by wilt inducing bacteria, disrupt the flow of fluids in the xylem vessels, which, though may result in death of plants, are not toxins as their action is not direct on the protoplast. Similarly pectolytic and cellulolytic enzymes, which do not kill protoplasts, are not oxins. In fact, cellulase enzymes are used to produce protoplasts. However, some other enzymes are toxins. The toxin of *Clostridium welchii*, lecithinase, even though an enzyme, is a toxin because of its disruption of cell membranes. Such enzymes are toxins. Lumsden and Bateman (1968) have reported a phosphatide degrading enzyme—phosphatidase, associated with pathogenesis in *Phaseolus vulgaris* infected with *Thielaviopsis basicola*. In general, enzymes are equal to "aggressins" of mammalian pathology, which are defined as metabolites that enhance virulence of pathogens but are not directly toxic to the host. Degradative enzymes, e.g., pectolytic enzymes certainly promote invasion of the host by breaching host barriers. Graniti (1970) proposed the term "phytoaggressins" for these enzymes.

Classification of Toxins

The terminology is in a confused state. The following terms are frequently used in phytopathological literature.

(1) **Phytotoxin.** Any compound produced by a microorganism which is toxic to plants, is a phytotoxin. Phytotoxins are non-specific, incite few or none of the symptoms incited by the pathogen and, as happens in most cases, show no relation between toxin production and pathogenicity.

Lycomarasmin and alternaric acid are examples of phytotoxins.

(2) **Vivotoxin.** Dimond and Waggoner (1953), who coined the term defined vivotoxin "as a substance produced in the infected host by the pathogen and/or its host, which functions in the production of disease but is not itself the initial inciting agent of disease" They laid down certain criteria which a substance must fulfil before being called a vivotoxin. Their requisites of a vivotoxin are:

(*i*) it must be isolated from diseased plant but should not be present in the healthy host;

(*ii*) it must be characterized chemically, and

(*iii*) when introduced in pure form into healthy host it must produce the symptoms of disease or a portion of the syndrome. Fusaric acid, and piricularin are typical examples of vivotoxins.

Pathotoxin

Pathotoxins play a causal role in disease production and produce, in susceptible plants, symptoms characteristic of the disease. Following criteria have been laid down for a chemical to be called a pathotoxin (Wheeler and Luke, 1963):

(*i*) the toxin, applied at concentrations which could be reasonably expected in or around the diseased plant, produces in a susceptible host all the symptoms characteristic of the disease,

(*ii*) the pathogen and the toxin exhibit similar host specificity,

(*iii*) the ability of the pathogen to produce the toxin varies directly with its ability to cause disease, and

(*iv*) a single toxin is involved.

Wheeler and Luke were conscious of the difficulties in fulfilling all the criteria—specially the last step which "required to meet the demands of *Occum's* razor".

Thus pathotoxins are 'chemopathogens' of biological origin which by substituting the pathogen provide tools for investigating the nature and course of pathogenesis.

Victorin, wild fire toxin and toxin produced by *Periconia circinata* are pathotoxins.

Host Specific Toxin

Pringle and Scheffer (1964) coined and defined 'host specific toxin' as a metabolic product of a pathogenic microorganism which is toxic only to its host. Pathotoxins of Wheeler and Luke are host specific toxins of Pringle and Scheffer.

Victorin (produced by *Helminthosporium victoriae*), toxins of *Periconia circinata*, and *Helminthosporium carbonum* and some species of *Alternaria* e.g., *A. kekuchiana* are host specific toxins.

Scheffer and Pringle (1967)—divided toxins into two broad categories.

(1) **Primary determinants** (toxins) are essential for pathogenicity and are largely the host specific toxins but may also include some non-specific toxic substances.

(2) **Secondary determinants** (toxins) are not required for pathogenicity but control the virulence. Scheffer and Pringle treat wildfire toxin, secreted by *Pseudomonas tabaci*, as a secondary determinant. Mutants of the bacterium that do not produce the toxin can still be pathogenic though they fail to produce the halo, typical of the toxin producing strains.

Graniti (1970) has advocated the use of only two terms:

(*i*) **Phytotoxins** ***sensu stricto***, for microbial products that interfere with or alter the metabolism of plant cells. These are produced in the host, on its surface or even outside the host, for example in soil.

(*ii*) **Vivotoxins**. Phytotoxins whose role in pathogenicity has been demonstrated. "Host specific toxins" and pathotoxins are only host specific vivotoxins.

Victorin

The toxin victorin is produced by ***Helminthosporium victoriae***, the causal organism of "Victoria blight" disease of Oats. The fungus is soil and seed borne and causes basal stem and root necrosis, in addition to the conspicuous red striping of the leaves. Litzenberger (1949) demonstrated that the culture filtrate of the fungus contained a toxin that could reproduce the 'leaf striping', typical of the disease. Wheeler and Luke (1954) gave the name "Victorin" to the toxin and called it (1963) a pathotoxin. The toxin has not been detected *in vivo* but extracts from diseased leaves do show the presence of the toxin.

Chemical Nature

Active preparations are obtained from culture filtrates of the pathogen. A seedling root growth bioassay is used to assay the relative toxicity. The toxin is a low-molecular weight peptide (mol. wt. 867), linked to a base—a nitrogen containing sesquiterpene moiety, called "Victoxinine". The linkage, being unstable, is broken by mild hydrolysis.

Role of Victorin in Disease Development

The following five evidences suggest victorin's key role in development of the disease:

(1) Only those strains that produce the toxin are pathogenic, non-producers are non-pathogenic.

(2) Plants that are susceptible to the fungus are susceptible to

the toxiu also. Resistant host plants resist the toxin action. Cultivars intermediate in scale of resistance and susceptibility show intermediate sensitivity to toxin also.

(3) Spores contain the toxin which is released during germination. A trace of toxin is shown to be present at the point of contact of the germinating spore with the host. Further, if a drop of toxin is added to the infection drop the non-pathogenic strain behaves like a pathogenic strain. This indicates involvement of the toxin in initial stages of infection.

(4) The toxin shows extremely high biological activity. In bioassay tests of toxicity, ·0002 μg/ml of most active preparations gave complete inhibition of susceptible seedling root growth. Resistant roots tolerated at least 400,000 times higher concentrations with no effect on growth.

(5) The toxin causes the same biochemical changes in the host as are caused by the pathogen.

Mode of Action

The primary effect of the toxin is on the permeability of the host plasma membrane. Samaddar and Scheffer (1968) detected leakage of electrolytes within 2 minutes of toxin treatment of host tissue (Fig 9.1). The slope of the curve suggested that the effect was instantaneous.

Another strong evidence suggesting plasma membrane as the lesion site came from observations of Scheffer and Sammadar (1970) and Samaddar and Scheffer (1971). They observed that plasma membranes of isolated protoplasts break after a brief exposure to the toxin. Table 1 shows the differential toxicity to hosts and non-host plants.

TABLE 1. EFFECT OF VICTORIN (0.16 μG/ML) ON SURVIVAL OF ISOLATED PROTOPLASTS (AFTER SCHEFFER AND SAMADDAR, 1970)

	Percent survival after 1 hour	
	Control	*Treated*
Susceptible Oats	93	0
Resist. Oats	89	92
Corn	96	94
Sorghum	92	95

Other responses like increased rate of respiration, decreased up take and utilization of phosphates and amino acids, decrease in amylase activity of aleurone layer in oat seeds and increased dark carbon dioxide fixation are secondary effects. The toxin has no effect on isolated mitochondria (Black and Wheeler, 1966) or permeability: the oxygen and phosphorus uptake (P O ratio) remain unaffected. Wheeler and Black (1962) suggested that salts, organic acids and other materials, leaking from the vacuole due to damaged permeability of the protoplast, stimulate mitochondria to increased activity by processes similar to "salt respiration"[1] described by

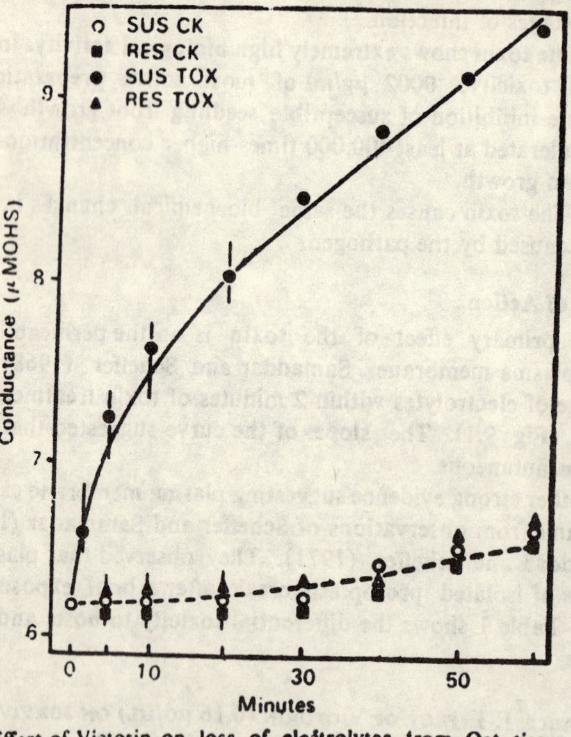

Fig. 9.1. Effect of Victorin on loss of electrolytes from Oat tissue (suc—susceptible tissue: Res=resistant; CK=control; Tox=Toxin treated; 'minutes' refers to time after exposure of tissue to toxin).

Lundegardh (1955) and Bennet Clark and Bexon (1943).

Thus the primary and most initial event of toxin action is damage to the membranes which brings in its trail several other metabolic derangements as secondary effects.

[1]Lundegardh and Burstrom (1933) observed that the rate of respiration incr ases when a plant or tissue is transferred from water to a salt solution. The increased respiration is called salt respiration

Resistant varieties of oat inactivate the toxin. Thus while 40 per cent toxicity was recovered from the susceptible tissues exposed to victorin solution, no toxicity could be recovered from the resistant tissues. The resistant cells appear to lack the receptor site, as such cells do not respond in any observable way. One plausible explanation could be the lack of a specific host enzyme necessary to bind the toxin to the cell membrane. The drug isonicotinic acid hydrazide is dependent on host peroxidase for fixation in the bacterial cell. The enzyme is present only in susceptible strains of bacteria. This is yet to be ascertained for victorin.

Lycomarasmine

Lycomarasmine (produced by *Fusarium lycopersici*, causing wilt of tomato) was the first toxin isolated in pure form from the culture filtrate of the fungus by Plattner and Clauson Kass (1945). The toxin is unique in that it is produced only by *F. lycopersici* Though the parasite produces disease only in tomato plants, the toxin is aspecific.

Chemical Structure and Mode of Action

It is a dipeptide with a mol. wt. of 277.

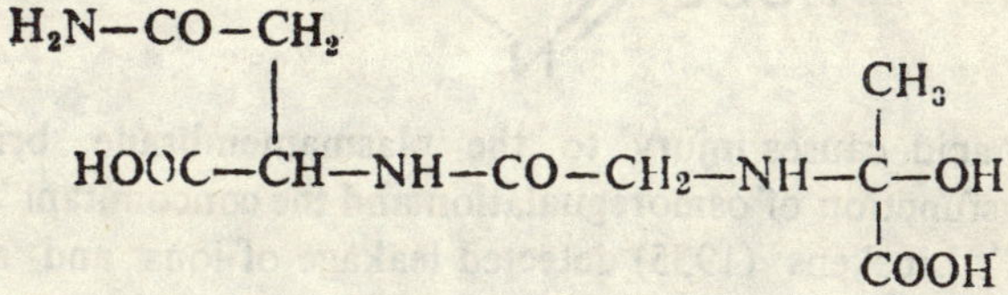

A solution of lycomarasmine is capable of causing wilting and necrosis of excised tomato leaves at minimum concentration of 5×10^{-4} molar.

It is a structural analogue of the yeast growth factor—strepogenin, an oligopeptide containing glutamic acid against aspartic acid of lycomarasmine. To this has been ascribed the toxin action. The toxin action is reversed by adding strepogenin externally. Thus, it is believed that lycomarasmine causes wilting action by interfering with strepogenin metabolism, resulting in its deficiency.

Another aspect of study, mainly by Gäumann's school, suggested that lycomarasmine chelates iron ions or dislodges them from essential sites and transports the toxin—Fe complex to the leaves, where the unstable complex is decomposed under the influence of light. Fe accumulates in toxic concentrations in the leaves which results in damage to the permeability of the leaf cells. Excessive transpira tion results and flaccidity of leaves occurs.

Fusaric Acid

Gaumann et al (1952) isolated a second toxin from culture filtrates of *Fusarium lycopersici* and called it fusaric acid. Lakshmi-narayanan and Subramanian (1955) reported fusaric acid as a vivotoxin. Unlike lycomarasmine, fusaric acid is produced by several genera and species of fungi. Some of these fungi also pro-duce toxic metabolites, closely related to fusaric acid or different from it. This is one of the reasons that despite numerous studies, the significance of fusaric acid in pathogensis is not yet clear. It is known to play a pathogenic role in wilt of tomato, cotton and bakane disease of rice.

Chemistry and Mode of Action

It is 5, *n*-butyl pyridine-2 carboxylic acid

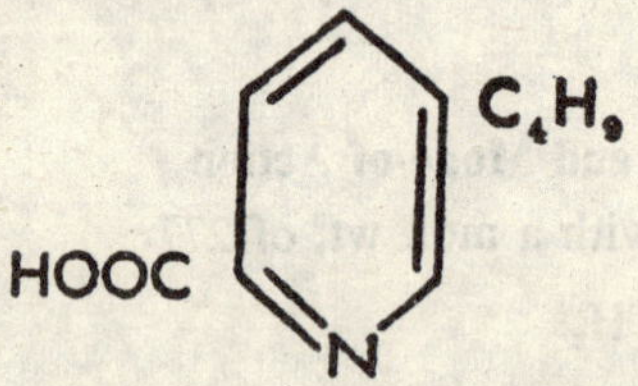

Fusaric acid causes injury to the plasmamembrane, bringing about a dysfunction of osmoregualation and the concomitant "ionic imbalance". Linskens (1955) detected leakage of ions and amino acids from fusaric acid-treated tomato leaf-tissues. The water per-meability of protoplasts and the water balance of whole plant is altered. It also increases, apparently, the rate of respiration and inhibits polyphenol oxidase activity. Like lycomarasmine, fusaric acid chelates iron causing its deficiency.

The two parts of the toxin molecule are essential for toxicity. The carboxyl group at α-position is associated with the inhibition of respiration rate and polyphenol oxidase enzyme. The aliphatic side chain in β-position is necessary for damage to water perme-ability of cells; severity of damage increasing with the length of the chain from methyl to butyl. Of the two, the first appears to be more critical in pathogenesis A volatile decarboxylation product of fusaric acid, 3 - *n*-butyl pyridine, is more injurious to water per-meability, but has not been detected *in vivo* or in the atmosphere around diseased plants.

Fusaric acid has been detected in the rhizosphere (Kalyanasundaram, 1958) and their detoxification by several fungi too is known. The producers of the toxin are more active detoxifiers. Fusaric acid is a product of active metabolism, probably connected with the TCA cycle.

Wild Fire Toxin

It is a bacterial toxin produced by *Pseudomonas tabaci*, causal agent of the 'wild fire' disease of tobacco.

The toxin resembles Victorin in that it reproduces all the symptoms of the disease, yet differs in being nonspecific. Though not detected in host tissues, there are strong evidences suggesting its *in vivo* production. Wheeler and Luke (1963) include it among pathotoxins. Scheffer and Pringle (1970), however, call it a secondary determinant, as it does not condition pathogenicity. Non-toxin-producing mutant strain of the pathogen continues to be pathogenic even though fails to produce the yellow halo, typical of the toxin producing strain.

Chemistry and Mode of Action

It has been characterized as a new type of amino acid with the formula $C_{10}H_{16}O_6N_2$ (molecular weight 260), which appears to be a structural analogue of methionine. Causing methionine deficiency by actively competing for active enzyme sites, appears to be the *modus operandi*. Reversal of toxic action by methionine supplementation, though shown for *Chlorella vulgaris*, is ineffective with tomato tissues. If methionine could undo the toxin damage to tomato tissues, wild fire toxin would be more readily acceptable as a pathotoxin.

α-picolinic acid, Piricularin and Pyriculol

α-picolinic acid ($C_6H_5NO_2$) and piricularin ($C_{18}H_{14}N_2O_2$), were implied by Tamari and Kaji (1954) in the blast disease of rice caused by *Piricularia oryzae*. Both toxins have been isolated from culture filtrates in crystalline form and detected in extracts from infected tissues. Both cause symptoms of disease.

α-picolinic acid interferes with respiration by chelating Fe^{+++} of respiratory enzymes; its specific effect on catalase is made good by adding Fe^{+++} externally.

Piricularin ($C_{18}H_{14}N_2O_2$) also inhibits respiration and several respiratory enzymes. A point of considerable importance is that the fungus secretes a protein in the culture medium, which binds piricularin. Piricularin, in bound form is innocuous to the fungus but retains its toxicity to host tissue.

Surprisingly, since 1954, there have been no confirmation of these substances as the toxins of *P oryzae*. However, a third phytotoxic substance 'pyriculol' ($C_{14}H_{16}O_4$) was isolated from cultures of *P. oryzae* (Iwasaki et al, 1969 and Sato and Kozaka, 1970), which causes the natural 'blast' lesion and also inhibits the growth of rice seedlings. The compound is a heptaketide (Turner, 1971). Rao and Suryanarayanan (1971), who earlier failed to detect piricularin and α-picolinic acid, were able to identify 'pyriculol' in the culture filtrates of 6 isolates of *P. oryzae*. They observed great variability in pyriculol production under different cultural conditions. The isolates on repeated subculturing lost the ability to produce the toxin.

In addition to the above 3 toxins, two more toxic substances recovered by Iwasaki et al (1972) viz.. 3, 4 dihydro 3,4,8-trihydroxy-1 (2H) naphthalenone and Tenuazonic acid. The latter substance has been isolated from infected tissues also, and thus, is a vivotoxin. Rao and Suryanarayanan (1974) detected 'pyriculol' *in vivo* only in infected *Brachiaria mutica*.

It is not yet possible to ascribe the role of the different toxins in the blast disease syndrome.

All the toxins described above involved facultative parasites. Do obligate parasites too produce toxic substances? The answer can be given only with some reservations. Some mild toxic substances have been implicated with a few obligate parasites but the results are inconclusive and need further elucidation. On the contrary, data indicate that in the initial stages of infection, these parasites provide a stimulatory effect on the growth of the host. The absence of typical toxins in such diseased tissues, easily agrees with their mode of life. Obligate parasites—the elite among pathogens, establish a relation of 'peaceful coexistence' as Gaumann would call it; and in such an atmosphere there is no scope for any violent disruptive mechanism of pathogenesis. In the case of facultative parasite, where good health of the host is not of much concern to the pathogen, elaboration of highly toxic substances is commensurate with their destructive mode of life.

The toxins produced by the pathogens are not accidental products of metabolism but products of an inherent metabolic feature,

Table 9.2. Summarises Chemical Nature and Action of Important Toxins

Toxin/Organism		Chemical nature and structure	Mode of action
1 Victorin	*Helminthosporium victoriae*	Peptide linked to sesquiterpene moiety.	1. Primary effect is on plasma membrane, damaging permeability. 2. Increases respiration and affects several other metabolic systems as secondary effect.
2 Lycomarasmine	*Fusarium lycopersici*	Dipeptide 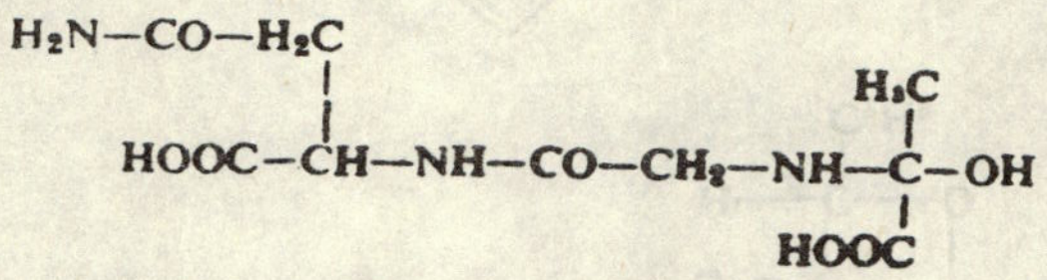	1. Causes strepogenin deficiency. 2. Chelates, Fe^{+++}, free and bound, injures permeability of leaf cells.
3. Fusaric acid	*Fusarium heterosporium* *F. lycopersici* *F. vasinfectum* *F. cubense* *F. moniliforme* *Nectria cinnabarina*	5-*n*-butyl picolinic acid	1. Affects permeability and water balance. 2. Depresses respiration, polyphenoloxidase and other enzymes. 3. Chelates Fe.

Toxin/Organism	Chemical nature and structure	Mode of action
4. Wild fire toxin/*Pseudomonas tabaci*	A new type of amino acid 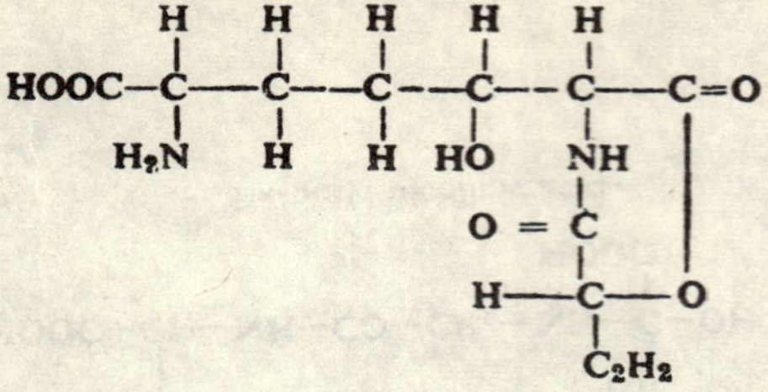	1. Interferes methionine metabolism.
5. α-picolinic acid/*Piricularia oryzae*	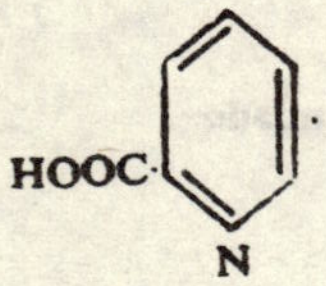	1. Affects catalase activity. 2. Chelates iron from respiratory enzymes.
6. Piricularin/*Piricularia oryzae*	$C_{18}H_{14}N_2O_3$	1. Depression of respiration and inhibition of several respiratory enzymes.
7. Pyriculol/*Piricularia oryzae*	$C_{14}H_{16}O_4$	Unknown.

developed during courśe of evolution and which has become established as an essential character determining inter-relations between host and the parasite.

REFERENCES

BENNET CLARK, T.A. and BEXON, D., 1943, "New Phytologist," 42: 65-92.

BLACK, H.S. and WHEELER, H.E., 1966, "Biochemical effects of victorin on oat tissues and mitochondria," *Amer. J. Bot.*, 53: 1108-1112.

BROWN, W., 1915, "Studies in the physiology of parasitism. I. The action of *Botrytis cinerea*," *Ann. Botany* (London), 29: 313-48.

BUDDENHAGEN, I. and KELMAN, A., 1964, "Biological and physiological aspects of bacterial wilt caused by *Pseudomonas solanacearum*," *Ann. Rev. Phytopathol*, 2: 203-230.

DE BARY, A., 1886, "Uber einige Sclerotinien und Selerotien Krankheiten," *Botan. Ztg*, 44: 377-387, 393-404, 409-426, 433-441, 449-461 and 465-474.

DIMOND, A.E. and WAGGONER, P.E., 1953, "On the nature and role of vivotoxins in plant diseases," *Phytopathology*, 43: 663-669.

GRANITI, A., 1972, "The evolution of toxin concept in plant pathology," in *Phytotoxins in plant disease*. R.K.S. Wood *et al* (eds.,) Academic Press, pp. 1-18 (London, N.Y.).

GÄUMANN, E., NAEF-ROTH S. and KOBEL, H., 1952, Uber Fusarinsaiire im 3 weites Welketoxin des *Fusarium lycopersici*, Sacc.," *Phytopath. Z*, 20: 1-38.

———, 1954, "Toxins in plant disease," *Endeavour* B, 198-204.

IWASAKI, S., NOZOE, S., OKUDA, S., SATO, Z., and KOZAKA, T., 1969, "Isolation and structural elucidation of a phytotoxic substance produced by *Pyricularia oryzae*. Cavara," *Tetrahedron Letters*, 45: 3977-3980.

———, MURO, H., NOZOE, S. OKUDA, S. and SATO, Z., 1972, "Isolation of 3, 4-dihydro-3, 4, 8-trihydroxy-1 (2H)-naphthalenone and tenuazonic a id from *Pyricularia oryzae*. Cavara," *Tetrahedron Letters* 1: 13-16.

KALYANASUNDARAM, R., 1958, "Production of fusaric acid by *Fusarium lycorpersici* Sacc. in the rhizosphere of tomato plants," *Phytopath. Z.*, 32: 25-34.

LAKSHMINARAYANAN, K. and SUBRAMANIAN, D., 1955, "Is fusaric acid a Vivo toxin ?" *Nature*, London, 176: 697-698.

LINSKENS, H.F., 1955, "Der Einfluss der toxigenen welke auf die Blattausscheidunger der formatenpflanze," *Phytopath. Z.*, 23: 89-106.

LITZENBERGER, S.C., 1949, "Nature of susceptibility to *Helminthosperium victoriae* and resistance to *Puccinia coronata*," *Phytopathology*, 39: 300-318

LUMSDEN, R.D. and BATEMAN, D.F., 1968, "Phosphatide degrading enzymes associated with pathogenesis in *Phaseolus vulgaris*– infected with *Thielaviopsis basicola*," *Phytopathology*, 58: 219-227.

LUNDEGARDH, H., 1955, "Mechanisms of absorption, transport, accumulation, and secretion of ions." *Ann. Rev. Pl. Physiol.* 6: 1-24.

PLATTNER, P.A. and CLAUSON-KASS, N., 1945, "Uber ein Welke erzeugendes stoffwechsel produkt Non *Fusarium lycoperisici* Sacc," *Helv, Chim Acta*, 28: 188-195.

PRINGLE, R.B. and SCHEFFER, R.P., 1964, "Host specific plant toxins," *Ann Rev. Phytopathol.*, 2: 133-156

RAO, A.N and SURYANARAYANAN, S., 1976, "Studies on the toxins of *Pyricularia*. I. Detection of Pyriculol in cultures of *Pyricularia*," *Proc. Ind. Acad. Sci* (B) 80(1) 31-40.

SAMADDAR, K.R. and SCHEFFER, R.P., 1968, "Effect of the host specific toxin in *Helminthosporium victoriae* on host cell membranes," *Plant Physiology*, 43: 21-28.

———, and SCHEFFER, R.P., 1971, "Early effects of *Helminthosporium victoriae* toxin on plasma membrane and counteraction by chemical treatments," *Physiol, Plant Pathol*, 1: 319-328.

SATO, Z. and KOZAKA, T., 1970, "New toxic substance produced by *Pyricularia oryzae*," *Ibid.*, 36: 153 (Abstract).

SCHEFFER, R.P. and SAMADDAR, K.R., 1970, "Host specific toxins as determinants of pathogenicity," *Recent Advances in Phytochemistry*, 3: 123-142.

———, and PRINGLE, R.B., 1967, "Pathogen produced determinants of disease and their effects on host plants," in *The Dynamic Role of Molecular Constituents in Plant Parasite Interaction*, C.J Mirocha and I. Uritani (eds.), St. Paul, The Bruce Publishing Co., Minnesota.

SMITH, R.E., 1902, "The parasitism of *Botrytis cinerea*," *Botan. Gaz.*, 33: 421-436.

TAMARI, K. and KAJI, J, 1954, "On the biochemical studies of *Piricularia oryzae* Cavara, the causative mould of the blast disease, local name, Imochi disease of rice plants. Part 2. The Isolation of picolinic acid and Pirucularin, the toxic substances produced by *Piricularia oryzae*," *Niigata Univ. Fac Agric. Bull.*, 5: 33-42.

TURNER, W.B, 1971, *Fungal Metabolites*, Academic Press, London and New York.

Wheeler, H.E. and BLACK, H.S., 1962, "Changes in permiability induced by victorin." *Science*, 137. 983-984.

WHEELER, H E and LUKE, H.H., 1954, "Studies of a toxic agent produced by *Helminthosporium victoriae*." *Phytopathology*, 44: 334.

———, 1963, "Microbial toxins in plant disease," *Ann. Rev., Microbiol* 17: 233-242.

10

DISEASE RESISTANCE IN PLANTS—I PROTECTION

DESPITE the enormous number and ubiquity of micro-organisms with bewildering pathogenic potentials, the majority of plants remain healthy. Losses due to individual diseases of some crops are rather negligible. Thus, resistance is the rule and disease an exception. Were it not so the survival of plants would have become difficult. To withstand the hostile environment of pathogenic micro-organisms plants have evolved several effective mechanisms of resistance. There is 'built in' resistance that prevents entry of pathogens into the plants and provides a general protection. In case it fails and the pathogen gets a foothold and enters inside, the host plant employs more vigorous methods to check any further growth of the pathogen. The plant triggers several metabolic systems resulting in formation of defense chemicals which are aimed at localizing and killing the pathogen.

The question of specificity of pathogenicity and resistance as expressed below, is of considerable significance.

	Plant A	*Plant B*
Fungus α	+	—*
Fungus β	—	+

There exist several resistance factors in plants and we can not pin-point resistance or susceptibility to a single factor: much in the same way as we can not assign pathogenicity to a single attribute of the pathogen.

Disease resistance can be categorized into:

(*A*) Protection, and

(*B*) Defense.

(*A*) *Protection* (Preformed resistance, passive resistance). The resistance factors that prevent the entry of the pathogen into the

*The sign + represents susceptibility and — represents resistance.

plants are included under protection. It is brought about by structural and chemical features that exist in the host prior to infection. This is passive resistance. The surface layers provide a physical barrier to penetration and the toxic chemicals, present on the surface or inside the host cells, provide an inhospitable environment that prevents the establishment of the pathogen.

(*B*) **Defense** (Induced resistance). It is a dynamic event, which starts after the pathogen has entered the host tissue and a physiological contact has been established. Thus it is induced resistance. The formation of new structural barriers and synthesis of defense compounds, that kill or restrict the growth of the pathogen, are defense reactions.

PROTECTION

Factors that protect plants against infection can be grouped under following heads: (*a*) structural protection; (*b*) chemical protection; (*c*) absence of nutrients, and (*d*) absence of common antigens.

Structural Protection

The pathogen first comes in contact with the surface layers of the plant where it germinates and multiplies. To establish infection the pathogen has to get an access to the host tissues. Entry is made either by directly penetrating the surface layers or through natural openings like stomata, lenticels and wounds.

The various structural features of host that participate in protection are as follows: (*i*) structure of epidermis, (*ii*) structure of stomata, and (*iii*) mechanical tissues. The first two features prevent germination and penetration by the pathogen while the mechanical tissues restrict spread inside the host tissues.

(*i*) **Structure of epidermis.** The waxy cuticle and the hairs above the epidermis being hydrophobic prevent water drops from adhering to the leaf surface. This prevents deposition and germination of the pathogens. The cuticle limits the diffusion of nutrients needed by the pathogen during the initial stages of infection. However, the cuticle does not make any significant contribution to resistance. Even a well defined cuticle may not serve as an effective barrier.

A thick epidermal wall makes direct penetration difficult. The resistance of Barbery to black rust (*Puccinia graminis*) has been related to the thickness of the epidermis (Melander and Craigie, 1927). If the epidermis is ruptured the resistance is lost. Similarly, Japanese pear and flax varieties, resistant to *Alternaria kekuchiana* and *Melampsora lini*, have thick epidermal layers. It is a common observation that puncturing of the epidermis in artificial inoculations allows the weak pathogens to infect the host, which otherwise fail to parasitise on an intact surface. The epidermis in such cases is a decisive hurdle to the pathogen.

(*ii*) **Structure of stomata.** Infection by a few bacterial pathogens is greatly dependent on stomatal morphology. The resistance of the mandarin variety of Citrus to *Xanthomonas citri* is attributed to the narrow pore of the stomata which prevent entry of water, and thus the bacteria, which require a film of water for entry, are unable to infect. A cuticular ridge protrudes over the stomata that greatly narrows down the pore. (Fig. 10.1)

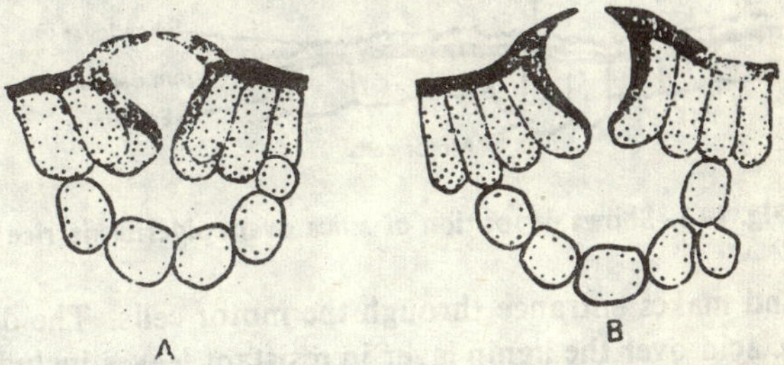

Fig. 10.1. Showing stomatal pore of (*A*) resistant, (*B*) susceptible varieties of *Citrus*.

Though most fungi are able to push through the closed stomata the germ tubes of rust uredospores enter only through open stomata. Hart (1926) observed that the stomata of rust-resistant wheat remain closed till late in the morning. By the time they open, the germ tubes of uredospores, formed earlier in the dew, get killed due to evaporation of the water. She termed such resistance as "functional resistance".

(*iii*) **Mechanical tissues.** Well developed mechanical tissues in the Sonem variety of wheat have been implicated with its high resistance to *Puccinia graminis*. The susceptible varieties, on the other hand, have only feeble development of such tissues. The sclerenchymatous tissues act as mechanical barriers to the growth of the pathogen. The black, long pustules of rice leaf smut (*Entyloma oryzae*) and "angular leaf-spots" are limited to areas in between the veins. They fail to cross the mechanical tissues overlying the veins.

Chemical Protection

(*i*) *Chemicals of the surface layers.* Waxes and cutin, present over the epidermal surface, provide a chemical barrier in the sense that they are resistant to enzymes of the pathogens. Wax extracts from apple leaves, resistant to powdery mildew (*Podosphaera leucotricha*), when deposited on susceptible leaves makes them resistant.

Lignin and silicic acid of the epidermal cells of rice leaves provide protection against 'blast' disease. The spores of *Piricularia oryzae* are unable to penetrate the epidermis anywhere except at the motor cells. The epidermal cells are covered with lignin and chlorogenic acid except over the motor cells which are covered by pectic material (Fig. 10.2). The pathogen degrades the pectic material

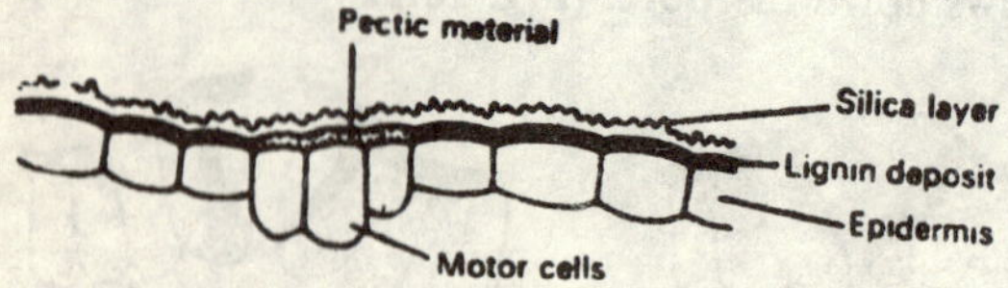

Fig. 10.2. Shows deposition of silica over epidermis in rice leaf.

easily and makes entrance through the motor cells. The deposition of silicic acid over the lignin layer in resistant leaves includes motor cells also and thus increases the resistance by blocking entry through the motor cells.

(*ii*) *Exudation of toxic chemicals.* Exudates from leaves and roots of resistant plants have been shown to inhibit growth of pathogens. Two examples will be taken.

Walker (1923) observed that whereas the onion plants with coloured bulbs were resistant to smudge disease, caused by *Colletotrichum circinans*, those with white bulbs were easily attacked. The resistant factor was present only in the dry outer scales of the coloured bulbs which when removed rendered them susceptible. In

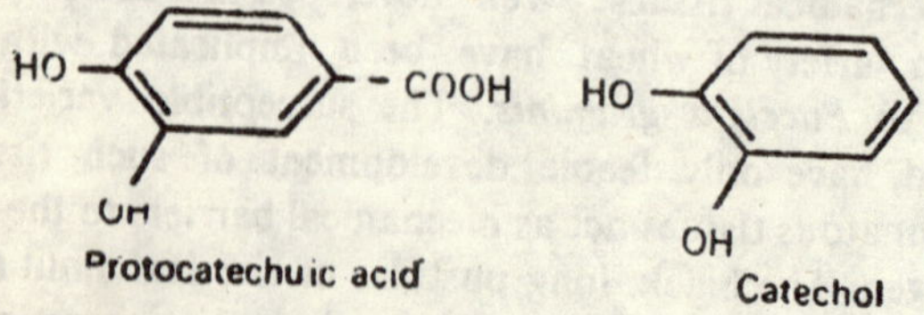

addition to the red pigments present in all the scaly leaves of coloured bulbs, the dead dry outer scales were demonstrated to contain two water soluble phenols. Link *et al.* (1922) and Link and Walker (1933) identified the phenols as Protocatechuic acid and catechol.

The two substances are toxic to the onion smudge fungus. Being water soluble, these diffuse out into the infection drop and prevent germination of the spores. The toxicity is not specific to this fungus.

Another classical example is that of the resistant varieties of flax (Timonin, 1940) which were observed to excrete Hydrocyanic acid (HCN) in the rhizosphere. HCN was found to be toxic to several organisms including *Fusarium lini*. Interestingly HCN does not affect *Trichoderma viridae* an antagonist of *Fusarium.*

(iii) Toxic chemicals present in plant cells. Cells of resistant plants show higher amounts of toxic chemicals which generally exist as phenols. Antifungal substances have been detected in juices of several plants. In the scab disease of potato, caused by *Streptomyces scabies*, the resistant hosts show higher amounts of chlorogenic acid. A similar higher level of chlorogenic acid occurs in the roots of potato plants resistant to *Verticillium* wilt. Patil *et al* (1966) observed a decline in total phenol content after 5 weeks of sprouting and the resistance was also lost by that time. There are several such reports of resistant varieties having more phenolic content than the susceptible plants.

The effect of naturally occurring polyphenols on virus infection is of minor significance. However some cases are known that implicate phenols in inhibiting virus infection. Stanley (1935) confirmed the earlier observation of Allard (1918) that tannic acid—a phenol—present in tobacco plants inhibited the infectivity of tobacco mosaic virus (TMV). It was also shown that the removal of tannic acid restored the infectivity of the virus. In comparison to other class of inhibitors of viruses, the role of phenols is insignificant (Bawden, 1955).

Absence of Nutrients

Plants lacking the nutrients essential for the pathogen will be saved from the disease. Mutant strains that need the growth factors, absent in host cells, lose their virulence. Thus the host specificity is explained by the availability of the particular nutrient(s) only in the specific host.

There are but few examples involving lack of nutrients in the host as a means of protection. In *Rhizoctonia* disease of seedlings, appresorium formation takes place only on the susceptible host.

Appresorium is not formed in pure culture either. However, if extracts from susceptible hosts are added to the culture medium, appresorium formation is triggered. Thus the factor for appresorium formation and growth is found only in the susceptible host. A deficiency of this factor in the host makes it resistant.

Nutritional deficiency in a host plant may be due to:

(*a*) *Lack of adequate amount of the nutrient* (turnips with a low level of histidine are resistant to three histidine-requiring mutants of *Erwinia amylovora*).

(*b*) *Inhibition of uptake of the nutrient by other compounds* (certain turnips have enough histidine but are resistant because other amino acids antagonise histidine uptake.)

(*c*) *Lack of compounds needed to facilitate the uptake of the nutrients present in enough amounts* (avirulent tryptophan —requiring mutants of *Pseudomonas* become virulent on tobacco if other amino acids except tryptophan are added.)

Thus, any of the above three conditions may deprive the fungus of its nutrition and the growth.

Absence of Common Antigens

Animals produce antibodies against antigens (foreign proteins) injected into the body by pathogens. But in case the antigen resembles a host protein (common antigen), antibody will not be formed and the disease will go undefended.

Though antibodies are not formed in plants against pathogens, 'common antigens' are demonstrated by serological methods to exist in the compatible host-pathogen systems. The existence of common antigens have been demonstrated in susceptible flax, cotton, sweet potato and their respective virulent pathogens. Absence of common antigens have been observed in resistant hosts.

REFERENCES

ALLARD, H.A., 1918, "Effects of salts, acids, germicides, etc. upon the infectivity of the virus causing the mosaic disease of tobacco," *J. Agric. Res.*, 13: 619-637.

BAWDEN, F.C., 1955, "Inhibitors and plant viruses," *Advance. Virus Res.*, 2: 31.

CRAIGIE, J.H., 1927, "Discovery of the function of the pycnia of the rust fungi," *Nature*, 120: 765-767.

HART, H., 1929, "Relation to stomatal behaviour to stem-rust resistance in wheat," 39: 929-948.

LINK, K.P., ANGELL, H.R. and WALKER, J.C., 1929, "The isolation of protocatechuic acid from pigmented onion scales and its significance in relation to disease resistance in onions," *J. Biol. Chem.*, **81**: 369-375.

———, and WALKER, J.C., 1933, "The isolation of catechol from pigmented onion scales and its significance in relation to disease resistance," *J. Biol. Chem.*, **100**: 379-383.

MELANDER, L.W. and CRAIGIE, J.H., 1927, "Nature of resistance of *Berberis sp.* to *Puccinia graminis*," *Phytopathology*, **17**: 95-114.

PATIL, S.S., ZUCKER, M. and DIMOND, A.E., 1966, "Biosynthesis of chlorogenic acid in potato roots resistant and susceptible to *Verticillium alboatrum*," *Phytopathology*, **56**: 971-974.

STANLEY, W.M., 1935, "Chemical studies on the virus of tobacco mosaic. IV. Some effects of different chemical agents of infectivity," *Phytopathology*, **25**: 899-921.

TIMONIN, M.I., 1941, "The interaction of higher plants and soil microorganisms. III Effect of by-products of plant growth on activity of fungi and actinomycetes," *Soil. Sci.*, **52**: 395-413.

WALKER, J.C., 1923, "Disease resistance to onion smudge," *J. Agric. Res.*, **24**: 1019-1040.

11

DISEASE RESISTANCE IN PLANTS—II DEFENSE

RESISTANCE to entrance seldom protects plants completely against disease and it is generally overrun by pathogens. Thus when protection fails and the pathogen succeeds in getting into the host, the more effective and dynamic mechanisms are pressed into operation. Cruickshank's (1966) observation that most plants are susceptible to most fungi but are resistant to most diseases, explains the situation well.

The defense mechanisms can be arranged in following categories:

A. Exclusion of the pathogen by new histological barricades (histological defense)
B. Accumulation of toxic substances (Phenols)
C. Hypersensitive defense reaction
D. Defense through growth substances and enzymes
E. Modification of substrates to resist pathogens' enzymes
F. Detoxification
G. Tolerance to disease.

In all these endeavours, except the toleranee, the speed of accomplishment is of critical importance.

EXCLUSION OF PATHOGEN BY DEFENSIVE BARRICADES (HISTOLOGICAL DEFENSE)

The injury caused to host cells, during establishment of the pathogen, induces cell division. The host 'walls off' the pathogen by (*a*) demarcation of infected tissues by cork layers, abscission cells, tyloses and gums, and (*b*) callus like swelling on the wall which forms a sheath around the hyphae.

Demarcation of Infected Tissues

(*i*) *Formation of cork layers around the infected area.* It is a common response against all kinds of pathogens (Fig. 11.1). The cork layers, often several layers deep, form an impervious wall that prevents further spread of the pathogen. It stops diffusion of toxic metabolites into the healthy zone and of nutrients into the lesion The pathogen is thus 'cut off' from rest of the tissues.

Such 'cicatrical' layers are formed in several diseases *e.g.* Anthracnose disease of grape, citrus scab and rot of sweet potato

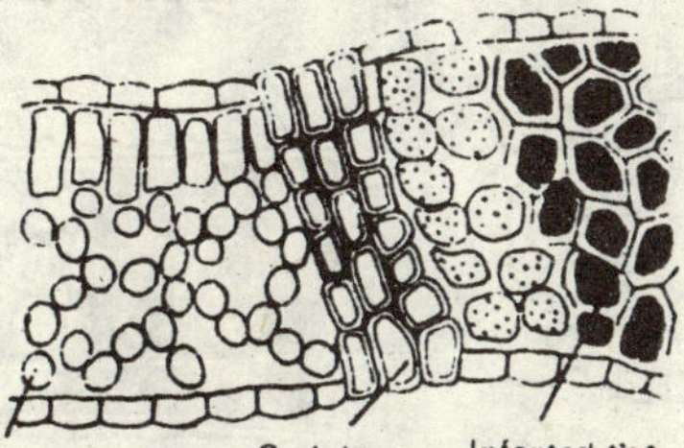

Fig. 11.1. Formation of a cork layer around infected tissue.

Deposition of lignin makes these barriers chemically more effective. Susceptible varieties fail to form the cork barrier, and the pathogen grows unhampered.

(*ii*) *Formation of abscission cells.* Layers of cells around the lesions separate apart by dissolution of the middle lamella. This results in detachment of the invaded cells from the adjoining tissues; which dry and fall out leaving a shot hole (Fig. 11.2).

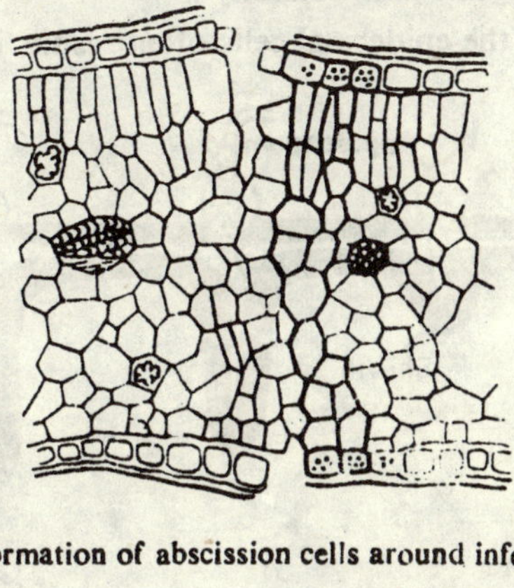

Fig. 11.2. Formation of abscission cells around infected zone.

Abscission cells are formed readily on leaves of stone fruits especially on young leaves.

(*iii*) *Formation of tyloses and gums.* Tyloses are balloon like extensions of parenchymatous cells into the lumen of vessels (Fig. 11.3A). These are frequently formed in vessels of diseased plants especially in wilt diseases. Toxic products of invaded cells, or of the pathogen, serve as stimulants. Gum accumulation in the interstices of the tyloses as well as on the walls of the vessels blocks the transport of water and nutrients and also the accompanying pathogens. Spread of pathogens to upper portions is prevented.

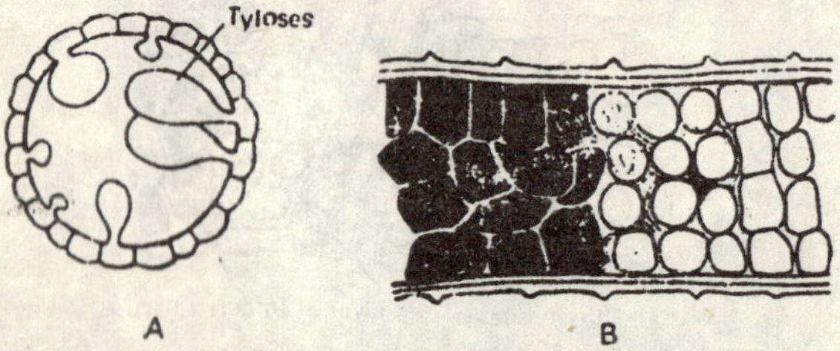

Fig. 11.3. Formation of A. Tyloses, and B. Gums, in 'blast' resistant rice leaf.

The effectiveness of this mechanism depends on the speed of formation of the tyloses and gums. If the blockade is completed when the pathogen is still in the young roots, the dispersal of the pathogen would be checked. In some diseases secretion of gums by cells around the lesions serves to ward off the pathogen. Rice varieties resistant to 'blast' disease and *Helminthosporium* leaf-spot, trap pathogens in the gums secreted by the adjoining cells (Fig. 11.3B).

Swelling of Cell Walls and Callosity

Penetration of the epidermal cells often results in swelling of the

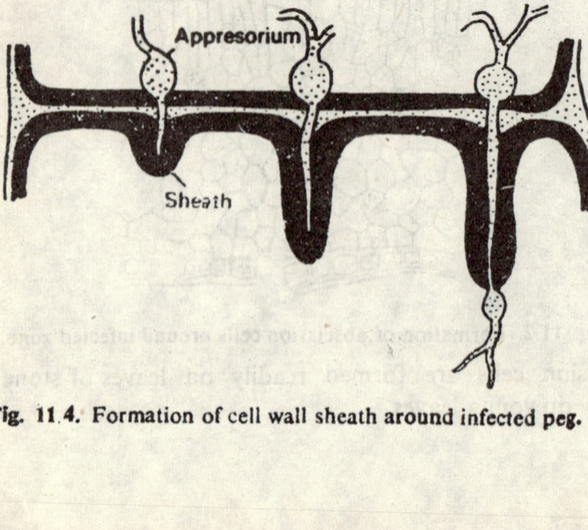

Fig. 11.4. Formation of cell wall sheath around infected peg.

cell wall. When tomato leaves are infected by conidia of *Piricularia oryzae*, the epidermal wall in contact with the appresorium gets swollen and lignified and prevents further invasion (Fig. 11.4). The effectiveness of such swellings in checking penetration is only of little value.

Sometimes, a callus is formed on the inner side of epidermal cells which grows around the intruding hyphae of fungi and forms a kind of sheath around it (Fig. 11.5). The nature of the sheathing material whether cellulose or callus substance, is not known. In some cases, granules, originating from the host cytoplasm accumulate to form the sheath. Frequently pathogens outgrow such sheaths and their effectiveness is of little significance.

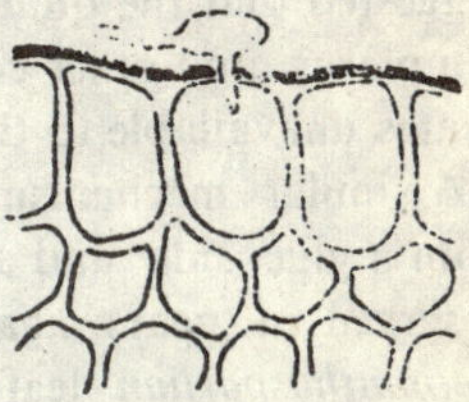

Fig 11.5. Callus sheath around infection peg

ACCUMULATION OF TOXIC SUBSTANCES

The plants respond to infection and also to mechanical injury by accumulation of toxic chemicals, which are mostly phenolics. The increase in phenolics may be due to release of phenols from non-toxic glucosides, migration from healthy tissues towards diseased areas or by augmented synthesis.

(*i*) **Release of phenols from non-toxic glucosides.** Kuc' (1963) and his associates have observed this type of resistance in apple scab disease (= *Venturia inaequalis*).A β-glycosidase, produced by

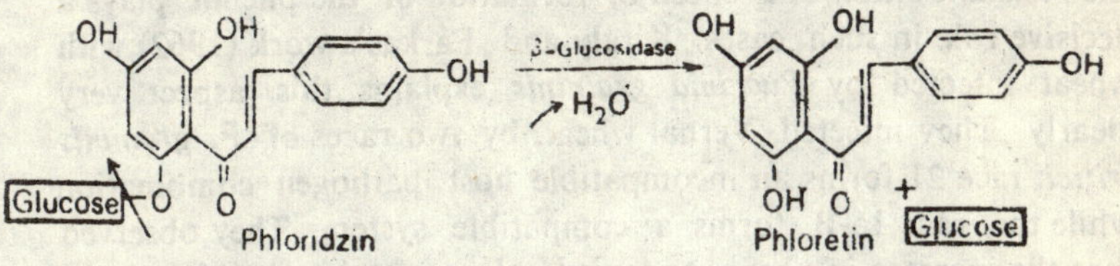

Fig. 11.6. Release of phlorietin from phloridzin by β-glucosidase action.

pathogen or the host, acts on the glucoside phloridzin, to release phloretin and glucose (Fig. 11.6). The phloretin is oxidized by a phenol oxidase to a quinone, which is highly toxic to the scab fungus. It is interesting that the glucoside phloridzin is present in susceptible varieties also, but the chain of defense reactions is triggered only in the resistant variety. The reason for such behaviour is, however, not known.

Quinones act against microorganisms by interrupting some vital enzymic processes. The toxicity may be due to: (*i*) binding of quinone nucleus to SH or NH_2 groups in the fungal or bacterial cell, or (*ii*) by disrupting the electron transport system. Kuc' (1963) suggested that the quinones bind with the pectin-cellulose-protein complexes and other cell wall materials, making the adaptive substrates unavailable to the hydrolytic enzymes of the pathogen.

A similar mechanism of resistance, involving splitting of phenol from a glucoside and a follow up oxidation to quinone, has been observed in pears against fire blight disease (*Erwinia amylovora*), *Helminthosporium* leaf-spot disease of rice and tomato against *Fusarium lycopersici*.

(*ii*) **Synthesis of phenols**. In cotton, infected by *Verticillium*, level of phenolic compounds increases in the roots and stems with a concomitant fall in the leaves. This suggests transport of phenols from leaves to roots and stems. But the increase is so high that it can not be accounted for by migration only. Biosynthesis must start and the evidences for this are well documented in several diseases.

Increase in phenol contents is reported in several important diseases, some of which are given in Table 11.1.

The list can be extended to a great extent, which suggests strongly that increased phenol level is a general response of plants against infection.

Phenol accumulation occurs in susceptible varieties as well. In some cases the amouht in susceptible varieties is higher than in the resistant ones. The speed of formation of the phenol plays a decisive role in such cases. Kiraly and Farkas's work (1962) with wheat infected by *Puccinia graminis* explains this aspect very clearly. They infected Vernal wheat, by two races of *P. graminis tritici*: race 21 forms an incompatible host pathogen combination while the race 15-B forms a compatible system. They observed that the increase in phenol to toxic level was attained much quicker in the incompatible combination. The increase is slow in the compatible combination even though the final level is higher than that in the resistant combination. The low level of phenol in the critical phase of infection fails to check the pathogen in the compatible combination and the disease occurs. The rate of synthesis thus makes all this difference.

Host/Pathogen	*Phenol accumulated*	*Structure*
Potato/*Phytophthora infestans*	Chlorogenic acid 2-3 fold increase (Sokolova *et al*, 1958)	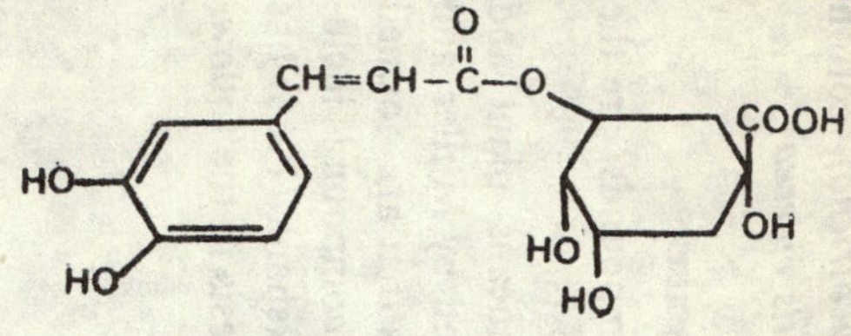 Chlorogenic acid
	Scopoletin 10-20 fold increase (Hughes and Swain, 1960)	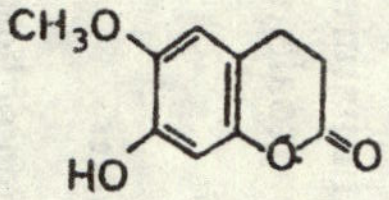 Scopoletin
Potato/*Streptomyces scabies*	Chlorogenic acid (Johnson and Schaal, 1952)	
Apple/*Venturia inaequalis* *Podosphaera leucotricha*	Scopoletin (Barnes and Williams, 1960)	
Sweet potato/*Ceratocystis fimbriata*	Chlorogenic acid, Caffeic acid Umbelliferone (Uritani, 1963)	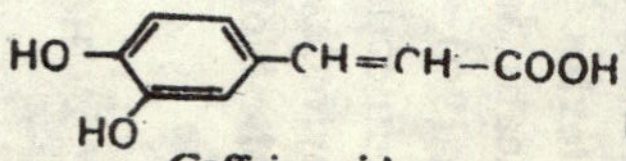 Caffeic acid 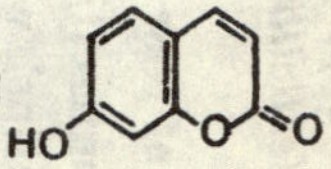 Umbelliferone
Tomato/tomato spotted wilt virus	Scopoletin (Best 1936, 1944)	
Tobacco " " "	—do—	

Similar rapid accumulation has been observed in the resistant varieties of apples (against *V. inequalis* and *P. leucotricha*), cotton (against *Verticillium*) and cabbage (against *Botrytis cinerea*).

Synthesis of New and Abnormal Phenols – Phytoalexins

A very interesting group of such abnormal compounds are the "phytoalexins', which are not found in healthy hosts.

Phytoalexins. The term phytoalexin (*phyton* means plant and *alexin* means warding off compound) was coined by Muller and Borger in 1940, for toxin or toxin like substances that are formed only post-infectionally. Except ipomeamarone, all compounds included in phytoalexins are phenolic in nature. Cruickshank (1966) has outlined the biological steps of phytoalexin synthesis in the following scheme.

Host plant surface + water

↓ Diffusion of host nutrients.

Host lechate + fungal spores.

↓ Germination and growth of fungus.

Mycoantigen + viable host cell

↓ Stimulation of intermediary host metabolism

Phytoalexin.

Fig. 11.7. Steps in phytoalexin synthesis.

Müller and Borger (1940) observed that inoculation of resistant potato plants by zoospores of *Phytophthora infestans* resulted in a rapid death of cells. The hypersensitive reaction caused the formation of a fleck of dead cells. Müller wondered as to what caused this hypersensitive reaction and postulated the formation of fungitoxic substances. The hypothesis, now known as the phytoalexin theory, is summarized by Cruickshank (1966) as follows:

(1) A principle designated as "phytoalexin" which inhibits the development of the fungus in hypersensitive tissue is formed or activated only when the host cells come into contact with the parasite.

(2) The phytoalexin is non-specific in its toxicity towards fungi.

(3) The basis of differentiation between resistant and susceptible hosts lies in the speed of formation of phytoalexin.

(4) The resistant state is not inherited. It is developed after the

fungus has attempted infection. The sensitivity of the host cell, which determines the speed of the host reaction, is specific and genetically determined.

Several 'phytoalexins' have been isolated and characterized. Some of them are described in Table 11.2 (*see* pages 116 and 117).

Some salient features of phytoalexins are as follows:

(*a*) Phytoalexins are broad based antifungal compounds. In a few cases bacteriostatic property has been detected.

(*b*) The stimulant has no role in determining the chemical nature of the phytoalexin; it is a characteristic property of the host.

(*c*) The biosynthesis of the phytoalexins is modified by several environmental factors. The quantity produced may be influenced by the factors like age of the host, inoculum concentration, temperature and aeration.

(*d*) In nature, the microbial agents serve as the main inducers but in laboratory conditions several chemicals, especially heavy metal ions, metabolic inhibitors and some amino acids, have been shown to be able to induce the phytoalexin synthesis. Uhera (1963) has suggested the use of heavy metals of some fungicides to bring about systemic resistance in plants.

(*e*) **Phytoalexins are generally not translocated and scarcely persistent. Its role as a protectant is doubted.**

HYPERSENSITIVE REACTION

It is a common defense mechanism incited by non-pathogens. No sooner the pathogen comes in physiological contact with the hypersensitive host, the adjoining cells collapse resulting in a necrotic fleck of tissue. Thus extreme resistance is accomplished by extreme susceptibility. The pathogen is killed and "cut off" from the healthy cells. The hypersensitive reaction (HR) has been extensively studied in rust fungi and *Phytophthora infestans*.

The size of the lesions depends on the degree of resistance of the host. The more resistant the variety, the smaller the area involved. Some immune varieties of wheat involve killing of only a few cells and it is difficult to locate the lesions by visual observation.

The hypersensitive reaction (HR) is brought about by a rapid accumulation of phenols in the invaded cells and their oxidation to

TABLE 11.2. SHOWS SOME IMPORTANT PHYTOALEXINS, THEIR ORIGIN AND CHEMICAL NATURE

Phytoalexin	*Host*	*Organism*	*Chemical nature+mol. wt.*	*structure*
Ipomeamarone (Hiura 1943)	*Ipomea batata*	*Ceratocysis fimbriata* and also by *Thielavia basicola* *Helicobasidium mompa* *Fusarium sp.*	Sesquiterpenoid $C_{15}H_{22}O_2$ (M. wt. 230)	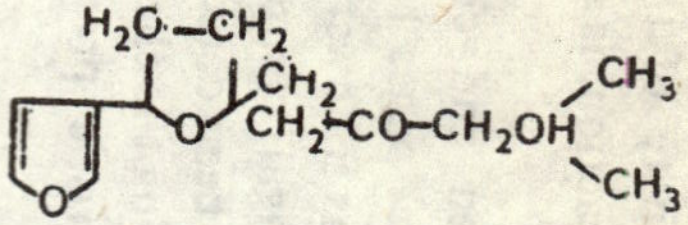
Orchinol (Gäumann *et al*) (1950)	*Orchis militaris*	*Rhizoctonia repens* and also by *Didymella exitiales* *Fusarium solani* *Ophiobolus graminis* *Orcheomyces hircinis*	Dihydrophenanthrene $C_{16}H_{16}O$	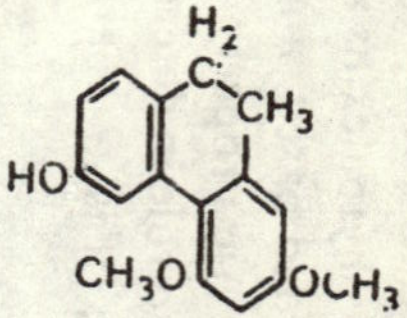
Isocoumarin [Condum & Kuc (1960)]	Carrot root	*Ceratocystis fimbriata* and *Helminthosporium carbonum*, *Fusarium lycopersici*, *Thielaviopsis basicola*, *Rhizopus stolonifer*	Isocoumarin $C_{11}H_{12}O_4$ (M. wt. 208)	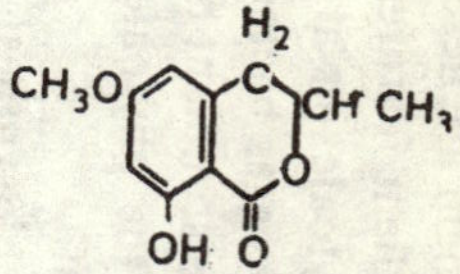

Pisatin (Cruick-shank and Perrin 1963)	*Pisum sativum*	*Monilina fructicola* and several other fungi	Chromanochroman (phenolic ether) $C_{17}H_{14}O_6$	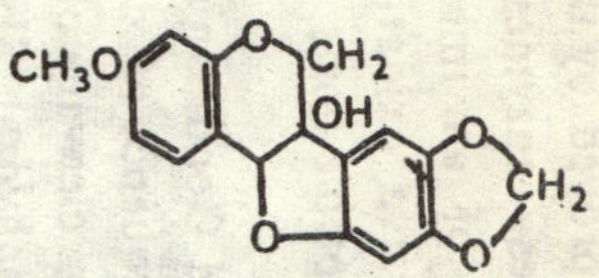
Phaseolin (Cruickshank and Perrin 1963)	*Phaseolus vulgaris* (French bean)	*Monilia fructicola*	Chromanochroman (Phenol)	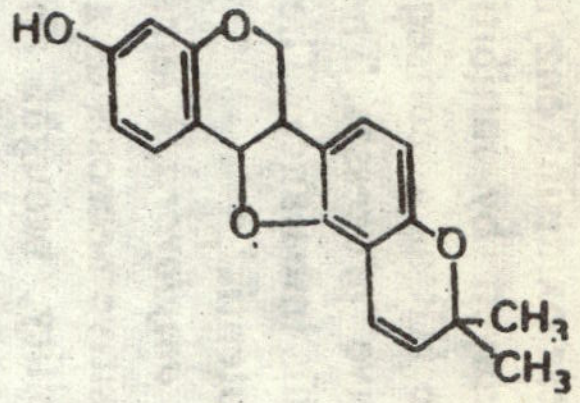
Trifolirhizen (Hietala, 1960)	(red clover) *Trifolium pratense*	Non pathogens	Isoflavnoid (Glucoside) $C_{22}H_{22}O_{10}$	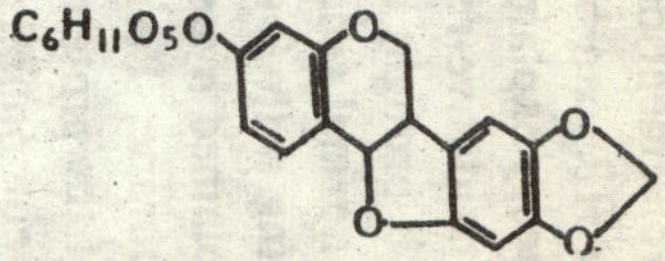

toxic quinones by polyphenol oxidase and peroxidase enzymes. The involvement of these oxidising enzymes is proved by the fact that the HR can be delayed by inhibiting these enzymes. The redox-theory explains the HR as a consequence of an imbalance in the oxidative-reductive processes. The excessive oxidation of phenols is brought by this imbalance. The susceptible hosts maintain the normal redox potential.

Working with *Erwinia amylovora*, Klement and Goodman (1967) have explained hypersensitive necrosis as a consequence of damaged membrane. Permeability, brought about by cleavage of S-S-bonds of the proteins Fig. (11.8). Synthetic sulphahydral (SH) containing compounds induced HR. This suggests S-S bonds as the site for attack of the 'inducer' of hypersensitive necrosis.

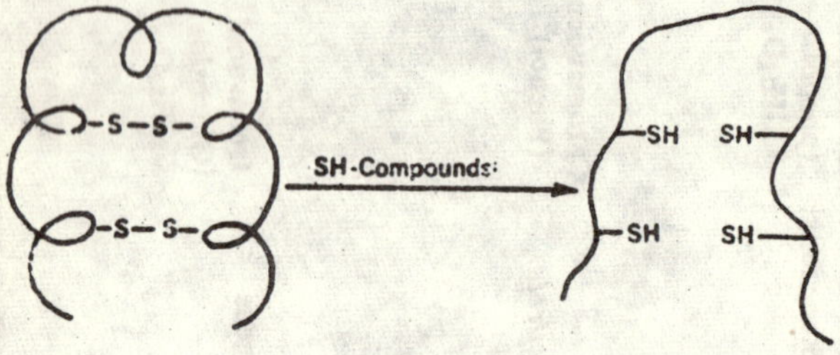

Fig. 11.8. Showing cleavage of S-S bonds of membranic proteins.

Loss of cell-permeability may result in leakage of phenols and mixing with their oxidative enzymes.

The salient points of HR can be summarised thus:

(*a*) Only living micro-organisms and viruses are able to induce HR.

(*b*) It operates only in incompatible host-parasite combination.

(*c*) The rate of HR appears quicker in more resistant plants than in susceptible ones.

(*d*) The necrosis is brought about by accumulation and excessive oxidation of phenols as a consequence of disturbed redox potential.

(*e*) The inducer of HR acts probably by damaging the membrane permeability by disrupting the S-S bonds.

Non-phenolic Defense

(*a*) *Defense through growth substances and enzymes*. Infection often cause increase in auxin level in host tissues. Auxins in general inhibit fungi. IAA may provide resistance by controlling the direction and rate of migration of defense compounds. Fungi produce IAA-oxidase to inactivate the auxin while the host in turn inhibits the *IAA oxidase* through formation of polyphenols like chlorogenic acid and caffeic acid. Further, polyphenols increase the auxin level by converting tryptophan to IAA.

Kinins and Gibberellins too have a role in disease resistance. Kinetin and benzimidazole maintain rust resistance in detached cereal leaves by preventing protein breakdown. Powdery mildew fails to grow on cucumber leaf discs floated in kinetin solution; the resistance is lost by transferring them to water. Selman (1964) working with tomato spotted-wilt virus observed reduction in lesion number by a four hour pretreatment, plus a three day post-inoculation treatment with kinetin. The effect depends upon the concentration of the kinetin used.

The role of ethylene in inducing various enzyme systems is well known. Ethylene is regarded as a hormone and, like auxins and cytokinins, activates some specific sites of plant DNA.

Enzymes. The important role of phenol oxidising enzymes has been described earlier. Peroxidase provides resistance to beans against *Pseudomonas phaseolicola*. The resistant varieties produce higher amount of peroxidase than the susceptible ones. The virulent strains inactivate it by catalase production. In the contest, production of peroxidase at a level that can withstand the inhibitory effect of catalase, will ensure resistance.

Dehydrogenases and amino acid oxidases are part of defense reaction of cabbage against *Botrytis cinerea*.

(b) Defense by modification of substrates of pathogen's enzymes. Polyvalent cations, especially calcium, have been reported to increase the resistance of bean tissues against *Rhizoctonia solani* (Bateman, 1969). Cations are known to release PME from host cell walls. The PME demethylates the pectin—the position of methyl groups is taken by calcium ions forming calcium pectate, which is resistant to hydrolysis by hydrolytic pectic enzymes. Thus the modified substrate brings in resistance.

Similarly protein-pectin-polyvalent cation complexes in green apple tissues provide resistance to hydrolytic enzymes of the pathogens. Potassium in general bestows resistance to disease. Resistant varieties in several diseases have been shown to have higher potassium content than the susceptible varieties. A wilt resistant flax contained more potassium and also showed greater uptake of this element, (Dastur and Bhatt, 1964). The role of potassium is associated with maintenance of membrane permeability, and is of great importance in disease resistance.

(*c*) *Defense by detoxification of toxins.* In principle antitoxic defense reaction neutralizes the toxins and other necrogenous substances. There are but few reports of biochemical antibiotic defense in plants.

Fusaric acid is detoxified by resistant cotton and tomato plants which metabolize the toxin and convert it to a nontoxic compound *N-methyl fusaric acid amide*.

Romanko (1959) observed that intact tissues of oat varieties resistant to *Helminthosporium victoriae* were able to incapacitate the toxin victorin. When cut stems of oat plants were exposed to victorin solution (40%) toxicity was recovered from susceptible tissues while none from resistant tissues. This could be due to inactivation or mobilization of the toxin by the host tissues. The other explanation of non-recovery of toxicity could be the lack of a receptor site or the enzyme that could bind it to the receptor site.

TOLERANCE TO DISEASE

As early as 1894, Cobb observed that some heavily rust infected wheat varieties produced fairly good crop of grains. Disease and damage are not synonymous in this context. It is frequently seen that equal severity of disease does not result in equal damage to different varieties. Those that give fairly good yield of the economically important product are said to be 'tolerant' to the disease. Tolerance is usually shown in cases where disease symptoms appear on parts other than the desired product. In rusts symptoms appear on the leaf but the grain remains unaffected. However, the same is not true for smuts which mostly involve replacement of the grains by the fungus. The disease is measured by damage done to the grains; here disease is synonymous to damage, and tolerance in such cases is unlikely.

To qualify as tolerant a plant must show susceptibility equal to a less tolerant one. Mathews (1970) described tolerance in virus diseases as a situation where the virus multiplies and spreads widely in the host but produces mild or negligible symptoms. Dropkin and Nelson (1960) have included "tolerance" in their useful chart describing the various types of host pathogen interaction for nematodes. Tolerant host shows good growth despite good development of the pathogen.

		Host Growth	
		Good	Poor
Parasitic growth	Good	Tolerant	Susceptible
	Poor	Resistant	Intolerant

There are several definitions for tolerance in the literature. According to Schafer (1971) "Tolerance may be defined as that capacity of a cultivar resulting in less yield or quality loss relative to disease severity or pathogen development, when compared with other cultivars or crops."

Like other forms of resistance, tolerance is a relative phenomenon occurring in different amounts in various hosts. It may also occur with other forms of resistance.

REFERENCES

BARNES, E.H. and WILLIAMS, E.B , 1960, "A biochemical response of apple tissues to fungus infection," *Phytopathology*, 50: 844-846.

BATEMAN, D.F., 1969, "Alteration of cell wall components during pathogenesis by *Rhizotonia solani*," in *Role of molecular constituents in plant parasitic interaction*." Mirocha and Uritani (eds.).

BEST, R.J , 1936, *Aust J. Expt. Biol Med. Sci*, 14: 199.

———, 1944., *Ibid.*, 22: 251.

BREDENBERG, J.B. and HIETALA, P.K., 1961, "Investigation of the structure of trifolirhizen, an antifungal compound from *Trifolium pratense*. L". *Acta. Chem. Scand*, 15: 696-99.

COBB, N.A., 1894, "Contributions to an economic knowledge of Australian rusts (*Uredineae*). *Chap*," 10. *Agr. Gaz.* N.S.W., 5: 239-250.

CONDUN, P. and KUC, J., 1960, "Isolation of a fungitoxic compound from carrot inoculated with *Ceratocystis fimbriata*.," *Phytopathology*, 50: 267-270.

———, 1962, "Confirmation of the identity of a fungistatic compound produced by carrot root tissues," *Phytopathology*, 52: 182-183.

CRUICKSHANK, I.A.M., 1963, "Studies on phytoalexins. VIII. The isolation and general properties of phaseolin and viciatin, the phytoalexins of *Phaseolus vulgaris*," *Life Sciences* 2: 680.

CRUICKSHANK. I.A.M. and PERRIN, D.R., 1960, "Isolation of a phytoalexin from *Pisum sativum* L.", *Nature*, 187: 799-800.

CRUICKSHANK, I.A.M. and PERRIN, D.R., 1966, "Defence mechanisms in plants," *World Rev. of Pest. Control*, 5(4): 161-175.

DASTUR, R H and BHATT, J.G., 1964, "Relation of potassium to *Fusarium* wilt of flax," *Nature*, 201: 1243-1244.

DROPKIN, V.H. and NELSON. P.E., 1960, "The histopathology of root knot nematode infections in soybeans," *Phytopathology*, 50: 442-447.

GÄUMANN, E.R, and BRÄUN, R. and BAZZIGHER, G , 1950, "Uber induzierte

Abwehreaktionen bei Orchideen.," *Phytopath. Z.*, 17: 36-62.

HIETALA, P.K., 1960, "A counter current distribution method for separation of chemical compounds," *Ann. Acad., Sci. Fennicae Serr.*, A 11: 100, 1-69.

HIURA, M., 1943, "Studies in storage and rot of sweet potato," *Sci. Rep. Gifu Coll. Agric.*, 50: 1-5.

HUGHES, J.C. and SWAIN, T., 1960, "Scopolin production in potato tubers infected with *Phytophthora infestans*," *Phytopathology*, 50: 398-400.

JOHNSON, G. and SCHAAL, L.A., 1952, "Relation of chlorogenic acid to scab resistance in potatoes," *Science*, 115: 627-629.

KLEMENT, Z. and GOODMAN, R.N., 1967, "The hypersensitive reaction of infection by bacterial plant pathogens," *Ann Rev. Phytopath*, 5: 17-44.

KIRLAY, Z. and FARKAS, G.L., 1962, "Relation between phenol metabolism and stem rust resistance in wheat," *Phytopathology*, 52: 657-64.

KUBOTA, T. and MASTURA, T., 1953, "Chemical studies on the black rot disease of sweet potato," *J. Chem Soc. Japan Pure Chem. Sect.*, 74: 248 251

KUC, J., 1963, "The role of phenolic compounds in disease resistance," in *Perspectives of biochemical plant pathology*, S. Rich (eds.), *Bull 663 Conn. Agric. Expt. St.*, pp. 20-30.

MATTHEWS, R.E.F., 1970, *Plant Virology*, pp.771. Academic Press, New York.

MULLER, K.O. and BORGER, H., 1940, "Experimentelle Untersuchungen uberdie *Phytophthora*—Keristenz der kartoffel," *Arb. Biol.*, Reichsanstalf Landu. Fortwirtsch, Berlin. 23: 189-231.

PERRIN, D.R. and BOLTOMELY, W., 1962, "Studies on Phytoalexins. V. The structure of pisatin from *Pisum sativum* L.," *J. Am. Chem. Soc.*, 84: 1919—1922.

PERRIN, D.R., 1964, Tetrahedron letter No. 1, p. 29.

ROMANKO, R.R., 1959, "A physiological basis for resistance of Oats to victoria blight," *Phytopathology*, 49: 32-36.

SCHAFER, J.F., 1971, "Tolerance to plant disease," *Ann. Rev. Phytopathol*, 9:235-252.

SCHELLENBAUM, M., 1959, "Isolierung und knostitutionsaufklarung ais orchinols (Dissertation Eidgenossis chen Technischen Hochsehule, Zurich, Prom. Nr. 2977

SELMAN, I.W., 1964, "The effect of kinetin on infection of Petunia and tomato leaves with tomato spotted wilt virus," *Ann. Appl Biol.*, 53: 67-76.

SOKOLOVA, V.E., SOVELYEVA, O.N. and RUBIN, B A., 1958, "The character of conversions of chlorogenic acid in potato tubers infected with *Phytophthora infestans*," C.R *Acad Sci.*, U.R.S.D 123, 335.

SONDHEIMER, 1961, "Possible identity of fungistatic compounds from carrot roots," *Phytopathology*. 51: 71-72.

UHERA, K., 1963, *Bull Hiroshima. Agric. College*, 2: 41.

URITANI, I., 1963, "The biochemical basis of disease resistance induced by infection," in *Perspectives of biochemical plant pathology*," S. Rich (ed.), *Bull. 663. Conn. Agric. Expt. St.*, New Haven, U.S.

12

GENETICS OF HOST-PARASITE INTERACTIONS

PATHOGENICITY and resistance are heritable characteristics of the pathogen and the host, respectively. Since Biffen's elucidation (1905) of the inheritance of the resistance in simple Mendelian fashion, spectacular progress has been made in our understanding of the genetic aspects of parasitism and disease resistance. The mechanisms of variability that make the pathogens versatile in their behaviour and host range are now well known. The genetic basis of host specialization has been elucidated in most elegant experiments of Flor. We shall examine some of these aspects here.

The study of the inheritance of pathogenicity or resistance is done by hybridization of species, race or strains and examining the pattern of distribution of this character in the progeny. The character that appears in the F_1 generation is the dominant character and the ratio in which the dominant character is distributed in the F_2 generation tells whether the character is regulated by a single gene (3:1 ratio) or two genes (15:1 ratio).

Inheritance of Pathogenicity

Virulence is mostly inherited as a recessive character. Haliskly (1965) crossed race 6 (avirulent) and race 7 (virulent) of *Ustilago avenae*, and found that all the individuals of the F_1 generation were avirulent as shown by their failure to infect the host. This suggested that avirulence was dominant. In the F_2 generation the ratio between the avirulent and virulent individuals was 3:1. Thus in this case virulence was governed by a single gene with avirulence being the dominant character

Flor, working with Flax rust, found that in the cross between the virulent, (race 22) and avirulent (race 25), all the F_1 offsprings were avirulent on the host. The F_2 progeny showed 3:1 ratio between the avirulent and virulent individuals. Thus the inheritance of pathogenicity was monogenic, with avirulence as the dominant character.

In wheat stem rust also, avirulence was found to be dominant and controlled by a single gene.

Similar pattern is known for wheat stem rust (*Puccinia graminis tritici*). However, there are cases where the F_1 individuals showed virulence greater than the parents. This is called "transgressive segregation". This was found for races belonging to *Tilletia caries*, and *Tilletia foetida*. Again, crosses between *Sphacelotheca sorghi* and *S. cruenta* (*Sorghum* smuts), the F_1 progeny showed intermediate pathogenicity.

Cytoplasmic Inheritance

The appearance of the character in F_1, depending on its association with the female partner, indicates that the character is transmitted through the cytoplasm and not the chromosomes.

Race 7 and Race 11 of the wheat stem rust were hybridized. The F_1 offsprings showed pathogenicity of the female parent. If race 7 was taken as maternal partner, the F_1 individuals behaved as race 7. When this was reversed and race 11 served as female parent (and race 7 as male), F_1 offsprings were all similar to race 11 in infectivity.

Inheritance of Resistance in the Host

Unlike pathogenicity, resistance is usually inherited as a dominant character. One, two or more genes may govern the resistant character. Accordingly, resistance is monogenic, bigenic, oligogenic or polygenic. The single dominant resistant gene is called major gene, while components of the polygenic control are called minor genes. The inheritance of resistance in flax against *Melampsora lini* (Flax rust) is the best studied example. Henry in 1930 first examined the inheritance by crossing the immune varieties (viz., Bombay, Ottawa 770 B and Argentine Selection) with the susceptible varieties (viz., Saginaw or Winona). He invariably found all F_1 individuals to be immune against rust infection. This indicated the dominant nature of the gene governing resistance. The F_2 indviduals showed the presence of the resistance and susceptible individuals in the ratio of 3:1 in some crosses and 15:1 in others. This indicated that resistance was monogenic as well as bigenic in the different varieties.

Resistance as a dominant character has been found with wheat and barley against powdery mildew, wheat against rusts and potato against *Phytophthora*. But always resistance is not dominant.

Monogenic resistance (=Major gene resistance) is more stable and provides greater resistance to specific races of pathogen than the polygenic resistance, which provides resistance against large number of races, though with lesser efficiency.

Genetic Aspects of Host-Parasite Interaction

The genetic basis of host-parasite interrelationships has been most thoroughly studied in flax rust. Flor (1955), after a very detailed study, gave a 'gene for gene' hypothesis to explain the genetic basis of host specificity, which is now found to be true for several other host-pathogen systems. No area of knowledge is so much identified with one man as is the undertanding of genetics of host-parasite interaction with H.H. Flor.

Nature of Resistance and Virulence in Flax Rust

Flor found that resistance in flax against the rust (*Melampsora lini*) was inherited as a dominant character. while the virulence of the pathogen was controlled by a recessive gene. From over 200 existing varieties of flax, 16 were selected to be used as differential varieties for the identification of races. By inoculating all the available rust collections on these differential varieties, 24 races of the rust were identified. These races were studied for the homozygous or heterozygous nature of virulence. Out of the 24 races tested, two races, (viz., race 22 and race 24) were homozygous for virulence.

Flor identified 27 dominant genes for resistance (*R* genes) which occur on 5 loci as multiple alleles (about 5 *R* alleles for each locus). These loci were designated as *K*, *L*, *M*, *N* and *P*. Three of these loci (*K*, *L* and *M*) were inherited independently while two (*N* and *P*) were linked. Thus in any plant five genes for resistance can be incorporated. He also found that the virulence, a recessive character, was inherited independently; not linked and there was no evidence of multiple allelism in *V*-genes.

Several other studies have confirmed that (*i*) resistance is inherited as a dominant character whereas the virulence is recessive, and (*ii*) that resistant genes are allelic while the virulent genes show no allelism.

The 'Gene for 'Gene' Relationship

The 'gene for gene' relationship develops as follows:

The introduction of a resistant new gene (*R*) in the host favours the survival and spread of resistant hosts. Those which lack the *R* gene are eliminated from the host population. This change, which increases the number of resistant plants has an impact on the populations of pathogens as well. Those that can survive (that have virulent gene *V*) on the resistant host, though initially in minority, increase their number by reproduction. Natural selection allows spread only of the virulent pathogens having the *V* gene. Thus by this shift, the population of pathogens comprises those which have the *V* gene. Now the host population is open to attack by virulent pathogens, the value of *R* gene in the host diminishes and ultimately falls to zero when the pathogenic population comprises of individuals having the *V* gene only. In the process, the resistant hosts become susceptible by the effects of *V* gene. The *R* gene thus becomes useless and is no more spread by selection. It is well known that several rust resistant wheat varieties are rendered susceptible by the development of virulent rust races and new resistance genes are needed.

Thus the selection of the *R* and *V* genes are interdependent. The *R* gene is selected and spread when *V* gene is rare and *vice versa*. This continues in a cyclic manner. The interaction between *R* and *V* loci generate a cycle of selection pressures that leads to maintenance of a balanced polymorphisms in the two interacting populations.

By each addition of a resistant gene, the number of parasitic regenome is reduced to half. How? We shall examine. When host resistance is dominant and allelic, as in flax rust, a heterozygote, say $R_1 R_2$ will show wider range of resistance than their two homozygotes, R_1, R_1 or R_2,. R_2. Thus in populations, breeding by outcrossing, the heterozygote superiority maintains the *R* allele polymorphism. The selected heterozygote host population will bring forward two related genes and not one. Since virulence in the pathogen is a recessive character, changes at a single host locus will affect gene frequencies at two loci in the pathogen. Also, replacement of either of the two alleles at the host locus by a third resistance gene will affect the gene frequencies at a third locus in the parasite. This can best be explained with an example The resistant host genotype $R_1 R_1$ will bring forward virulent genes $V_1 V_1$, $V_2 V_2$ in the pathogen and ultimately become susceptible; a new resistant gene R_3 will replace R_1 or R_2 to confer resistance. This will make the plant to filter out $V_1 V_1$, $V_2 V_2$ pathogens in favour of any individuals having $V_3 V_3$ genes for virulence. Since *R* genes are polymorphic in the host, only those pathogens will survive on a resistant sector of a susceptible population, which will have virulence genes at three different loci, at least.

Flor found that usually each host locus has five alleles. This makes it necessary, then, that the pathogens must have five related virulent genes at five different loci. The five changes from susceptibility to resistance, that are possible in the host, will each require a matching response at a new *V* locus, which till then was not contributing to the virulence of the pathogen.

Person (1966, 67) gave a model to explain the above situation (Fig. 12.1).

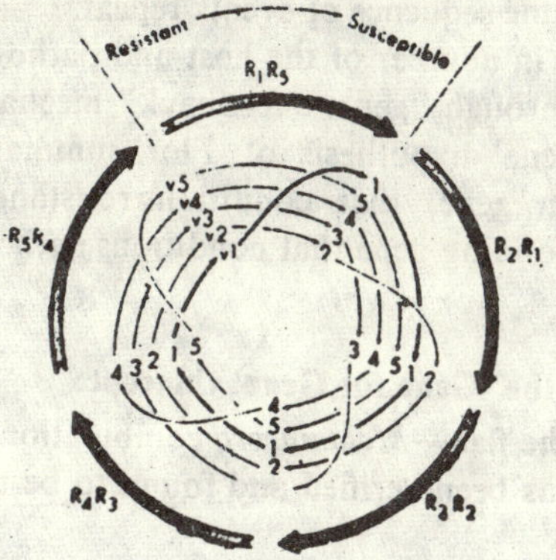

Fig. 12.1. Shows interaction between the five *R*-alleles (for resistance and their five related *V*-loci (virulence) in a system of balanced polymorphism.

The figure shows that the five *R* alleles at a single locus and their five related *V* loci may interact in a system of balanced polymorphism. The figure does not illustrate the relative frequencies of the different *R* alleles or of the alleles for avirulence at the five *V* loci.

The outer circle shows that a single locus in the host, the five *R* alleles appear in sequence R_1 to R_5. The five inner circuits represent the change in the frequency (abundance in population) of *V* alleles at the five loci in the parasite. The *V* allele moves outward rapidly from centre with the introduction of a new *R* allele and falls slowly inward with the replacement of the *R* allele by another.

The figure illustrates what happens in the five successive conversions of host population from resistance to susceptibility.

As will be evident, at each of the five steps in the cycle the incoming *R* gene is the one whose related *V* gene is at lowest frequency. It shows that the incoming *R* allele has high selective value. The resistance allele is one against which pathogenic population has least individuals (lowest frequency) with virulence gene to survive the resistance. With the introduction of the *R* allele, the frequency of the related *V* allele in the pathogen increases and reaches maximum when all pathogens in the population are those which have the virulent *V* allele. By this time, the resistance of the host has yielded place to susceptibility. When the *R* allele becomes ineffective, the frequency of the allele declines—and reaches the level that allows reintroduction of the new *R* allele and the second step starts. The same sequence of events repeat.

The fluctuations in number of the host and pathogen are brought into equilibrium through "genetic feedback" mechanisms

The 'gene for gene' hypothesis of Flor, summarises the above situation. For each gene, that conditions resistance in the host, there is a corresponding gene that conditions pathogenicity in the parasite.

Current Status of the 'Gene for Gene' Concept

In addition to the flax—*Melampsora* combination, the 'gene for gene' hypothesis has been verified and found to be true for several other systems.

Wheat

Rusts: *P. graminis tritici: P. striiformis* and *P. recondita*
Smuts: *Tilletia caries, T. controversa* and *Ustilago tritici*
Powdery Mildew: *Erysiphe graminis tritici*

Avena: Puccinia graminis avenae and *Ustilago avenae*
Solanum: Phytophthora infestans, Synchytrium endobioticum
Helianthus: Puccinia helianthi
Coffee: *Hemileia vestatrix*
Malus: Venturia inaequalis
Lycoperiscon: Cladosporium fulvum
Hordeum: Erysiphe graminis tritici.

The hypothesis is applicable to most host-pathogen systems in which resistance is conditioned by major (vertical resistance) genes; virulence increases in step wise fashion. Hatchett and Gallun (1970) have reported a similar 'gene for gene' relationship between plants and insects. Flor (1971) suggests the possibility of similar relationship between other insects, nematodes, bacteria and viruses and their host plants.

REFERENCES

BIFFEN, R.H, 1905, "Mendel's laws of inheritance and wheat breeding," *J. Agric. Sci. Camb*, 1: 4-48.

FLOR, H.H., 1955, "Host parasite interaction in flax rust, its genetics and other implications," *Phytopathology*, 45: 680-685.

———, 1956, "The complementary genic systems in flax and flax rust," *Adv. Genetics*, 8: 25-54.

———, 1971, "Current status of the gene for gene concept," *Ann. Rev. Phytopathol*, 9: 275-296.

HALLISKY, P.M., 1965, "Physiologic specialization and genetics of the smut fungi. III," *Bot. Rev.*, 31: 114-150.

HATCHETT, J.H. and GALLUN, R.L., 1970, "Genetics of the ability of the Hessian fly *Mayetiola destructor*, to survive on wheats having different genes for resistance," *Annals. Ent. Soc.*, America 63: 1400-1407.

HENRY, A.W., 1930, "Inheritance of immunity from flax rust," *Phytopathology*, 20: 707-721.

PERSON, C., 1966, "Genetic polymorphism in parasitic systems," *Nature*, 212: (5059) 266-267.

———, 1967, "Genetic aspects of parasitism," *Can. J. Bot.*, 45: 1193-1204.

13

PHYSIOLOGY OF THE DISEASED PLANTS

DISEASE has been defined as a malfunctioning process. The sick plant suffers. Its all metabolic processes get upset, the net effect of which appears in the form of visible external symptoms. A plant pathologist never knows when the plant got sick, how rapidly it worsened and whether it is recovering and then at what rate. The plant never 'tells' its suffering; and to use Horsfall's phrase, it never groans, grumbles, never has convulsions, fever, no pulse rate, no blood pressure and has no red cells to count. Inspite of this, plant pathologists have done well in elucidating the crippled metabolism of the sick plant.

Any process that is looked into is found deranged. Osmoregulation, photosynthesis, respiration, nitrogen metabolism and phenol metabolism are the best studied. Some of these disturbances are 'primary' while others are secondary effects of the former. Osmoregulation and respiration are found in all kinds of diseases to be the first affected processes. The host-parasite relationship involves a chemical warfare between the host and the pathogen. Pathogenicity is determined by the successful production of chemicals inside the host plant tissue, which alter the host metabolism in a way that is beneficial to the pathogen. The obligate parasites do it in a 'subtle' way without making much structural damage to the host tissue. They don't cause death of cells. They establish a relation of "peaceful co-existence" (Gäuman's phrase). Rusts, powdery mildews and downy mildews are obligate parasites, which are recognised by fungal structures and not other 'symptoms'. Thus we see conidiophores and conidia in the 'powdery' and 'downy' growth. In rusts, we see fungal sori with least damage to the host tissue. The facultative parasites are 'unsophisticated' that way. They run down the host tissue, secrete toxic chemicals and enzymes that degrade cell walls and organelles with the result that the tissues die. In fact, most of these pathogens, in a way, live saprophytically 'eating' dead food on the living host, unlike the 'elite' obligate parasites. No wonder, then, that all the joints of the metabolism are disarticulated.

The host, on its part, struggles, fights to lose or to win. For this they have ready made 'chemicals' toxic to the fungi, or, synthesize them after coming in contact with the pathogen. Phenols are the chemicals which, more than any, are involved in resistance. These are secondary metabolites, not needed in normal physiology.

OSMOREGULATION

Breakdown of osmoregulation occurs due to damage to the membranes. This brings about loss of turgor, leakage of ions and other important substances, which form the most important 'mileu interior'. The loss of ions disturbs the ionic environment which is needed for the functioning of several enzymatic reactions of the cells. The 'ionic imbalance' sets in. The damage to membranes of the roots disturbs the uptake of nutrients and water and in other tissues it disturbs their retention and accumulation.

Much work has been done with toxins. For victorin and some other host specific toxins it has been conclusively proved that membrane damage is the first act of the pathogen. Involvement of toxins in breaking osmoregulation of host cells has been examined in chapter 9. It is now being believed that permeability damage is a common initial event in disease incited by all kinds of pathogens. It precedes changes in respiration, which is usually supposed to be the first expression of disease. It has been proved with victorin that permeability alterations are the cause of respiratory disturbances.

Thatcher (1943) offered a "nutrition hypothesis" of parasitism, based on permeability damage of infected plants. Increased permeability of host tissue membranes aids in the establishment of the pathogen. On the opposite, a decreased permeability offers resistance to infection. Permeability damage, as indicated by leakage of ions, has been detected in the early stage of infection in several diseases caused by fungi, bacteria and viruses. Cook and Stall (1968) found loss of electrolytes from leaves of susceptible *Capsicum annum* infected with *Xanthomonas vesicatoria*. A causal relation between permeability changes in root tissues and wilting has been shown in tobacco pepper plants (*Capsicum frutescens*) infected with tobacco etch virus by Ghabrial and Pirone (1967). Sadasivan and his associates reported 'ionic imbalance' in *Fusarium*-infected cotton plants (Gnanam, 1956; Sadasivan and Saraswathi Devi, 1957). A similar "ionic imbalance" has been reported in *Verticillium* wilt of cotton and blast disease of rice by Dube (1971, 1973). All these workers have reported potassium and calcium to be lost maximally, which when supplied externally, provide resistance. Calcium determines the overall thickness of unit membrane; and of all elements only calcium deficiency results in membrane disruption. Bateman (1966) related resistance to calcium for beans against *Rhizoctonia solani*. Membrane damage has been observed in electron microphotographs of infected tissues.

Hall and Wood (1970) and Mount and Bateman (1970) have reported endo-PGTE (endo-polygalacturonate *trans*-eliminase), a pectic enzyme, to damage plasma membrane and cause leakage of ions. The rupturing of the plasma membrane occurs at plasmodesmata, as a result of excessive influx of water from the hypotonic solution (formed by enzymic dissolution of the middle lamella) into host cells. Mechanism of membrane damage by toxin action is discussed in chapter 9. In addition to toxins, phosphatidase and tyrosinase have also been implicated by Lumsden and Bateman (1968) and Cory (1967) respectively, in membrane damage. Dube and Srivastaya (1975) and Wheeler and Hanchey (1968) have reviewed studies on the role of permeability damage in plant disease.

PHOTOSYNTHESIS

The destruction of chloroplast and chlorophyll is a common feature in leaves showing chlorotic symptoms, which are most commonly encountered with viruses. Several bacterial and fungal pathogens too cause chlorosis. Very little information is available regarding disturbances in the photosynthesis of diseased plants. Chlorosis and necrosis of tissues, detachment of leaves and tissues and defoliation naturally affect the amount of the net photosynthate. In addition, the ability of the existing chlorophyll is impaired.

Chlorophyll Content

The enzyme chlorophyllase degrades chlorophyll into chlorophyllide (chlorin ring containing the magnesium) aad the phytol chain.

$$\text{Chlorophyll} \rightleftharpoons \text{chlorophyllide} + \text{phytol}$$

The chlorophyllide crystallizes outside the chloroplast, in the cell. Reduced chlorophyll content and increased chlorophyllase activity has been reported by Peterson and McKinney (1938) for mosaic diseases caused by four viruses. The chlorophyll degradation results in destruction of the lamellar arrangement in the chloroplasts. The formation of new plastids in the growing tissues is affected. However, in several cases, "green islands" appear due to increased chlorophyll content. Wang (1961) showed that in bean, infected by *Uromyces phaseoli*, the 'green islands', at the margin of the infection, were not only photosynthetically active but recorded greater starch content than the adjoining healthy tissues. In powdery mildew of wheat, Allen (1942) recorded reformation of chlorophyll after an initial destruction, in the 'green islands'. The rest of the leaf continued to be chlorotic. Allen also found that the photosynthetic activity of the existing chlorophyll molecules in powdery mildew-infected wheat leaf, was unaffected. Braun (1955) found that the chlorophyll, obtained from toxin treated tobacco leaf tissue (wild fire toxin produced by *Pseudomonas tabaci*), were spectrophotometrically similar to those obtained from healthy tissues.

Effect on Photosynthetic Reactions

Photosynthetic Phosphorylation and Hill Reaction. Decreased rate of 'Hill reaction' (photochemical splitting of water: evolution of oxygen by illuminated chloroplast preparations) has been observed in chloroplasts isolated from infected tissues e.g. in advanced sugar beet yellows–virus infections (Spikes and Stout, 1955). Zaitlin and Jagendorf (1960) observed decreased photosynthetic phosphorylation and 'Hill reaction' in TMV-infected tissues. Similar results are known for wilt disease of cotton caused by *Verticillium albo-atrum*. This results in shortage of NADPH and ATP, which are needed in the CO fixation cycle

CO_2 Fixation. Enzymes of the CO_2 fixation are not bound to any structural elements of the chloroplast; they lie in the chlorophyll free, water-soluble portion of the chloroplasts. Usually these enzymes are not affected by virus infections.

The 'key' reaction of the Calvin cycle *viz*. fixation of CO_2 to the Ribulose diphosphate is mediated by the enzyme carboxydismutase (ribulose, 1-5 diphosphate carboxylase). Carroll and Kosuge (cf. Strobel and Mathre, 1970) reported that in TMV-infected tissues, the carboxydismutase activity declined with the degradation of the chloroplasts, and ultimately became undetectable in advanced phase of the disease.

Decreased rate of CO_2 fixaticn has been reported for several diseases. Also, increased dark fixation of CO_2 has been observed in rust (*Uromyces phaseoli*) infected bean leaves. Dry weight of these infected leaves increased (Zaki and Mirocha, 1964). Liven (1964) also observed similar enhancement of CO_2 fixation in *Uromyces*-infected bean and safflower. Thus, the infection was beneficial. Liven also found that the increased CO_2 fixation was characteristic of non-infected leaves in the rust infected bean plants. It is known that in rust diseases carbohydrates are transported to infected tissues from non-infected areas. The increased CO_2 fixation could, then, explain the increased carbohydrate synthesis.

Starch Metabolism. Accumulation of starch around rust pustules has been demonstrated. The carbohydrates from healthy tissues get diverted to the site of infection, where starch synthesis and accumulation occurs.

Virus infected plants show either excessive accumulation or depletion of starch. TMV-infected leaves show decreased capacity of both synthesis and degradation of starch. Starch accumulation is characteristic of some virus diseases e. g. potato leaf-roll, sugar-beet yellows and barley yellow dwarf.

RESPIRATION

The respiration of infected plants has received much attention from plant pathologists and an impressive literature, including several reviews (Allen 1953, 1954, Millerd and Scott 1962, Uritani and Akazawa, 1959, Sadasivan, 1963), is available in this area.

In 1934, Yarwood showed that leaflets of red clover (*Trifolium pratense*) infected with *Erysiphe polygoni*, had a higher rate of respiration (O_2 uptake) than the healthy ones. Since then, increased respiration is known to be a general phenomencn in the physiology of diseased plants. This has been demonstrated in several host plants infected by fungi bacteria and viruses. Chemical and mechanical injuries, too induce an increase in the rate of plant respiration and, therefore, it is an unspecific reaction of the host. This is quite understandable because any kind of disturbance in the metabolism will immediately show up in respiration.

Studies on respiratory stimulation were confined mainly to

quantitative aspects of increase in respiration rate of infected plants, until 1953, when Allen postulated that the increased respiration in mildewed leaves was the result of action a toxin produced by the fungus which uncouples phosphorylation from electron transfer in host cells (uncoupling of oxidative phosphorylation). Subsequently Allen (1954) suggested that stimulation of anabolic processes in infected tissues could also cause an enhancement in respiration. The involvement of new pathways, not operative in healthy tissues, too have been implicated which we shall examine later in this chapter.

The increase in respiration depends not only upon the nature of the pathogen but also on the genetic make up of the host. With the same host-parasite combination the respiratory response varies with the variety of the host and depends upon the degree of its resistance or susceptibility.

MECHANISM OF RESPIRATORY INCREASE

The direct contribution of parasite in the increased oxygen uptake is negligible. To bring about the augmented respiration in diseased plants, three mechanisms have been suggested: (*i*) Uncoupling of oxidative phosphorylation; (*ii*) augmented synthetic process (acceleration of ATP utilizing systems), and (*iii*) induction of respiratory enzyme systems not involving phosphorylation.

Uncoupling of Oxidative Phosphorylation

2,4-Dinitrophenol (DNP) is a well known uncoupler of oxidative phosphorylation, and causes increased oxygen uptake. Infected plants are usually insensitive to DNP treatment, and this indicates that the fungus has already brought about the uncoupling reaction, perhaps through a DNP like metabolic product. The "uncoupling", by preventing the formation of ATP molecules, releases the control over electron flow, imposed by the formation of high energy state. The untrapped energy appears in form of heat as is evidenced by the higher temperature recorded in diseased tissues. Yarwood (1953) reported an increase of 0.3 to 0.7°C of temperature in mildew and rust diseases. However, conclusive evidence to indicate lesser efficiency of augmented respiration in disease is lacking and, therefore, the cause of the increased oxygen uptake may be searched elsewhere.

The uncoupling has been reported in the diseases caused by

facultative parasites like *Helminthosporium victoriae*, *Gibberella saubinetti*, *Cercospora personata* and obligate parasite, *Puccinia graminis* causing rust of wheat.

Augmented Synthesis as Cause of Accelerated Respiration

The increased oxygen uptake is frequently accompanied by an acceleration of ATP utilizing systems, such as those involved in protein, nucleic acid, carbohydrate synthesis, production of organic phosphorus compounds and protoplasmic work. Protein synthesis is the most important among the ATP utilizing systems. This results in an increased ADP level which, in turn, enhances respiratory rate. Activation of protein synthesis has been documented in the following diseases (Table 13.1).

TABLE 13.1. DISEASES IN WHICH ACTIVATION OF PROTEIN SYNTHESIS HAS BEEN STUDIED

Diseases	*Augmented synthesis*	*References*
Root rot of sweet potato/ *Ceratocystis fimbriata*	Protein synthesis	Akazawa and Uritani (1955)
Late blight of Potato/ *Phytophthora infestans*	Protein synthesis, production of peroxidase isoenzymes.	Tomiyama (cf Stahmann, 1963)
Safflower infected with *Puccinia carthemi*	Stimulated growth	Daly and Sayre (1957)
Rust of bean leaves/ *Uromyces phaseoli*	Marked changes in proteins of host. Changes in isoenzymes of its enzymes.	Staples (cf. Stahmann, 1963)
Halo blight of beans/ *Pseudomonas phaseolicola*	Changes in isoenzymes pattern of 3 enzymes	Rudolf and Stahmann, (1964)

Induction of Respiratory Enzyme Systems

Increased respiration reflects activation of oxidative enzymes, which has been reported in many diseases. It is believed that in the early stage of infection only the normal route of oxidation, viz., cytochrome electron transport system is geared up and in advanced infections, during disintegration of tissues, other "terminal" oxidases, like ascorbic acid oxidase, polyphenol oxidase, catalase and peroxidase enzymes become active. The association of polyphenol oxidases with increased respiration in local lesion formation in tobacco tissues infected with TMV, and absence in systemic infections, led Merrett and Sunderland (1969) to suggest their relation to the necrotic process, rather than to virus multiplication.

The role of these oxidases is obscure in the metabolism of plants and these enzymes are not coupled with oxidative phosphorylation. Thus the activation of these enzymes results in greater consumption of oxygen with no concomitant ATP gain This simulates the action of the DNP. On the other hand the observation of light reversible inhibition of respiration by CO in mustard hypocotyls of safflower by Daly and Jensen (1961) proves an important role of cytochrome oxidase in oxygen uptake.

Reports of activation of oxidative enzymes are given in Table 13.2.

Abolition of Pasteur Effect

The increased rate of respiration, brought about by whatsoever mechanism, results in increased concentration of ADP and inorganic phosphorus in infected tissues. Either of these events also brings about abolition of the 'Pasteur effect', the importance of this phenomenon was emphasized by Sempio (1950). Pasteur effect depicts efficient utilization during aerobic breakdown of glucose. Abolition of this effect means poor utilization (which occurs during glycolysis), even in the presence of oxygen. In effect, this amounts to aerobic fermentation. Thus respiratory efficiency of infected tissues declines due to poor utilization of glucose. An important gain to their plant in the abolition of the Pasteur effect, is the accumulation of intermediates of carbohydrate breakdown, which are utilized in the formation of phenolic compounds. Phenol synthesis is triggered in diseased tissues.

Shifts in Pathways of Respiration

Alterations in respiratory pathways is another feature of the respiratory metabolism of the sick plant. Such alterations are detected by observing the response of the plant to chemicals which specifically block enzymes of the respiratory pathways. Thus, insensitivity to sodium fluoride shows inaction of glycolysis. Malonic acid, a competitive inhibitor of succinic dehydrogenase, if fails to depress oxygen uptake, suggests that the Krebs cycle is inoperative. Operation of the HMP pathway is indicated by a decline in the C_6/C_1 ratio (less than 1), which is measured with isotopes or by increased activity of enzymes, characteristic of this pathway e.g.

TABLE 13.2. LIST OF DISEASES IN WHICH THE OXIDATIVE ENZYMES ARE REPORTED TO INCREASE

Enzyme	*Host/Pathogen*	*Reference*	*Remarks*
Ascorbic acid oxidase	Wheat leaves/*Puccinia triticina*	Kiraly and Farkas (1957)	Brings about electron transfer from $NADPH_2$ (generated in HMP pathway) to oxygen (O_2) suggested by Goddard (1951-52).
	Oats/*Helminthosporium victoriae*	Grimm and Wheeler (1963)	
	Cabbage/*Fusarium oxysporum*		
	Rice/*Piricularia oryzae*	cf. Goodman *et al*. (1969)	
	Tobacco/*Pseudomonas solanacearum*	Maine (1960)	
Polyphenol oxidase	Tomato/*Fusarium lycopersici*	Sanwal (1956)	
	Wheat leaf (resistant variety)/*Puccinia graminis*	Kiraly (1959)	Related to resistance and local lesion formation in virus infections. No increase in systemic infections.
	diseases caused by/*Fusarium* spp. /*Phytophthora, Cochliobolus, Ceratocystis, Thielavia*, and *Helicobasidium*	(cf. Farkas and Kiraly, 1962)	
	Tobacco/*Pseudomonas solanacearum*	Maine (1960)	
	Tumor tissue of tomato/*Agrobacterium tumefaciens*	Nagy *et al*. (1938)	
	Tobacco/Tobacco mosaic virus	Farkas *et al*. (1960)	
	Datura stramonium/Cucumber mosaic virus (CMU)	Farkas *et al*. (1960)	
Peroxidase	Potato/*Phytophthora infestans*	Tomiyama (cf. Stahmann, 1963)	
	Gall tissue of tomato/*Agrobacterum tumefaciens*	Nagy *et al* (1938)	
	Bean/*Pseudomonas phaseolicola*	Rudolf and Stahmann (1964)	Increased peroxidase favours resistance
	Cucumber/Cucumber mosaic virus	Menke and Walker (1963)	
	Sweet potato/vein clearing virus	Loebenstein and Lindsey (1963)	

glucose-6-phosphate dehydrogenase and 6-phosphogluconate dehydrogenase. Such change over to HMP pathway is recorded during mechanical injury also.

Changes in the operation of the Krebs cycle have been recorded in several plants by observing insensitivity to malonic acid treatment. Some examples are as follows: (*i*) Rusted and mildewed wheat; (*ii*) potato tuber infected by *Gibberella saubinetii*; and (*iii*) groundnut infected by *Cercospora personata*.

However, Uritani and Akazawa (1962) have reported a more vigorous glycolysis and Krebs cycle in black rot of sweet potato caused by *Ceratocystis fimbriata*.

A shift from glycolytic to HMP pathway has been detected in several diseases, caused by fungi, bacteria and viruses (only in local lesion type). Some examples are:

(1) Rust and mildewed wheat.
(2) TMV-induced local lesion in *Datura stramonium*.
(3) Potato tubers infected with *Phytophthora infestans*.
(4) Tomato tumor incited by *Agrobacterium tumefaciens*.
(5) Halo blight of Tobacco/*Pseudomonas tabaci*.

Respiration and Resistance

As early as 1912 Bach suggested that oxidative enzymes might protect the host by oxidizing the toxic substances of parasitic origin. His words were prophetic indeed. Sempio (1950) in his concept of metabolic resistance to parasitic attack emphasized that it is the ratio between synthetic reactions (like photosynthesis) and breakdown reactions (like respiration) that decides resistance or susceptibility. Tissues having higher rate of photosynthesis and low rate of breakdown will be resistant and *vice versa*. Thus, if a resistant and susceptible variety, both have same photosynthetic rate, the susceptible variety should respire faster than the resistant.

Andal (1959) and Lakshmanan(1959) found direct relation between susceptibility and rate of respiration. However, opposite conclusions were arrived at by Rubin and his co-workers in USSR (cf. Farkas and Kiraly, 1958), who observed that resistant varieties of potato, cabbage and citrus (against *Phytophthora*, *Botrytis* and *Penicillium* respectively) respired more vigorously than the susceptible plants. Swamy (1964) made similar observations with resistant varieties of groundnuts against *Cercospora personata*.

The pattern of respiratory rate changes with the advancement of disease. Samborski and Shaw (1956) observed that in wheat rust, the resistant variety recorded a steep rise followed by an equally steep fall of respiration, while the susceptible variety had a steady continuous rise of respiration. Similar observation was made by Millerd and Scott (1956) in barley leaves infected by powdery mildew.

Activation of polyphenoloxidase activity results in an active production of toxic substances (quinones) that arrest the growth of pathogen though at the cost of the death of host cells. Increased phenol synthesis during the "aerobic fermentation" operating in infected cells, is another important contribution of pathological respiration towards active resistance. The peroxidase enzymes have been implicated in resistance of bean against *Pseudomonas phaseolicola* by Rudolf and Stahmann (1964). Resistant varieties produced more peroxidase enzymes than the susceptible variety. Supporting evidence came from higher catalase production by the virulent isolates of the pathogen. The enzyme sapped the resistance of the host by suppressing its peroxidase activity.

It is an important observation that resistance of the host to fungal parasites can be reduced by inhibiting the oxidative enzymes. This proves that oxidative enzymes participate in 'active' disease resistance. The fact that the respiratory inhibitors also inhibit virus multiplication, suggests a link between the two.

NITROGEN METABOLISM

Changes in Protein Synthesis

Infected plants show change in the quality and quantity of their proteins. Some proteins are formed by the pathogens themselves in the infected cells. The adjacent non-infected tissues also show stimulation and degradation of specific proteins. Maximum changes in protein metabolism are brought about by viruses.

On entering the host cell, the infectious RNA induces the formation of RNA synthetase. Since this RNA is degradable by host RNase, the virus multiplies in the nucleus, which lacks RNases. The enzyme RNA synthetase (=RNA replicase or RNA dependent RNA polymerase) brings about the synthesis of viral RNA, or RNA template. It does not need DNA as primer and is insensitive to actinomycin-D. The RNA polymerase of the host cells, on the other hand needs DNA primer and is inhibited by actinomycin-D. The infecting virus RNA is single stranded but now there is evidence that during the replication the single strand forms a duplex, two stranded structure, called the replicative form (RF). The complementary strand (minus strand) serves as template for the production of the original plus RNA strand in presence of

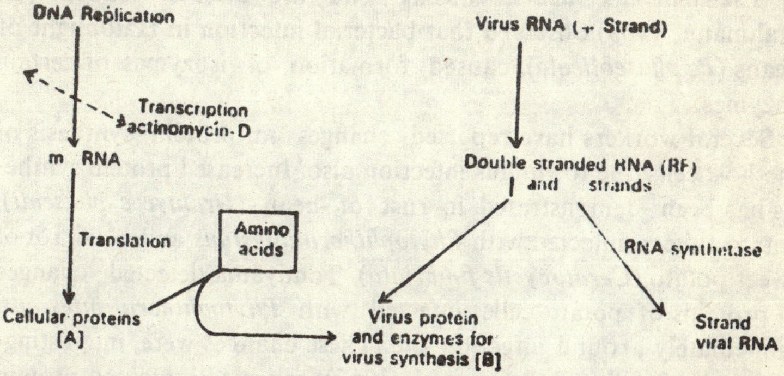

Fig. 13.1. Illustrates RNA and protein synthesis in healthy (A) and virus-infected (B) cells.

RNA synthetase. Thus the negative strand serves as the messenger RNA for viral RNA synthesis. A second enzyme is postulated to catalyze the production of the minus strand. The genetic information for the coat proteins resides in the plus strand.

Reddi (1964) observed degradation of the ribosomes in TMV infected tobacco tissues. Ribonucleoside moieties are utilized in the synthesis of viral RNA. The amount of TMV-RNA synthesized was approximately equal to the amount of RNA degraded. Reddi also found that cytoplasmic ribosomes were free from RNA template of the virus. This indicated that cytoplasm was not the site of virus-directed protein synthesis. On the other hand, the nuclei of the infected leaf tissues, in electron microphotographs, showed the presence of the virus particles. Reddi concluded that TMV is synthesized in the nucleus.

Penman *et al* (1963) working with Poliovirus infected Hela cells observed that the 'new' ribosomes were formed after degradation of old ribosomes on which the viral proteins are synthesized. This could be occurring in infected plants also.

The effects of TMV, as reported by Reddi, however, have not been reported in other virus diseases and a generalization about ribosome degradation can not be made at this stage. The virus infection does not make much change in the total protein content of infected leaves:

In crown gall disease, caused by *Agrobacterium tumefaciens*, the actively dividing tumor cells synthesize more proteins (3-6 times more) than the normal tissue. The opposite i.e. breakdown of proteins has been observed in tobacco tissues treated with the toxin of *Pseudomonas tabaci*, causing wild fire disease. Rudolf (cf. Stahmann, 1963) observed that bacterial infection in Haloblight of beans (*P. phaseolicola*), caused formation of isozymes of certain enzymes.

Several workers have reported changes in protein synthesis of the host cells due to fungus infection also. Increased protein synthesis has been demonstrated in rust of bean (*Uromyces phaseoli*), potato tubers, infected with *Phytophthora infestans* and black rot of sweet potato (*Ceratocystis fimbriata*). Tomiyama detected changes in proteins of potato cells infected with *Phytophthora infestans*, immediately around infection site. These changes were, interestingly, race specific. Pozsar *et al* (1966) reported increased protein content in rust infected beans and also found that the protein increase was due to protein synthesis of the fungus. Staples, working with bean rust *Uromyces phaseoli*, observed marked changes in the host proteins. He showed that the rust induces the formation of new host isozymes. The synthesis of some enzymes, which disappear as the leaf matures, was observed to continue as a result of infection.

Kiraly *et al* (1966) observed that the 'green islands' in bean rust disease surrounding the infection centre, show increased protein synthesis while the zone between the 'green islands' show protein degradation and low capacity of protein synthesis. Thus, the infected tissues show characteristic resemblance to tissues under senescence.

Changes in Nucleic Acid Metabolism

Alterations in the nucleic acid metabolism have been reported in several diseases. Changes in the nucleic acid level have been reported in diseases caused by all kinds of pathogens. The host RNA content in TMV-infected tobacco leaves increases shortly (eight hours) after inoculation but afterwards (70 hours), the viral RNA occurs in greater quantity, when the total RNA also shows an increase. Total RNA increased in infected tissues, while it decreased in the uninoculated control tissues.

The process of gall formation (*Agrobacterium tumefaciens*) involves an integration of the host and bacterial genomes. This transformation of the host cells is irreversible and growth control mechanisms go out of gears. A rise in RNA content of tumor tissues is also observed.

In rust-infected wheat, marked increase in RNA content has been observed, accompanied by enlargement of the nucleoli and the huclei. The DNA content also increases. The histones decrease. Since histones have a regulatory role, any change in its quantity is related to change in regulation of gene action. The increased RNA is contributed by both the host and the rust fungus. The nuclei after the initial swelling, collapse and get degraded. This swelling of nuclei, followed by degradation, has been observed in several fungal diseases. Swelling represents increase in RNA while collapse indicates proteolysis.

Reports of changes in nuclei and metabolism, in addition to studies on rust diseases, are available for facultative parasites also. An increased RNA, (and also DNA) has been reported in early stages of club-root galls of cabbage (caused by *Plasmodiophora brassicae*) and *Helminthosporium* leaf-blight of rice. The spurt in RNA results in increased protein synthesis.

Changes in Amino Acid Metabolism

Free amino acids usually increase in virus infected plants and it is suggested that some of the symptoms on leaves are due to changes in the amino acid concentration. Pipecolic acid an uncommon amino acid, is found in increased amounts in necrotic tissues of several diseases. Its accumulation occurs through other injuries also. It has been assigned a role in symptom production.

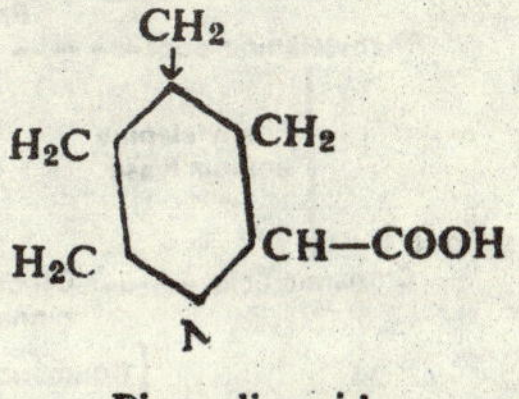

Pipecolic acid

Increased amount of free amino acids was detected in Halo-blight of bean (*Pseudomonas phaseolicola*) by Patel and Walker (1963). Some new amino acids, ornithine and histidine, which were not present in healthy tissues, were detected in diseased tissues.

PHENOL METABOLISM

Phenolic Compounds and their Biosynthesis

Phenols are compounds which contain one or more aromatic (benzene) rings with one or more phenolic hydroxyl groups. The simplest compound is phenol. Compounds having more than one hydroxyl group on the benzene ring are called 'polyphenols'. Phenolic compounds are widely distributed in higher plants and fungi anthocyanins, leucoanthocyanins, anthoxanthins, hydrobenzoic acid, glycosides, sugar esters of quinic and shikmic acids, esters of hydroxycinnamic acids and coumarin derivatives make up the vast group of phenolics. The earlier term "tannin" has been replaced by "polyphenols" to refer to any or a collection of these compounds.

Mostly they are polymerized terminal products of secondary metabolism, not important for life processes. Though simpler in structure than proteins, they offer a variety of structure which is not approached by any class of naturally occurring compounds. This variety of structure provides basis for a degree of specificity as antimicrobial agents and offers a partial explanation for the specificity of host-pathogen interaction.

Phenols are synthesized from carbohydrates via either the shikmic acid pathway, acetate pathway or modifications of both pathways. The shikmic acid pathway is more important (Fig. 13.2).

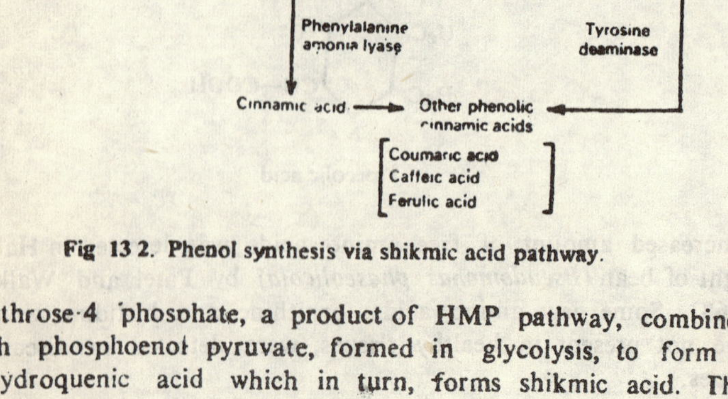

Fig 13 2. Phenol synthesis via shikmic acid pathway.

Erythrose-4 phosphate, a product of HMP pathway, combines with phosphoenol pyruvate, formed in glycolysis, to form a dehydroquenic acid which in turn, forms shikmic acid. The shikmic acid forms prephenic acid via several intermediate compounds. Prephenic acid gives rise to phenylalanine and tyrosine pools. The intermediates of these pools serve as precursors of various intermediate compounds. The deamination of phenylalanine by phenylalanine-amonia-lyase results in production of cinnamic acids, from which several phenolic cinnamic acids are formed by hydroxylation.

In the acetate pathway benzoid compounds are synthesized from polyacetic acids by head to tail condensation of acetate molecules. The synthesis of phenolics on infection would involve the accomplishment of the shikmic acid pathway. The increased rate of respiration (a general host response during pathogenesis) has been related to phenol production. The augmented respiration results from an increase in ADP concentration which is brought about by (*i*) acceleration of ATP utilizing processes like protein synthesis and cellular growth, (*ii*) by uncoupling of oxidative phosphorylation and (*iii*) by increase in respiratory enzymes, not involving phosphorylation. All these factors occur in diseased plants, resulting in an accumulation of ADP and concomitant speeding up of rate of respiration. The increased ADP synthesis results in the abolition of "Pasteur effect"[1] and accumulation of intermediate products of carbohydrate breakdown. The intermediate compounds C_3, C_4, C_5, C_6, etc. serve as precursors for phenol synthesis.

During infection, the pentose phosphate pathway is triggered to greater activity as is evidenced by an increase in C_6/C_1 ratio and also by detection of enzymes concerned with this pathway. This provides Erythrose-4-PO_4 and NADPH needed in phenol biosynthesis.The key enzyme *phenyl-alanine ammonia lyase*, and other enzymes of phenol biosynthesis e.g. *NADP reductase*, β-*glycosidase* are synthesised or activated. Ethylene, an important inducer of these enxymes and related metabolic changes, is formed vigorously in the infected and adjoining healthy tissues. Thus, the metabolic system for phenol synthesis becomes operative in diseased tissues. The speed of this accomplishment will govern the speed

[1]The suppression of fermentation in the presence of oxygen is known as 'Pasteur effect'. Fermentation being inefficient energy releasing process, more rapid consumption of glucose occurs in the absence of oxygen. In the presence of oxygen less carbohydrate will be consumed. This conservation of carbohydrate for the cell is Pasteur effect. Abolition of Pasteur effect in diseased plants results in less efficient aerobic breakdown of carbohydrates and increased rate of respiration tries to make up the loss.

of phenol synthesis, which is of critical importance in establishment or exclusion of the pathogen.

The abnormal phenolics i.e., phytoalexins are not induced by ethylene. Uritani and his group (1963) have demonstrated the *de novo* synthesis of various enzymes of ipomeamarone synthesis in sweet potato. The biosynthetic pathway is not functional in healthy tissues.

REFERENCES

AKAZAWA, T. and URITANI, I, 1955, "Respiratory increase and phosphorus and nitrogen metabolism in sweet potato infected with black rot," *Nature*, 176: 1071-1072.

———, 1962. "Pattern of carbohydrate breakdown in sweet potato roots infected with *Ceratocystis fimbriata*," *Plant Physiol.*, 37: 662-670.

ALLEN, P.J., 1942, "Changes in the metabolism of wheat leaves induced by infection with powdery mildew," *Amer. J. Botany*, 29: 425-435.

———, 1953, "Toxin and tissue respiration, *Phytopathology*, 43: 221-229.

———, 1954, "Physiological aspects of fungus disease of plants," *Ann. Rev. Pl. Physiol.*, 5: 225-248.

ANDAL, R., 1959, *Studies on foot rot disease of rice*, Ph. D. thesis, University of Madras.

BACH, A.N., 1912, (cited by G.L. Farkas and Kiraly, 1958).

BATEMAN, D.F. and MILLER, R.L., 1966, "Pectic enzymes in tissue degradation," *Ann. Rev. Phytopathol.*, 4: 119-146.

BRAUN, A.C, 1955, "A Study on the mode of action of the wild fire toxin," *Phytopathology*, 43: 659-664.

COOK, A.A. and STALL, R.E., 1968, "Effect of *Xanthomonas vesicatoria* on loss of electrolytes from leaves of *Capsicum annum*," *Phytopathology*, 58: 617-619.

CORY, J.E., 1967, "Evidence for a role of tyrosyl residues in cell membrane permeability," *J. Biol. Chem.*, 242: 218-221.

DALY, J.M. and JENSEN, S.G., 1961, "Evidence for functional cytochrome oxidase in tissues infected by obligate parasites," *Phytopathology*, 51: 759-765.

DALY, J.M. and SAYRE, R.M., 1957, "Relations between growth and respiratory metabolism in safflower infected by *Puccinia carthami*," *Phytopathology*, 47: 162-168.

DUBE, H.C., 1971, "Studies on permeability changes in cotton (*Gossypium hirsutum*) caused by *Verticillium albo-atrum*," *Proc. Indian Acad. Sci.*, 50(B): 185-193.

———, H.C., 1973, "Electrolyte changes in blasted rice," *Curr. Sci.*, 42 (7): 249-250.

———, H.C, and SRIVASTAVA, H.P., 1975, "Altered Permeability as a trigger for pathogenesis" in *Advances in Mycology and Plant Pathology*, published by Prof. R.N. Tandon Birthday Celebration Committee, New Delhi, pp. 271-275

FARKAS, G.L. and KIRALY, Z., 1958, "Enzymological aspects of plant diseases. I. oxidative enzymes," *Phytopath., Z*, 31: 251-272.

FARKAS, G.L, KIRALY, Z. and SOLYMOSY, F., 1960, "Role of oxidative metabolism in the localization of plant viruses," *Virology*, 12: 408-421.

FARKAS, G.L, and KIRALY, Z., 1962, "Role of phenolic compounds in the physiology of plant diseases and disease resistance," *Phytopath, Z.*, 44: 105-150.

FARKAS, G.L., LOVERKOVICH, L. and KLEMENT, Z., 1963, "Increased activity of glucose-6-phosphate dehydrogenase in tobacco leaves affected by *Pseudomonas tabaci*," *Natur wissenschaften*, 50: 22-23.

GHABRIAL, S.A., PIRONE, T.P., 1967, "Physiology of tobacco-etch virus induced wilt of tobacco peppers," *Virology*, 31: 154-162.

GNANAM, P., 1956, "Conductivity studies in cotton plants infected by *Fusarium vasinfectum*," *Proc. Indian Acad. Sc.*, 44 (B): 125-129.

GOODMAN, R.N., KIRALY, Z. and ZAITLIN, M., 1967, *The biochemistry and physiology of infectious plant disease*, Van Nostrand Co. Inc. N.Y.

GRIMM, R. and WHEELER, H., 1963, "Respiratory and enzymatic changes in victoria blight of oats," *Phytopathology*, 53: 436-440.

HALL, J. A. and WOOD, R.K.S., 1970, "Plant cells killed by soft rot parasites," *Nature* (London), 227: 1266.

KIRALY, Z. and FARKAS, G.L., 1957, "On the role of ascorbic acid oxidase in the parasitically increased respiration of wheat," *Arch. Biochem. Biophys*, 66: 474-485.

KIRALY, Z., POZSAR, B.I. and HAMMADY, M.E.L., 1966, "Cytokinin activity in rust infected plants: Juvenility and senescence in diseased leaf tissue," *Acta Phytopathol. Hung.*, 1: 29-37.

LAKSHMANAN, M., 1959, "Respiratory changes in Fusarium infected cotton," *Phytopath. Z.*, 36: 406-418.

LIVEN, A., 1964, "Photosynthesis in healthy and rust-affected plants," *Plant Physiol.*, 39: 614-621.

LOBESNSTEIN, G. and LINDSEY, N., 1963, "Effect of virus infection on peroxidase activity and C_6/C_1 ratios," *Phytopathology*, 53: 350. (Abstr.)

LUMSDEN, R.D. and BATEMAN, D.F., 1968, "Phosphatide degrading enzymes associated with pathogenesis in *Phaseolus vulgaris* infected with *Thielaviopsis basicola*," *Phytopathology*, 58: 219-227.

MAINE, E.C., 1960, (cited by Goodman *et al.* 1967, see above).

MENKE, G.H. and WALKER, J.C., 1963, "Metabolism of resistant and susceptible cucumber varieties infected with cucumber mosaic virus," *Phytopathology*, 53: 1349-1355.

MERRETT, M.J., and BAYLEY, J., 1969, "The respiration of tissues infected by virus," *Bot. Rev.*, 35: 262-392.

MERRET, M.J. and SUNDERLAND, D. W., 1969, "Respiration rate and adenosine triphosphate formation in tissues infected by Tobacco mosaic virus," *J. Expt. Bot.* 20 (65): 717-731.

MILLERD, A. and SCOTT, K.J, 1956, "Host parasite relations in powdery mildew of barley, II. changes in respiratory pattern," *Austr J. Biol. Sci.*, 9: 37-44.

———, 1962, "Respiration of the diseased plant," *Annu. Rev. Plant Physiol*, 13: 559-574.

MOUNT, M.S and BATEMAN, D.F., 1970, "Cellular injury caused by an endopolygalacturonate *trans*-climinase from *Erwinia amylovera*," *Phytopathology*, **60**: 577.

NAGY, R., RIKER, A.J. and PETERSON, W.H., 1938, "Some physiological studies on crown gall and contagious tissue," *J. Agr. Res.*, **57**: 545-555.

PATEL, P.N. and WALKER, J.C., 1963, "Changes in free amino acid and amide content of resistant and susceptible beans after infection with the halo blight organism," *Phytopathology*, **53**: 522-528

PENMAN, S, SCHERRER, K., BECKER, Y. and DARNELL, J.E., 1963, "Polyribosomes in normal and polio virus-infected Hela cells and their relationship to messenger RNA.," *Proc. Nat. Acad. Sci.*, **49**: 654-662.

PETERSON, P.D. and MCKINNEY, H.H., 1938, "The influence of four mosaic diseases on the plastid pigments and chlorophyllase in tobacco leaves," *Phytopathology*, **28**: 329-342.

POZSAR, B.I., KRISTEV, K. and KIRALY, Z., 1966, "Rust resistance induced by amino acids: A decrease of the enhanced protein synthesis in rust infected bean leaves," *Acta. Phytopathol. Hung.*, **1**: 203-208.

REDDI, K.K., 1963, "Studies on the formation of tobacco mosaic virus ribonucleic acid, II. Degradation of host ribonucleic acid following infection," *Proc. Natl. Acad. Sci. U.S.*, **50**: 75.

RUDOLF, K. and STAHMANN, M.A., 1964, "Interaction of peroxidases and catalases between *Phaseolus vulgaris* and *Pseudomonas phaseolicola* (Halo blight of bean)," *Nature*. **204**, 474-475.

SADASIVAN, T.S. and KALYANASUNDARAM, R.K., 1957, "Spectrochemical studies on the uptake of ions by plants, I. The Lundegardh flame technique of ash analysis of toxin/antibiotic invaded cotton plants," *Proc. Indian Acad. Sci.*, **43** (B): 271-275

———, and SARASWATHI DEVI, L.S, 1957, "Vivotoxins and uptake of ions by plants," *Curr. Sci*, **16**: 74-75.

SADASIVAN, T.S., 1963, "Respiration in plant under pathogenesis," *Indian Phytopath.*, Presidential address.

SAMBORSKI, D.J. and SHAW, M., 1956, "The physiology of host parasite relations, III the effect of *Puccinia graminis tritici* Eriks. and Henn. on the respiration of the first leaf of resistant and susceptible species of wheat," *Canad. J. Bot.*, **34**: 601-619.

SANWAL, B.D., 1956, "Polyphenoloxidase and ascorbic acid oxidase activity in tomato plants infected with *Fusarium lycopersici*," *Proc. Indian Acad Sci.*, 44 B: 257-270.

SCOTT, K.J., CRAIGIE, J.S, and SIMILLIE, R.M., 1964, "Pathways of respiration in plant tumours," *Plant Physiol*, 39: 323-327.

SEMPIO, C., 1950, "Metabolic resistance to plant diseases," *Phytopathology*, **40**: 799-819.

SPIKES, J. D and STOUT, M., "Photochemical activity of chloroplasts isolated from sugar beet infected with virus yellows," *Science*, **122**: 375-376

STAHMANN, M.A., 1963, "Perspectives of biochemical plant pathology," (ed. S. Rich). Bull. N. 663. The Connecticut Agri. Expt. Station, New Haven.

STROBEL, G.A. and MATHRE, A.E., 1970, *Outlines of Plant Pathology*, Van Nostrand Reinhold Co., New York.

SWAMY, R N., 1964, "Respiration in *Cercospora* infected groundnut tissues

I. Respiratory stimulation following infection," *Phytopath., Z*, 50: 227-234.
II. "Effect of some enzyme inhibitors and 2, 4-dinitrophenol on respiration," *Indian. J. Expt. Biol.*, 2: 193-197.

THATCHER, F.S., 1943, "Cellular changes in relation to rust resistance," *Can. J. Res.* Sect. 'C'., 21: 151-171.

URITANI, I. 1963, "The biochemical basis of disease resistance induced by infection," in *Perspectives of Biochemical Plant Pathology*, pp 4-14. Bulletin 663 of Connecticut Agric. Expt. Station, New Haven, U.S.

URITANI, I. and AKAZAWA, T., 1959, "Alteration of the respiratory pattern in infected plants," in *Plant Pathology*, vol 1, Horsfall, J.G. and A.E. Dimond, (eds.) Academic Press, New York.

WANG, J.D., 1961, "The nature of starch accumulation at the rust infection site in leaves of pinto bean plants," *Can. J. Botany*, 39: 1595-1604.

WHEELER, H and HANCHEY, P., 1968, "Permeability phenomenon in plant disease," *Ann. Rev. Phytopath.*, 6: 331-350.

YARWOOD, C.E, 1934, "Effect of mildew and rust infection on dry weight and respiration of excised clover leaeflts," *J. Agri Res.*, 49: 549-558.

———, 1953, "Heat of respiration and diseased leaves," *Phytopathology*, 43: 675-681.

ZAITLIN, M and JAGENDORF, A.T., 1960, "Photosynthetic phosphorylation and Hill reaction activities of chloroplasts isolated from plants infected with tobacco mosaic virus," *Virology*, 12: 477-486.

ZAKI, A.I. and MIROCHA, C.J., 1964, "Respiration, photosynthesis and dark fixation of $C^{14}O_2$ by healthy and rusted bean leaves," *Phytopathology*, 54: 913.

———, 1965, "Carbon dioxide fixation by rust infected bean plants in the dark." *Phytopathology*, 55: 1303-8.

ZEN, 1964, "Biochemical approach to the study of host-parasite relationship," in *Host Parasite Relations in Plant Pathology*, (eds) Kiraly and Ubrizsy, pp. 51-52

14

CONTROL OF PLANT DISEASES—I

THE knowledge gleaned from the science of plant pathology is put to use when methods are devised to save the damage to crops. It is here that plant pathology becomes a community aspect and the endeavour to treat plants for protection or cure—an art. The knowledge, obtained in laboratories, becomes power in the field. "The fungus is made," says Horsfall (1957) "to sit and beg."

The maxim, "prevention is better than cure" gets its fullest expression with plant diseases. Cure is costly and agriculture is an economic endeavour. It is in this respect that the health of an individual, unlike a human being, becomes unimportant. The control measures fall into two groups: (9) Prevention, and (2) Cure.

PREVENTION

Prevention of disease is achieved by:

(*a*) **Avoidance of the pathogen**: By planting at times and in areas where inoculum is absent, rare or ineffective.

(*b*) **Exclusion of inoculum**: Preventing the entry and establishment of pathogens to new areas.

(*c*) **Eradication of inoculum.**

(*d*) **Protection**: By preventing infection with the help of chemical barriers, and

(*e*) **Disease resistance**: By use of resistant hosts.

CURE

(1) **Therapy.** Reducing severity of disease in infected plants.

For satisfactory control more than one of the control measures have to be adopted. But before choosing the control measure, a knowledge of the characteristics of causal agent, the host, the cultural and climatic conditions in which the crop is grown, and the disease dynamics is essential.

Avoidance of the Pathogen

Diseases can be escaped by (*i*) Use of disease free planting materials obtained from reliable sources, and (*ii*) by modification of cultural practices. The latter can be done by choosing site and time of planting which ensure least chances of infection. Cabbage yellow, caused by *Fusarium*, occurs maximum at 26°C, which is most suitable for the pathogen. The disease can be avoided by growing cabbage during cooler months, at temperature below 17°C. The disease is insignificant at this temperature and the crop grows well. This is one example, but many more diseases can be avoided.

Exclusion of Inoculum

Pathogens may be kept away (excluded) by preventing entry of pathogenic propagules (by inspection and quarantine) and by eliminating insect vectors, which eliminate the chances of pathogens coming in contact with the hosts.

Preventing Entry of 'Non-resident' Pathogens. By growing together, a balance establishes between the host and the pathogen and the disease is kept at a minimum. Most of the epiphytotics (e.g. downy mildew of grapes in Europe, chest nut blight, and Dutch elm disease in USA)have occurred by the transportation of pathogens to disease free areas. The plants had not evolved resistance and, therefore, fell—one and all, causing catastrophe to people Laws were enacted to prevent such lapses and now quarantines are international obligation.

Plant Quarantines. The Food and Agriculture Organization (FAO) of the United Nations has contributed significantly towards crop protection. The international plant protection organization, under the auspices of FAO, keeps watch on plant diseases occurring in any part of the world. With the help of its six regional branches, it provides information and suggests control measures through its official plant protection bulletin. The six regional branches are:

(1) The European Plant Protection Organization (EPPO)
(2) The Inter-African Phytosanitary Commission
(3) The Plant Protection Committee for the South East Asia and Pacific region.
(4) Organisma International Regional de Sanidad Agroperclana.
(5) Convenio Interamericano de Protection Agricola
(6) Near East Plant Protection Commission.

In addition, every country has its own domestic quarantine and assists the international programme.

Modus Operandi of Quarantines for Healthy Plant Materials. *Inspection and certification* at the source of export and at the receiving end. Plant materials are detained to test for latent infections. For large consignments the diseases are checked by sampling method. For example the bacterial ring rot and brown rot diseases of potato are detected statistically by using the formula

$$W = \frac{\log P}{\log (1-p)}$$

where W=sample size, number of tubers examined
P=the probability of failing detection
p=the number of infected tubers.

The method detects disease if up to 3 per cent of tubers show symptoms.

Any restriction to trade, which of course quarantine procedures cause, should be done when it is a must; and for that it is necessary to confirm thatt

(1) The pest will be introduced if not regulated.
(2) If introduced, the pest will be harmful.
(3) That there are no natural methods of introduction of the pathogen (like wind dissemination).

Thus, for a pathogen which can not be carried through plant materials or will not be harmful in the climatic conditions of the receiving country or can come through wind, there is no justification for restricting trade by quarantine regulations.

Plant quarantines are not very effective because they, usually, are not established soon enough and also because of biological or physical limitations in implementation.

For scientific purposes, special permits are issued to import cultures of pathogens or diseased materials.

Plant Quarantine Organization in India

In 1914 the "Destructive Insects and Pests Act" (DIP Act) was passed by the Government of India, which forbids introduction of exotic pests and diseases into the country from abroad. The Agricultural Pests and Disease Acts of the various states prevent inter-state spread of pests within the country.

The plant quarantine stations work at the major sea and air-ports, which check the entry of forbidden diseased materials. The plant protection adviser issues certificates and import permits.

Eradication of Inoculum

Pathogens can be eradicated (destroyed) by the following methods:

(1) Biological control.

(2) Crop rotation.

(3) Removal of diseased plants and plant debris.

(4) Soil treatment.

(5) Removal (elimination) of the pathogen from infected/infested plant materials.

(1) **Biological Control.** Biological control brings about reduction in activity of a pathogen by another organism. This may be *biocidal* or *biostatic*. In biocidal biological control one organism kills the other, while in biostatic the organisms only inhibit the pathogen.

Biological control is defined by Garrett (1965) as "any condition under which, or practise whereby survival or activity of a pathogen is reduced through the agency of any other living organism (except man himself), with the result that there is a reduction in the incidence of the disease caused by the pathogen."

Sanford (1926) first suggested that control of potato scab (caused by *Streptomyces scabies*) was caused by increase in saprophytic bacterial populations, as a result of green manuring.

Mechanisms of Biological Control. 'Antagonism' which plays key role in biological control operates in three ways (Jackson, 1965; Park, 1965):

(*a*) **Antibiosis**—where one species secretes some chemicals which inhibit the growth of the other.

(*b*) **Exploitation**—where one is parasitic or predatory on other, and

(*c*) **Competition**—The organisms compete for a substance which is in short supply.

Garrett (1956, 1963 and 1970) has placed emphasis on competition. The "competitive saprophytic ability" of an organism, determines the extent of its success as a soil inhabitant. In the biological control, the soil is modified in such a way that the pathogen is overgrown by harmless saprophytic organisms. This is not easy. The soil is biologically buffered and any alteration does not last long as the microbial populations tend to regain the natural balance Nonetheless, success has been achieved in controlling some diseases by biological-control method.

The real value of the biological control does not lie in "natural" methods but in abstracting their underlying mechanisms and applying them.

(2) **Crop Rotation.** This method is useful against those pathogens which can not persist for long in the soil in the absence of the host viz. the soil invaders. Soil inhabitants cannot be overcome by this method.

In this method host plants and non-host plants are rotated in a way that reduces the inoculum size in the soil.

Success has been achieved in control of "take all" disease of wheat (caused by *Ophiobolus graminis*) by rotating wheat cultivation with oats. The other successful examples are Soyabean-potato rotation against root-rot of bean, caused by *Fusarium*,. *Rhizoctonia* and *Thielaviopsis*.

(3) **Removal and Destruction of Diseased Plants.** This method of control has been the most practised one. It includes the following:

(*i*) **Roguing diseased plants.** This removes the source of inoculum and prevents infection of adjacent healthy plants. On first appearance of smutted whips of sugarcane and the smutted wheat ears, the infected plants are removed.

(*ii*) **Elimination of alternate hosts and collateral hosts.** The removal of the alternate host, by breaking the life cycle of the pathogen, prevents the chances of formation of new races. In cases where the inoculum for infection of economically important host comes only from the alternate host, the removal of the latter may give complete control. Pine blister rust has been controlled completely by eradicating *Ribes* sp., the alternate host. The eradication of barberry bushes, prevents the chances of formation of new physiologic races by sexual recombination. Since the pathogen survives even in the absence of the alternate host on wheat plants on hills in India and southern wheat areas in America, the barberry eradication could not end the disease.

Some weeds act as collateral hosts and perpetuate the pathogen. Their destruction, if feasible, will control the disease.

Oversummering or overwintering crop residues serve as inoculum bearers and, therefore, must be destroyed.

(4) **Soil treatments.** The infested soil may be treated so as to inactivate or reduce the inoculum. The soil may be treated with chemicals, heat, flooding or fallowing. Most fumigants are non-selective and reduce population of most organisms. Root disease of *Citrus*, caused by *Armillarea mellea*, is controlled by fumigation with carbon-disulphide. This was explained by Bliss (1951) and later confirmed by Garrett (1958) to be due to survival of *Tricho-*

derma viridae, which later became dominant. Thus, it was biological control brought about by soil treatment.

Flooding of fields is done to control *Fusarium* wilt of banana in the tropics. Fallowing gives good control of certain diseases.

Elimination of the Pathogens

Planting materials, if contaminated or infected, may be used after removing the pathogen. Elimination of pathogens can be done by sorting or chemical treatment.

Sorting. It may be done mechanically by hand for diseases of bulbs etc., which show visible infection. Diseased materials may be removed in this way.

Lighter seeds of parasitic higher plants like *Cuscuta* and *Orobanche* can be separated by fanning or by immersing in brine water. The lighter seeds float on the surface.

Chemical Treatment. The presence of externally seed borne pathogens can be known by sowing a sample in a Petridish lined with moist filter paper. Contaminated seeds are treated chemically for disinfection. Some bacterial infections are detected by flourescence of infected areas in ultraviolet light. Detection of virus infection in potatoes is done by growing a bud or "eye" in an insect proof chamber.

The contaminated seeds are treated with chemicals to kill the surface contaminants or those that are not deep seated. This is done by steeping the seeds in a solution or suspension of chemicals or by dusting. The chemical should kill the pathogen without damaging the seeds.

For killing internally seed borne pathogens, three methods are available: (*i*) Hot water treatment; (*ii*) Exposure to hot air, and (*iii*) Immersion in chemicals.

The hot water treatment is used successfully for disinfecting wheat seeds against loose smut fungus *Ustilago nuda*, cabbage seeds infected with *Alternaria brassicae*, and several others.

The hot air treatment is used to eliminate viruses. Kunkel in USA (1936) and Kassanis in UK (1949, 1954) pioneered this method. Plants are heat treated in 'hot boxes' and the shoots produced during the heat treatment are disease free, which can be removed and propagated.

Virus free stocks are obtatined also by 'Tip Culture' method (Holmes, 1955) and by growing excised apical meristems (Kassanis and Varma, 1967). Prior heat treatment is helpful even in these methods.

The use of chemicals in eliminating deep seated pathogens is of recent origin and some breakthrough has been made in this direction.

PROTECTION

Plants are protected against infections by use of chemicals, "cross protection" and modification of cultural practices.

(1) **Chemical protection.** Chemicals are used as eradicants and therapeutants but it is only as protectants that they are used maximally. As a protectant, the chemical prevents the inoculum from entering the host and establishing infection.

Fungicides when present on the plant surface either prevent or delay spore germination, which results in failure of infection. Most protectant fungicides are in form of very fine powder which have low water solubility. Usually repeated applications of the protectant chemical are needed.

Organic fungicides are used as seed protectants which, when coated on the seeds, prevent the seedlings from infection by damping off and seed rot organisms.

(2) **Cross protection.** Mild viruses prevent infection by a more serious virus and protect against serious loss. This phenomenon called 'cross protection" is utilized in disease control. By cross-protection is meant the decrease in disease that would be expected from inoculation of plants with one pathogen, which is caused by earlier inoculation with another or the same pathogen or by simultaneous inoculation with another pathogen. Cross protection which originated with virus disease is now well known in wilt diseases, especially those caused by *Fusarium oxysporum*.

When a virulent pathogen is inoculated it initiates host resistance reactions that prevent its growth and later of the virulent pathogen. The virulent pathogen does not invoke the resistant response of the host and grows sufficiently to cause disease.

(3) **Protection by modification of host nutrition.** The relation of nutrition with susceptibility is well known. Nitrogen in excess generally causes susceptibility. Calcium endows resistance to plants against some wilt and soft-rot pathogens. Less nitrogen makes plants resistant. Diseases that result from deficiency of mineral elements may be avoided by foliage or soil applications of

the deficient element. Soil leaching may take away the excess elements; nonetheless, balanced nutrition does not provide protection against most of the diseases.

A low pH reduces apple scab and damping off in tree nurseries. Pathogens like *Fusarium*, which can tolerate wide range of pH, can not be controlled by this method.

USE OF DISEASE RESISTANT HOSTS

Use of resistant varieties is the best method of disease control. Resistant varieties frequently fall to new physiologic races and, therefore, breeding for resistance is a continuous programme. For many soil borne pathogens, use of resistant varieties is the only solution. It would be difficult to grow crops in some areas profitably in the absence of resistant hosts. Resistant varieties have been a great success with rust and smut diseases of cereals; fusarial wilts of cotton, tomato, water melon, sugarcane mosaic, curly top of sugar beets and so many.

Inheritance of Resistance

Biffen (1905) first proposed that resistance to yellow rust in wheat was inherited as a Mendelian character. Then it was not taken seriously. Now it is well known that inheritance of resistance to specific diseases in plants is controlled by the common genetic principles. It may be controlled by a single gene pair (monogenic) or it may be quantitatively controlled by few (oligogenic) or more genes (polygenic). The genes may be dominant or recessive. Diseases in which several races of the pathogen are known (e.g. wheat rusts, flax rust, covered smut of wheat, late blight of potato etc.) a particular host variety may contain genes specific for resistance against few races only. In any breeding programme, first task is to find out the races of the geographical area where the variety will be used and genes for resistance against all the prevalent races must be incorporated into the forthcoming variety. The physiologic races have specific allelic genes related to specific allelic resistance genes of the host. The relation between the genes of the host and the pathogen has been analysed extensively by Flor (1955) with flax rust (*Melampsora lini*). He proposed his "gene for gene" hypothesis which says that for each gene controlling resistance or susceptibility in the host, there is a corresponding gene for virulence or avirulence in the pathogen. This is now reported for a number of other host-pathogen combinations.

Since the plant has to be resistant against all diseases, resistance genes for more than one disease are incorporated. Difficult it is, but not insurmountable. Varieties of several important crop plants (e.g. barley, tobacco, tomato, cotton) are available that are resistant against several major diseases. It is known that monogenic resistance is more stable than the polygenic.

In breeding for resistance against one pathogen, varieties may be developed with high susceptibility against some other disease. This happened with the Victoria cultivar of oats, derived from the cross, Richland x Victoria. Richland was resistant against the stem rust and Victoria against crown rust. The progeny, resistant against the two rusts earned great popularity, which, before long, proved to be highly susceptible to a blight called Victoria blight, caused by *Helminthosporium victoriae*. The breeders had overlooked the fact that victoria parent was highly susceptible to this disease.

The resistant varieties, with whatsoever care they have been brought out, sooner or later, fall to some newly developed or immigrant pathogenic race. This happens more with varieties having monogenic than those with polygenic resistance. Some tobacco resistant varieties have been withstanding the bacterial wilt pathogen for over 20 years, and still going strong. However, it can never withstand for indefinite period. The famed wheat variety "Khapli" which faced tremors since 1908, ultimately gave up in 1950. Rust races, virulent on Khapli, had appeared.

Vertical and Horizontal Resistance

Vander Plank (1963) introduced the concept of vertical and horizontal resistance. When a variety is more resistant to some races of a pathogan than to others, the resistance is vertical or perpendicular. In horizontal resistance, the variety is resistant against all races. Horizontal resistance is usually polygenic in inheritance while vertical resistance is brought about by oligogenes Browning and Frey (1969) have added the word "population resistance" to describe the situation where the fungus population cannot increase and damage the well buffered host population.

Sources of Resistance

(1) **Selection and multiplication of resistant individuals from existing varieties.** Orton (1900) obtained his resistant cotton variety from this method. He observed that some cotton plants did

not show wilting in the heavily infected crop. He collected seeds from these plants and planted them in wilt infested soil. By several such plantings he ultimately got most resistant plants which grew well on most infested soils. Sorghum varieties, resistant against root disease (*Periconia circinata*), have been developed by this method in California.

(2) **Search for resistant genes in related species.** Resistance against a particular disease is more likely to be found in plants of that area where the disease actually belonged to. The germplasm of wild species often provides the source of resistance. In several of the presently grown resistant varieties, the resistance has been brought from wild species. Sometimes, the crosses do not succeed and the progeny becomes sterile (form no seed) or certain undesirable characters, linked with resistance, appear in the hybrid.

Developing Resistant Varieties—The Techniques

This is done by selection or breeding.

(1) Selection of surviving individuals from heavily infected fields and multiplying them.

(2) Crossing resistant species with susceptible hosts.

(3) By irradiating wild host plants. Crosses that are normally not compatible are made possible by irradiation of pollen grains. This method has great potentials but has been successful only in some cases. Leaf-rust resistance in *Aegilops umbellata* was successfully transferred to wheat.

(4) The environmental effect on the resistant variety must be checked before releasing for cultivation.

REFERENCES

BIFFEN, R. H., 1905, "Mendel's laws of inheritance and wheat breeding," *J. Argic. Sci., Camb.*, 1: 4-48.

BLISS, D.E., 1951, "The destruction of *Armillarea mellea* in Citrus soils," *Phytopathology*, 41: 665-683.

BROWNING, J.A. and FREY, K.J., 1969, "Multiline cultivars as a means of disease control," *Ann. Rev. Phytopathology*, 7: 355-383.

FLOR, H.H., 1955, "Host-parasite interaction in flax rust—its genetics and other implications," *Phytopathol* 45: 680-685.85.

GARRETT, S.D., 1956, *Biology of Root Infecting Fungi*, Cambridge University Press, London.

———, 1958, "Inoculum potential as a factor limiting lethal action by *Trichoderma viride*. Fr on *Armillaria mellea* (Fr.). Quel." *Trans. Brit Mycol. Soc.*,

41: 157-164.

GARRETT, S.D., 1963, "A comparison of cellulose decomposing ability in five fungi causing cereal root rots " *Trans. Brit. Mycol Soc.*, 46: 572-576.

———, 1965, "Toward biological control of soil borne plant pathogens," in *Ecology of Soil Borne Plant Pathogens*, K.F. Baker and W.C. Snyder (eds.), University of California Press, Berkley, California, 4-17.

———, 1970, *Pathogenic root Infecting fungi*, Cambridge University Press, London, 202.

HOLMES, F.O., 1955, "Elimination of spotted wilt from dahlias by propagation of tip cuttings," *Phytopathology*, 45: 224-226.

HORSFALL, J.G., 1957, "Mechanism of fungitoxicity," *Advances in Pest Control Research*, Vol. 1: 193-218, Interscience Publisher, Inc., New York.

JACKSON, R M., 1965, "Antibiosis and fungistasis of soil micro-organisms," in *Ecology of Soil Borne Plant Pathogens*, University of California Press, California.

KASSANIS, B., 1949, "Potato tubers freed from leaf-roll virus by heat," *Nature*, London, 16: 881.

———, 1954, "Heat therapy of vines-infected plants," *Ann. Appl. Biol*, 41: 470-474.

———, and VARMA, A., 1967, "The production of virus-free clones of some British potato varieties." *Ann. Appl. Biol*, 59: 447-450.

ORTON, W.A., 1900, "The wilt disease of cotton and its control," *Bull. Div. Veg. Physiol Path.*, U.S. Deptt of Agric., No. 27: 26.

PARK, D., 1965, "Survival of micro-organisms in soil" in *Ecology of Soil Borne Plant Pathogens*, K.F. Baker and W.C. Synder (eds), University of California Press, Berkley, California, 82-92.

SANFORD, G.R., 1926, "Some factors affecting the pathogenicity of *Actinomyces scabies*," *Phytopathology*, 16: 525-547.

VANDER PLANK, J.E., 1963, *Plant diseases: Epidemics and Control*, Academic Press, New York and London, p. 349.

15

CONTROL OF PLANT DISEASES—II
CURE (=THERAPY)

THE control measures that 'cure' the diseased plants by killing the pathogen inside the host plants or by ameliorating disease severity, comprise therapy.

Therapy may be by physical or chemical means.

Physical Therapy includes surgery and heat treatment.

(1) **Surgery.** In plant pathology it is a simple affair removing diseased limbs from pear trees infected with fire blight, excission of bark from 'bleeding canker' and 'stripe canker' of Cinchona and removal of infected wood in branch canker in tea. Besides this, surgery is practised in diseases forming galls, knots etc.

(2) **Heat treatment.** This is of particular significance in eliminating viruses from fruit-tree-propagating stock and in eliminating nematodes from bulbs.

Chemotherapy

It is based on the strategy of: (1) throwing the invader out, (2) inactivating their toxic products, and (3) altering the susceptibility of the host.

Chemotherapeutants act within the plant, so as to kill the established pathogens or protect plants from invasions by pathogens.

Requisites of a chemotherapeutant. A chemotherapeutant should be:

(*i*) mobile within the plant,

(*ii*) compatible with the plant,

(*iii*) persistent as a fungitoxic substance or as other functional entity.

(*iv*) should have low phytotoxicity, and

(*v*) low mammalian toxicity.

Classification of chemotherapeutants: Chemotherapeutants fall into two categories:

(*i*) the systemic fungicides, and

(*ii*) compounds that alter pathogenesis by action on pathogens' products or on host itself.

While upward translocation is common in plants, downward movement is infrequent. However, some materials like methylphenoxyacetic acid, when applied on foliage, show exudation through the roots.

Mechanism of Action

The chemotherapeutants act directly on the pathogen or indirectly by conferring resistance or changing metabolism in a way that the host, no more, remains a good substrate. Direct action involves :

(*i*) Fungitoxic or fungistatic effect, and (*ii*) Inactivation of toxic products of the pathogens.

Topical and Systemic Chemotherapy

Treatment of the injury of surface cells or a localized area is topical therapy. Systemic therapy eliminates infections through systemic action. Certain organic mercury fungicides eliminate primary infections in apple scab (topical therapy). Some bactericides have eliminated bacterial diseases like crown gall and fire-blight through systemic action. Cure of cherry leaf-spot by the antibiotic actidione is another example of systemic action.

Now large number of chemotherapeutants are known, including such diverse materials as 8-hydroxy quinoline, chlorinated phenoxyacetic acids, sulphonamides, pyridine 2-thiol, naphthaline acetic acid and 6-azauracil. Wain and Carter (1972) have listed the various chemotherapeutants and the diseases they control.

In 1957 Kuc', Williams and Shay showed the use of amino acids (alanine and phenylalanine) as chemotherapeutants, by increasing resistance of apple tissues against scab disease. This is an example of chemotherapy by altering the metabolic response of the host. Phenylalanine caused the accumulation of two toxic phenolic substances—phloretin and phloridzin. Since then, several amino acids and their analogues have been found to have chemotherapeutic activity. Usually, amino acids, other than those involved in nitrogen metabolism, are effective. Van Andel (1966) has discussed their mechanism of action.

Oils of several types have been used in chemotherapy. The sigatoka leaf-spot disease of banana has been controlled by oil. Certain pathogenic fungi grow well in mineral, synthetic and glyceride oils but the same oils control diseases caused by them. Oil is an unconventional compound in plant-disease control and can not be placed in any of the existing categories of materials that control diseases. Therapeutic action is involved in all the diseases controlled by oils which include some caused by fungi and several caused by viruses. Calpouzcs (1966) explains the therapeutic action of oils to be due to their physical rather than chemical properties.

REFERENCES

CALPOUZOS, L., 1966, "Action of oil in the control of plant disease," *Annu. Rev. Phytopathol.*, 4: 369-390.

KUC, J., WILLIAMS, P.B and SHAY, J.R., 1957, "Increase of resistance to apple Scab following injection of host with phenylthiourea and phenylalanine," *Phytopathology*, 47: 21-22.

VAN ANDEL, O.M., 1966, 'Amino acids and Plant diseases," *Annu. Rev. Phytopathol*. 4: 349-368

WAIN, R.L. and CARTER, G.A., 1972, "Nomenclature and definitions" in *Systemic Fungicides*, Marsh (eds.), Longman Group Ltd, 321 pp.

16

CONTROL OF PLANT DISEASES—III
CHEMICALS USED IN CONTROL MEASURES

THE 'arsenal' of plant pathology is now equipped with most sophisticated ammunitions to wage war on the unscrupulous pathogens, which are many and deceitful.

Planning for disease control involves strategy and tactics: demands a knowledge of the 'weak point' in the life cycle of the pathogen to hit hard and completely. The success will depend not on the amount of job done but on the job left unfinished, for the 'left over' may spring up again and next time be more horrible than ever before.

Prevost (1807) used a chemical to control disease—copper sulphate against bunt of wheat. Before him Hamberg (1705) had suggested the use of mercuric chloride as a wood preservative. Ward (1880) postulated that plants could be protected if chemicals were applied before infection. The Bordeaux mixture discovered by Millardet (1882) had the greatest impact in plant pathology and is still used against the powdery mildew of grapes in Europe. The discovery of dithiocarbamates in 1930's by Du Pont Company ushered in an era of organic fungicides. With the discovery of penicillin much hope was raised with antibiotics and now it is an important group in systemic fungicides.

Classification

The chemicals used for controlling plant diseases belong to diverse chemical groups and act in different ways. Nonetheless, they can be classified on the basis of.

(1) their chemical nature,

(2) mode of action, and

(3) on the basis of plant parts treated.

The classification on the basis of chemical nature is given in Table 16.1.

TABLE 16.1. CHEMICALS USED IN CONTROL MEASURES

Inorganic	*Organic*	*Miscellaneus*
Sulphur and its compounds e.g.	Dithiocarbamates	Oxanthiins
1. elemental sulphur,		
2. lime sulphur		
Copper and its compounds e.g.		Dinocap
1. Bordeaux mixture,	Heterocyclic nitrogen	
2. Burgandy mixture, etc.	compounds	
	Quinones	
Mercury and its compounds	Organo-mercurials	Oils
Mercuric chloride	Antibiotics	

On the basis of their local or systemic mode of action, chemicals are divided into systemic and non-systemic. Most of the chemotherapeutants are systemic in nature. Again these chemicals can be classified into: (*i*) Eradicants, (*ii*) Protectants, and (*iii*) Therapeutants.

Chemicals when act to reduce the inoculum at the source, are called eradicants. When they protect penetration and infection by forming a chemical barrier between host and the pathogen, the chemical provides protection and hence called protectant. Majority of fungicides are protectants. When the chemical enters the plant and kills the pathogen within the plant body it is therapy, and the chemical a therapeutant. Most of the therapeutants are systemic fungicides. The eradicants and protectants are non-systemic.

On the basis of plant parts treated, the chemicals can be designated as seed protectants, fruit protectants, foliage and blossom protectants and soil protectants. Since a chemical can effectively act on several parts, this classification is out of order.

Mechanism of Action

The chemicals, which are mostly fungicides, act on or interrupt the metabolic systems of an organism. The effectiveness of a chemical depends on its innate toxicity, and permeation. The toxicity depends on physical or chemical properties. Particle size, polarity, lipid solubility, adhesiveness on leaves, molecular size and shape are the physical properties that govern toxicity. The chemical properties include stability and capability to differentiate between cell constituents of the plant and the pathogen. This is not the case with insecticides. There, the chemical is aimed at the insect—so different from the plant. We deal with two plants, one to kill without damaging the other. Nevertheless, the chemicals that are basically toxic to all organisms, at certain concentrations, attack selectively and do not injure the crops. Thus a judicious choice of concentration may help in killing the pathogen and not the host.

The action of chemicals may involve:

(*a*) indiscriminate reaction with the functional groups of the cell, or

(*b*) specific reaction with certain cellular constituents.

The majority of fungicides belong to the first group and act by four general types of reactions.

(1) Oxidation-reduction,

(2) Secondary substitution,

(3) Chelation, and

(4) Complementary reactions.

The cellular reactants usually are thiol, amino and keto acids, chelators and metal containing compounds.

The specific toxicants react selectively and often the toxic chemical is recovered intact from the fungus cells, where they act. Often, the compounds compete with substrates for sites on enzymes —thus act as antimetabolites and inhibit enzyme activity. Interruption with protein and nucleic acid synthesis are known for several fungicides. Lukens (1971) has given a comprehensive review of the work done on the chemistry of fungicidal action.

The differential toxicity of a specific dose may be due to the fact that:

(1) the chemical penetrates the membranes of the fungus cells more rapidly than those of the host.

(2) the fungus cannot metabolize the toxic chemical to less toxic compounds while the host can, and

(3) the fungicides inactivate certain critical intermediate mesabolic compound in the fungus which is not of much significance for the host.

Criteria of a Good Fungicide

(1) A good fungicide should be able to resist degradation by environmental factors.

(2) It should have good coverage of plant surface. Coverage is improved by adding wetting agents and spreaders. However, coverage reduces adherence of the chemical.

(3) It should not be toxic to plants and should not leave residue, which is toxic to human beings or animals. In the recent years, the dangerous effects of residues have come to light.

Dosage Response Relations

A knowledge of the concentration of the chemical that is inhibitory to the fungus is important. By plotting percentage inhibition of spore germination (expressed in probit units) against log dosage of the toxicant, usually a linear dosage-response curve is obtained (Fig. 16.1). The dosage that kills 50 per cent of the spore population is called LD 50 and the dosage that inactivates or inhibits 50 per cent of the population is called ED 50 value. A knowledge of these values is essential in deciding the dosage to be employed in the control measure. The ratio of the ED 50 of the pathogen to that of the ED 50 of the host plant is used as an index of the safety margin for that pathogen and crop.

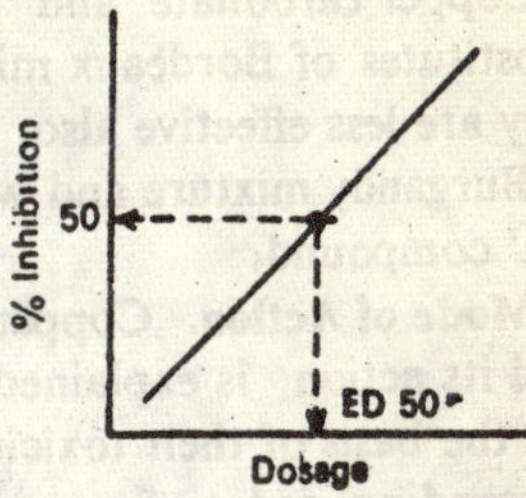

Fig 16.1. A typical dosage-response curve showing (*i*) relation between concentration of fungicide and per cent inhibition of fungus spores, and (*ii*) method to find ED 50 value of dosage.

The chemicals. (Fungicides, bactericides, viricides and nematocides.)

Inorganic Substances

Sulphur and its compounds. Elemental sulphur powder is an ancient fungicide and its toxic property has been exploited for the last two centuries. Elemental sulphur is best used as dust, however, wettable sulphur, used as paste or powder is becoming more popular. Lime-sulphur, being phytotoxic, is used as 'dormant spray' at low concentrations for controlling fruit diseases.

Mode of Action. The sulphur-sensitive fungi convert sulphur to hydrogen sulphide which is toxic. Horsfall (1955) suggested that sulphur competes with oxygen for protons and electrons in fungus respiration. Tweedy (1964) suggested that sulphur takes electrons between cytochromes b and c of the electron transport system. According to Owens (1960) the S_8 molecular form acts as a free radicle and denatures enzymes.

Copper and its compounds. The best known copper fungicide is the Bordeaux mixture, which consists of copper sulphate, lime and water in the ratio of 4:4:50. It controls many fungus and bacterial leaf-spots, blights, cankers and downy mildews. However,

Bordeaux mixture is phytotoxic and causes burning of leaves and fruits. The phytotoxicity, which is due to the copper, is reduced by increasing the ratio of lime in the mixture.

Copper carbonate and copper oxychloride are less phytotoxic substitutes of Bordeaux mixture and are called 'fixed coppers'. But they are less effective also. Copper sulphate is mixed with Na_2CO_3 in Burgandy mixture and with ammonium carbonate in the 'chestnut' compound.

Mode of Action. Copper is one of the most toxic metal cations and its action is explained to be due to inactivation of enzymes. On the basis of their toxicity, the metals fall into three groups, in descending order of activity. Group I includes Ag, Hg and Cu; group II includes Cd, Cr, Ni, Pb, Co and Zn and group III, Fe and Ca.

Mercury. Mercuric chloride is a well known surface disinfectant. Potato tubers and other propagative materials are dipped in 1:100 mercuric chloride solution for few minutes and washed before planting.

Organic Substances

Dithiocarbamates. Dithiocarbamates are, by far, the most important bactericides and fungicides. The discovery of dithiocarbamate in 1931, is hailed as one of the outstanding achievements of the past thirty years as it ushered in the era of organic fungicides. Dithiocarbimic acid, from which these compounds are derived, is unstable in nature, but by reacting with a metal or another molecule it forms stable and highly active compounds.

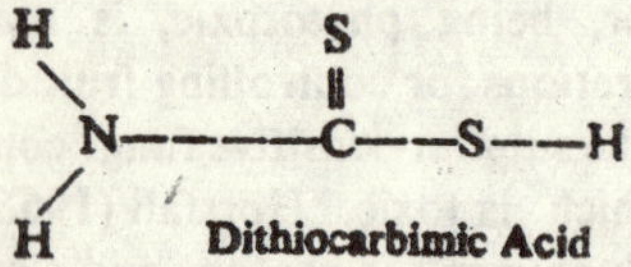

Dithiocarbimic Acid

In 1943, Dimond *et al* introduced ethylene bis-dithiocarbamates—another group of derivatives of this acid which are very important fungicides.

The dithiocarbamates, on the basis of their chemical structure and biological action, can be divided into two groups

Dithiocarbamates

Dialkyl-dithiocarbamates | Monoalkyl-dithiocarbamates

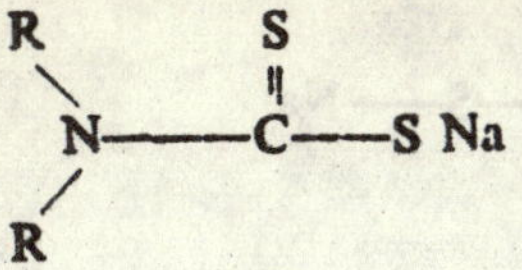

I

Example. Thiram, Ferbam and Ziram

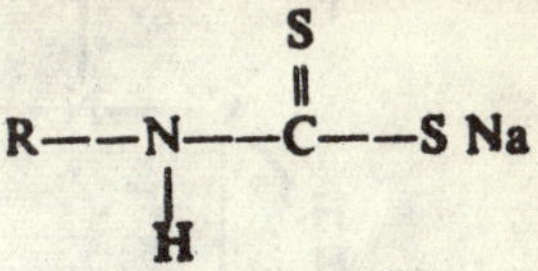

II

Example. Nabam, Maneb and Zineb

Dialkyl Dithiocarbamates

1. Thiram

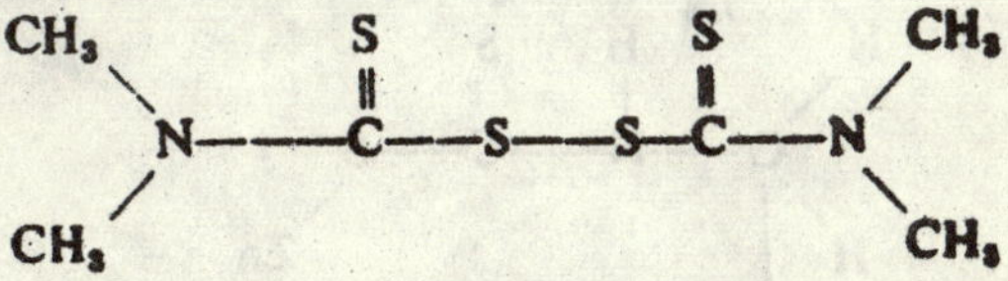

2. Ferbam

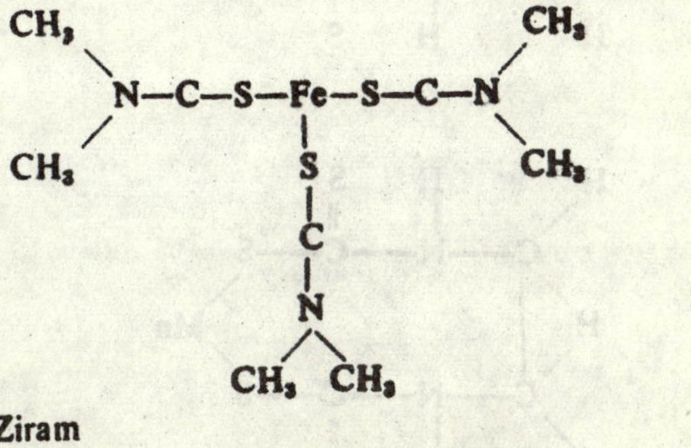

3. Ziram

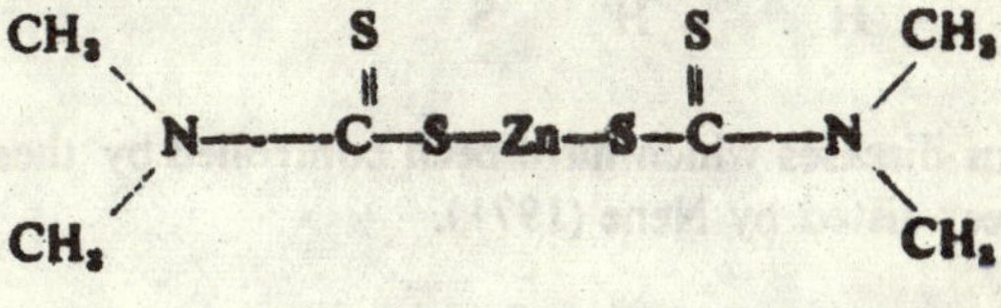

Monalkyl-dithiocarbrmates

1. Nabam

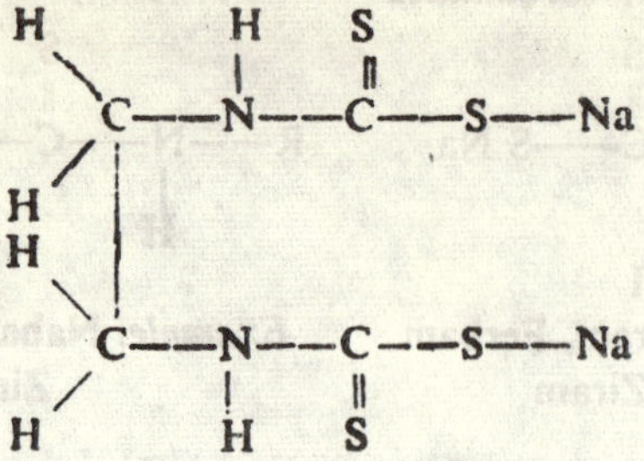

2. Zineb

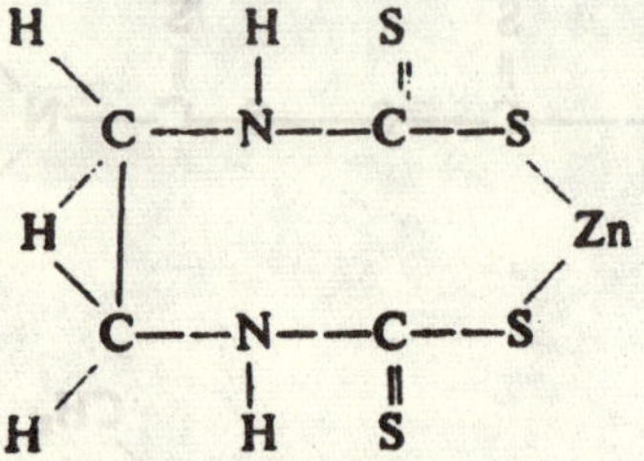

3. Maneb

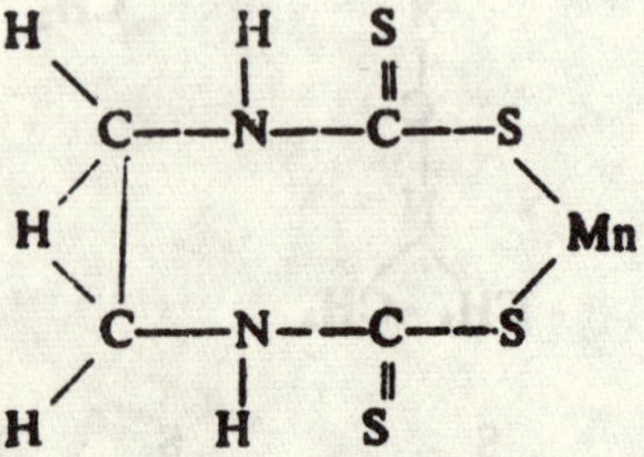

The various diseases which have been controlled by these fungicides have been listed by Nene (1971).

Mode of Action

How these substances permeate fungus cells is in dispute. Thorn and Richardson (1964) suggested that these substances are converted by the fungus to dithiocarbamate ion. This is disputed by Owens and Hayes (1964). The importance of the conversion into the anion is yet not known. Most of these compounds interfere with the enzymes of Krebs cycle. The monoalkyl dithiocarbamates—nabam, zineb, maneb etc., are non-toxic without previous exposure to air (Morechart and Grossan, 1965). Perhaps air degrades these substances into ethylene mono and disulfide, carbon disulfide and sulfides. All these are invoked to explain the toxicity of ethylene bis-dithiocarbamates.

Heterocyclic Nitrogen Compounds

The development (in 1952) of trichloromethylthiodicarboximides by Kittleson is an important event. Of the several heterocyclic nitrogen compounds recently introduced, Captan and Glyodin are the best known.

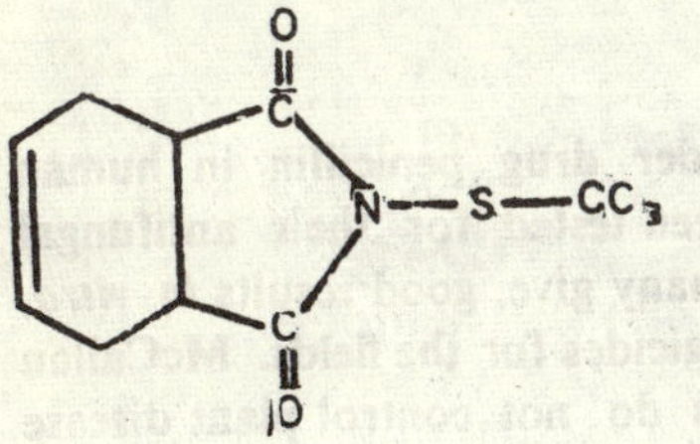

Captan

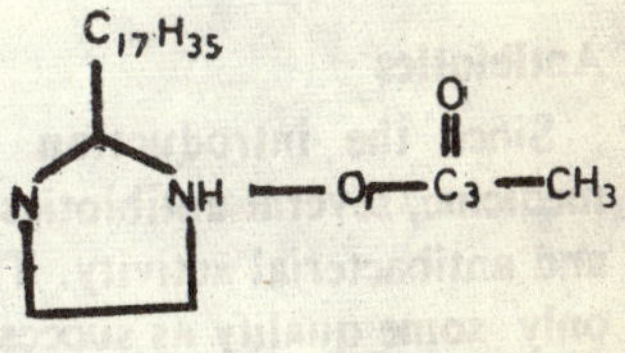

Glyodin

Captan is non-toxic to animals and controls a wide range of diseases. It is one of the outstanding fungicides of recent times.

Mechanism of Action

Glyodin acts as an antimetabolite and competes with guanine and xanthine. The toxicity of captan is unexplained. Reaction with thiol is suggested by Lukens *et al* (1965) to be responsible for toxicity.

Quinones

Quinones are well known antimicrobial chemicals. In natural plant resistance, these are formed by oxidation of phenols. On polymerization they form melanin.

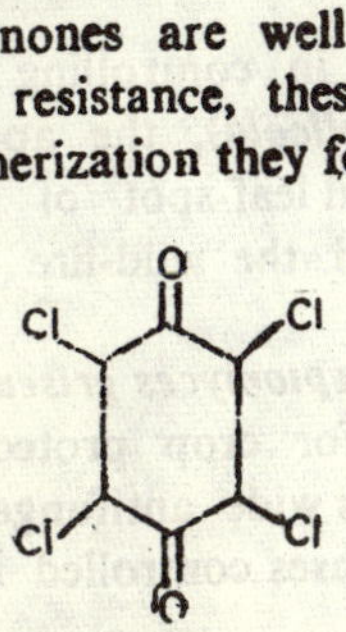

Chloranil

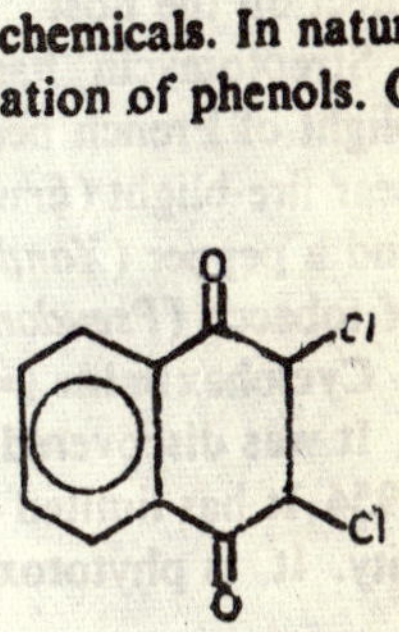

Dichlone

Chloranil and dichlone are the only two quinones which are commercially used as fungicides. Both are usually used as seed protectants.

Mode of Action. Inactivate enzymes containing S-H or amino groups.

ORGANO-MERCURY COMPOUNDS

Since mercury is toxic to human beings, its use is restricted and is used mostly for seed treatment. Only in very few cases they are used for foliage spray. Mostly, these are phenyl mercury derivatives e.g., phenyl mercury acetate, phenyl mercury chloride, phenyl mercury salicylate etc. Others are more complex, e.g., phenyl mercury dimethyldithiocarbamate.

Antibiotics

Since the introduction of wonder drug penicillin in human medicine, several antibiotics have been tested for their antifungal and antibacterial activity. Though many give good results *in vitro*, only some qualify as successful fungicides for the fields. McCallan (1969) remarked thus—"Antibiotics do not control plant disease very well and we have lost interest." Antibiotics are generally translocated to become systemic in the plants.

Of all the antibiotics tested, only streptomycin, cycloheximide (Actidione) and Blasticidin have been successful.

Streptomycin: source: *streptomyces* sp.

It is *in vivo* systemically active against a number of fungi of the order Peronosporales. This is a puzzle because streptomycin is ineffective on fungi *in vitro*. Since downy mildews are obligate parasites, *in vitro* effect of streptomycin can not be seen. The *in vivo* effect on the pathogens is believed to be an indirect effect, through effect on the host.

Streptomycin has been very successful in controlling hallow blight of French bean (*Pseudomonas phaseolicola*), the apple and pear fire-blight (*Erwinia amylovora*), bacterial leaf-spot of tomato and a pepper (*Xanthomonas vesicatoria*), and the wild-fire disease of tobacco (*Pseudomonas tabaci*).

Cycloheximide (=Actidione) source: *Streptomyces griseus.*

It was discovered in 1946 and released for crop protection in 1954. It has limited systemic activity but has wide antifungal activity. It is phytotoxic and mutagenic. Diseases controlled include cedar rust of apple, powdery mildew of beans and rose, rust of wheat (leaf and stem rusts), safflower and mint, leaf-spot of cherry, and bunts of wheat. However, it is costly.

Aureofungin: source: *Streptomyces cinnamomeus* var. *terricola*

Broad spectrum systemic antibiotic. It controls citrus gummosis (*Phytophthora citrophthora*), *Diplodia* rot of mango and *Alternaria* rot of tomato.

Griseofulvin. source: *Penicillium griseofulvum*, *P. janczewskii*

Several other species of *Penicillium* too produce griseofulvin. It is antifungal but is not translocated in the fungal hyphae. It attacks hyphal regions at the place of contact.

Blasticidin. Produced by *Streptomyces griseochromogenes* (Fukunaga *et al*, 1955), it is active against both bacteria and fungi and is selective in its antifungal activity. It is used in control of blast disease of rice (*Piricularia oryzae*).

Benzene Compounds: Some benzene compounds are toxic to micro-organisms and have been used as fungicides viz., dinitro-o-cresol, pentachloronitrobenzene (PCNB), dichloran dinocap and dexon. PCNB is well known as a soil fungicide and controls several soil-borne diseases.

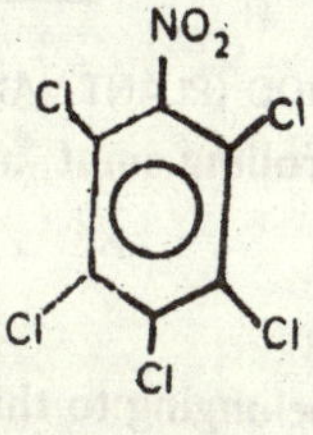

PCNB

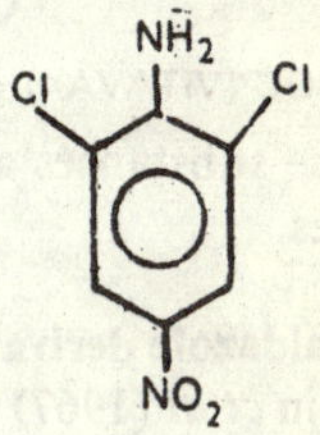

Dichloran

Miscellaneous Compounds

Some highly effective chemicals do not fall in any of the above categories, these are:

Dinocap (Carathane): 2-(1-methy-n-heptyl)-4,-6-dinitrophenyl crotonate. This substance is highly toxic against powdery mildews.

Oil. *Oils* of several types are now used for controlling plant diseases, particularly virus infections. The advantages of oil are that they are effective, cheap, non toxic and have good sticking and spreading properties. Their action is therapeutic.

Systemic Fungicides or Bactericides

Systemic fungicides or bactericides are chemotherapeutants that are translocated in plants to the site of infection, where they act. They act directly on the pathogen rather than by interference with pathogenic processes.

Oxanthiins and benzimidazole are two important groups of systemic fungicides.

Oxanthiins

The report of systemic fungicidal activity of oxanthiins by Von Schmeling and Kulka in 1966, was a big step ahead in chemotherapy. The two derivatives developed by these scientists are: (*a*) DMOC, 5,6-Dihydro-2-methyl-1, 4-oxanthiin-3-carboxyanilide and (*b*) DCMOD, 2,3-dihydro-5 carboxyanilido-6-methyl-1, 4 oxanthiin-4, 4-dioxide. They are known as vitavax and Plantvax (trade names).

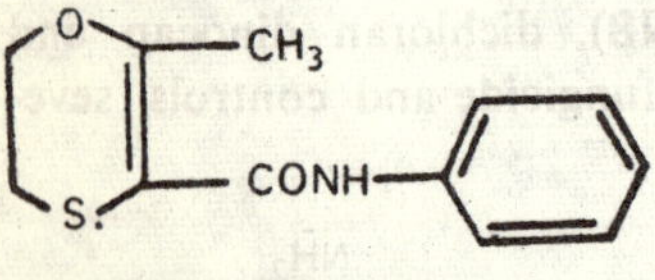

DMOC (VITAVAX)

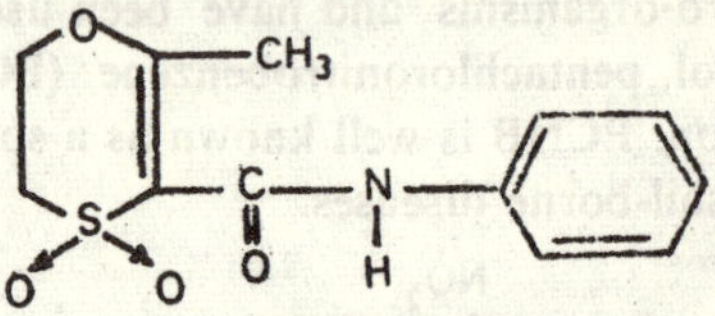

DCMOD (PLANT VAX)

These substances are very effective in controlling smut and rust diseases.

Benzimidazole derivatives

Erwin *et al* (1967) found two compounds belonging to this group to effectively control *Verticillium* wilt of cotton. These substances are Benlate and TBZ (trade names).

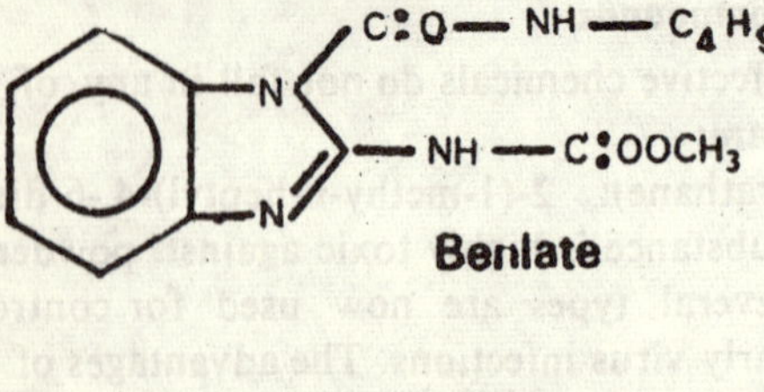

Benlate

REFERENCES

DIMOND, A.E., HEUBERGER, J.W. and HORSFALL, J.G., 1943, "A water soluble protectant fungicide with tenacity," *Phytopathology*, 33: 1095-1097.

ERWIN, D.C., ZENTMYER, G A., GALINDO, J. and NIEDERHAUSER, J.S., 1963,"Variation in the genus *Phytophthora*," *Annu. Rev. Phytopathol.*, 1: 375-396.

FUKUNAGA, K., MISATO, T. and ASAKAWA, M., 1955, "Blasticidin, a new anti-phytopathogenic fungal substance," *Bull. Agri Chem. Soc. Japan*, 19: 181.

HORSFALL, J.G., 1956, "Principles of fungicidal action," *Waltham* (Mass.): Chronica Botanica Co.

LUKENS, R.J., 1971, *Chemistry of Fungicidal Action*, Chapman and Hall Ltd., London. 136 p

———, RICH, S. and HORSFALL, J.G., 1965, "Role of the R group in the fungitoxicity of R-SccL compounds, *Phytopathology*, 55: 658-662.

MACCALLAN, S.E.A., 1969., "A perspective on plant pathology," *Ann. Rev. Phytopathol.*, 6: 1-12.

MILLARDET, P.M.A., 1885, (In French), English translation by Schheiderhn in *Phytopathol., Classics* 3 (1933).

MORECHART, A.L. and CROSSAN, D.F., 1965, "Studies on the ethylene bis-dithiocarbamate fungicides," *Delaware Univ. Agric. Expt. Stat. Bull.*, 357: 26.

NENE, Y.L., 1971, *Fungicides in Plant Disease Control*, Oxford and IBH Publishing Co., Delhi, 385 pp.

OWENS, R G , 1960., "Effects of elemental sulphur, dithiocarbamates, and related fungicides on organic acid metabolism of fungus spores" in *Developments in Industrial Microbiology*," 1: 187-205.

———, and HAYES, A.D., 1964, "Biochemical action of thiram and some dialkyl dithiocarbamates," *Contrib. Boyce Thompson. Inst.*, 22: 227-240.

PREVOST, B., 1807, "Memoire sur la cause immediete de la carie ou charbon des bles et de plusieurs autres maladies des plantes, et sur les preservatifs de la caire," *Phytopathology Classic*, 6: 1-4 (Trans. by Keitt, G.W., 1939).

THORN, G.D. and RICHARDSON, L.T., 1964, "Decomposition of ferbam," *Phytopathology*, 54: 910.

TISDALE, W.H. and FLENNER, A.L., 1942, "Derivatives of dithiocarbamic acid as pesticides," *Ind. and Erg. Chem*, 34: 501-502.

———, and WILLIAMS, T.. 1934, "Disinfectant," U.S. Patent 1972, 961.

TWEEDY, B.G , 1964, "A possible mechanism for the reduction of elemental Sulphur by *Monilina fructicola*," *Phytopathology*, 54: 910.

VON SCHMELING, B. and KULKA, M., 1966, "Systemic fungicidal activity of 1 4-oxanthiin Derivatives," *Science*, 152: 659-660.

PART TWO

PLANT DISEASES

17

ROTS AND DAMPING-OFF

PLANT diseases leading to rots and damping-off are caused by different types of microorganisms. The invasion of tissues involves the degradation of the cell wall constituents (mostly pectic substances and cellulose) of the host. Pathogens capable of forming the pectolytic and cellulolytic enzymes (especially the former) are, therefore, associated with such diseases.

Damping-off disease results in collapse and death of young seedlings. Generally fungi like *Pythium, Phytophthora, Sclerotinia, Thielavia, Glomerella, Fusarium, Rhizoctonia, Phoma* and *Ozonium* are responsible for such diseases. Sometimes, species of *Botrytis, Cylindrocladium* and *Diplodia* have also been found associated with damping-off diseases. The *modus operandi* of different organisms varies. Species of *Pythium* initiate infection in the tiny rootlets whereas *Rhizoctonia solani* causes necrosis of the succulent cortical tissue at the base of the stem. Infection of seedlings at such a delicate stage generally results in collapse and death of the plant. In case they survive, the growth remains stunted. The young seedlings contain substantially large amounts of pectic substances. Several species of *Pythium* and *Phytophthora* are known to produce different types of pectic enzymes. While the involvement of pectic enzymes is well documented, it is doubtful if cellulase plays any significant role in enzymatic destruction of cortical tissue of seedlings during invasion by these fungi (Husain and Kelman, 1959). Both cellulolytic and pectic enzymes are, however, effective during damping-off caused by *Fusarium moniliforme* (Venkataram, 1956) and *Rhizoctonia solani* (Kohlmeyer, 1956). Besides tissue disintegration, the damping-off fungi may also affect the seedlings by toxin action. Some such toxic principles have been recorded in *Pythium irregulare*—a common parasite of sugarcane (Bradenburg, 1950) and three species of *Pythium* (Damle, 1952). Damping-off at pre-emergence stage kills the seedlings before their emergence over the soil surface. There is complete rotting of radicle and plumule when they come out of the seed. High moisture content of the soil accompanied by high temperature usually favour such diseases.

The damping-off, which we are familiar with, is actually the one which manifests its symptoms at post-emergence stage. Infection is usually initiated at ground level and soft water-soaked areas appear on the young stems. With the advancement of disease, the stem rot gets constricted and finally collapses. Besides "the vegetable crops (tomato, brinjal, chillies and cucurbits etc.) and ornamental plants, the seedlings of forest trees are also affected in nurseries. When predisposing factors like high moisture and excessive organic manuring are available, these diseases take a heavy toll.

The control measures mainly comprise the use of seed-protectants to avoid pre-èmergence infection. Proper aeration of soil and good hygenic conditions are helpful in reducing the post-emergence infections.

DAMPING-OFF OF TOBACCO (*Nicotiana tobaccum* L.) SEEDLINGS

Pythium debaryanum Hesse, ***P. aphanidermatum*** (Eds.) Fitz., ***P. ultimum*** and ***P. arrhenomanes*** Dreschler have been found to be associated with this disease but ***P. debaryanum*** is usually more common.

Patches of infected plants are quite conspicuous in seed beds. In post-emergence phase, the affected seedlings show a girdle of brown decaying cortex extending upward and downward from the ground level, the seedlings subsequently collapse and topple down due to weakening of tissues at the base of the stem.

The hyphae of *Pythium debaryanum* penetrate the cell walls of the hypocotyle and ramify within and between the cells of the cortical parençhyma. Sporangia and oospores of the fungus are produced in the parenchymatic tissue, especially of the cotyledons. Sporangia are produced profusely; they get easily detached and are carried by wind or water to other locations where they germinate directly or indirectly (by zoospore formation) to cause infection. Disease is severe in ill-aerated and ill-drained soils.

Pre-emergence damping-off is easily controlled by treating the seeds with suitable seed protectants. Soaking the seeds in 0.2 per cent solution of Blitox-50 reduces incidence of the disease.

Periodical soil treatment of the nursery beds is also essential. It can be accomplished by heavy spray of formalin (1:50) over loose soil. For post-emergence damping-off, drenching the soil with Bordeaux mixture, 0.5 per cent Perenox, 0.2 per cent Fytolon is useful. In addition, overcrowding of the plants should be avoided. Cultivation should be carried out in light soil. Provision for proper irrigation and regular drainage should be made in the fields. Rotation of crops is also useful in certain cases.

ROTS

Rotting of plant parts is caused by a large number of fungal and bacterial pathogens. Fruits, bulbs, tubers and roots are the main targets of attack. On the basis of symptoms expressed over the infected tissue, the rots are distinguished in two types: (*i*) Dry rots, and (*ii*) Soft rots.

(1) **Dry Rots** are caused by few specialized fungi like *Fusarium* and *Cladosporium*. The infection is generally confined to storage organs of plants and usually there is no oozing of any fluid from the infected tissues, which collapse and shrink. Dry rot of potato and arvi (*Colocasia esculentum*) caused by *Fusarium coeruleum* is known from different parts of the world. Cucumber scab disease caused by *Cladosporium cucumerinum* is another example of dry rot. Since the exosmosis of liquid from the host cells usually reflects injury to plasma membrane, it is presumed that dry-rot causing organisms are possibly less toxigenic than the soft-rot ones. Tissue disintegration is much less as compared to soft-rots. Pierson and Walker (1954) observed that cellulose in the cell walls of cucumber (infected by *Cladosporium cucumerinum*) was considerably altered. Husain and Rich (1958) reported the production of PG and cellulase in culture by this fungus. Slow advancement of fungus is attributed to lack of PME activity.

(2) **Soft Rot** diseases are very common in fruits and vegetables. Usually they are prevalent in storage and transit. The basic principle involves the disintegration of parenchymatous tissue of the host leading to oozing of liquid on the host surface. A number of bacterial and fungal pathogens are capable of causing soft rot diseases. Culture filtrates of these organisms contain a number of pectic enzymes which are capable of tissue maceration by degrading the pectic substances of the host.

13 The exact mechanisms by which cells are killed by these plant

pathogens is not clearly understood. Brown (1955), Tribe (1955) and Fushtey (1957), however, are of the view that both killing and maceration are accomplished by pectic enzymes. It is difficult to visualize how pectic enzymes alone can to act as poisonous material for living protoplasm. Possibly, these enzymes make the plant cells more susceptible to other toxic metabolic products by rendering cells accessible to toxic molecules that do not enter the intact cells freely. Yet the other possibility is that the cells are killed before the enzymes act. This possibility rules out the involvement of pectic enzymes in causing the death of the cells. It is suggested that *Sclerotium rolfsii* (Higgins, 1927) and *Sclerotinia sclerotiorum* (Ovrell, 1952) kill the cells by oxalic acid that they produce. Evidences with more pathogenic organisms will be needed before reaching any decisive conclusions.

Rhizome Rot of Ginger

The disease is of wide occurrence. During initial phase of infection the leaves turn chlorotic from tip downwards, ultimately resulting in their withering and premature death. Subsequently, the foot and rhizome regions turn pale and the portion immediately above the ground becomes water soaked. In severe infections rhizome decomposes completely.

A large number of species of *Pythium* like *P. aphanidermatum*, *P. myriotylum*, *P. monospermum*, *P. gracile* and *P. butleri*, have been found to be associated with the disease. The pathogenicity tests, however, support the importance of first two species only. These species generally survive in soil or decayed rhizomes in the form of oospores. The disease is, thus, both soil-borne as well as rhizome-borne.

Waterlogging should be avoided and healthy rhizomes should be used for cultivation. Spraying Bordeaux mixture (5: 5: 50) on 2-3 weeks old plants also reduces the infection.

Pythium aphanidermatum is also known to be responsible for rhizome and root rot of turmeric. The disease is reported from Ceylon and certain regions of south India. Another species of *Pythium* viz., *P. graminicolum* is also reported to be associated with this disease (Middleton, 1943). It is also soil as well as rhizome-borne disease. Soil drenching with common fungicides like 1.0 per cent Bordeaux mixture, 0.1 per cent Ceresan or 0.3—0.5 per cent Blitox-50 controls the disease to a great extent.

Bud Rot of Areca Palms

Bud rot of Areca palms is very common in Western and Eastern India. The disease is popularly known as "Koleroga" and it is caused by *Phytophthora palmivora*.

Nuts are the main target of attack. During initial phase of infection, water-soaked areas appear at the base of the nuts. The infection gradually spreads to entire nuts which finally get detached and fall to the ground. In case of severe infections, the nuts are covered by whitish felty mycelium. Sometimes, the entire crown of a tree may get affected resulting in withering of leaves and young branches. Under such conditions, there is a total loss of the crop. Younger nuts are more susceptible. Six months old nuts are generally able to resist the infection.

Heavy rainfall favours spread of the disease. Oospores have not been observed in nature and the disease is perpetuated by mycelium which oversummers on tree tops. Fallen fruits, which lie in the vicinity, provide another rich source of inoculum.

Mass-scale and organized destruction of diseased tree tops, nuts and other affected parts is the most effective method of controlling koleroga. Periodical application of fungicides is difficult because of the height of the plant. Spraying with 1 per cent Bordeaux mixture before nut development and after break of monsoons is quite effective.

Bud Rot of Toddy and Coconut Palms

Phytophthora palmivora is responsible for the bud rot of toddy palm (*Borassus flabbelifer* L.) and coconut palm (*Cocos nucifera* L.). The disease is generally restricted to south India. At the initiation of infection, the leaves and leaf bases exhibit discolouration. Subsequently, they become yellow and dry up. Sometimes the entire growing points get affected which results in death of the plants. The disease is perpetuated by oospores or mycelium present in the dead plant parts. Propagation of the disease is carried out by sporangia.

Protective spraying with 5 : 5 : 50 Bordeaux mixture gives effective control.

Stem Rot of Papaya (*Carica papaya* L.)

The disease is widespread and has been reported from different papaya growing regions of the world. Infection is generally initiated during the monsoons. Water soaked patches appear on the stem, usually above ground level (Fig 17.1). Due to profuse rotting these areas turn dark-brown or black. The disease may spread from stem down to the roots. Leaves become yellow and may defoliate prematurely. Fruit development is reduced and they may fall before attaining normal size. Like other soft rots, there is disintegration of parenchymatous tissues of the stem which ultimately result in collapse of the entire plant.

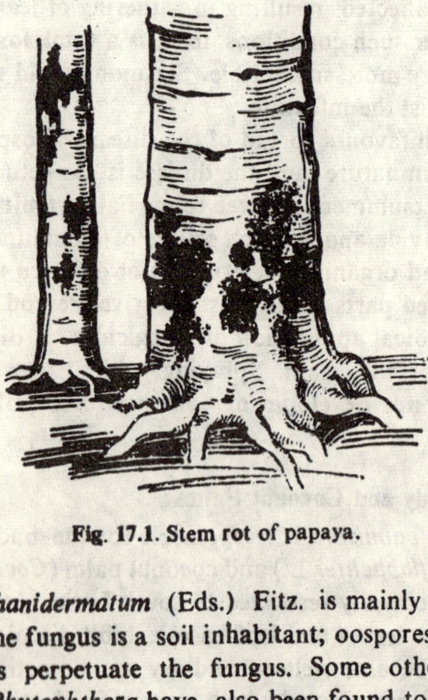

Fig. 17.1. Stem rot of papaya.

Pythium aphanidermatum (Eds.) Fitz. is mainly responsible for this disease. The fungus is a soil inhabitant; oospores produced in decaying parts perpetuate the fungus. Some other species of *Pythium* and *Phytophthora* have also been found to be associated. Consistent association of *Phytophthora parasitica* with stem and foot rot of papaya is known from some parts of the world.

Water-logging should be avoided and drainage near the plantation should be good. Regular destruction of diseased plants is also recommended. Soil drenching around the stem and spraying the stem with 6: 6: 50 Bordeaux mixture and 0.2 per cent Esso fungicide 406 considerably reduces the incidence of infection. Seed treatment with Agrosan GN or Ceresan has also been recommended.

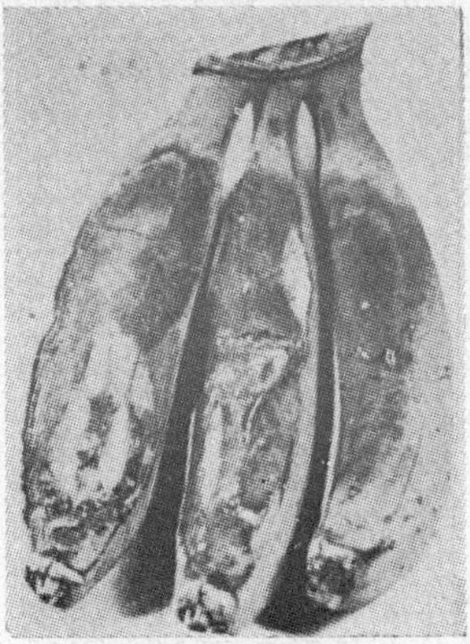

Fig. 17.2. Soft rot of banana caused by *Botryodiplodia theobromae*.

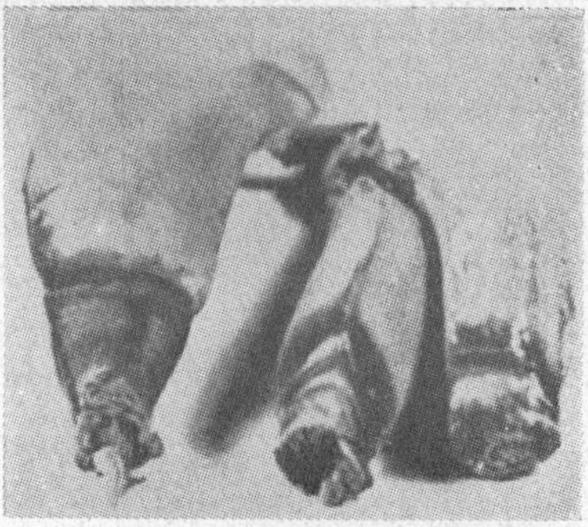

Fig. 17.3. 'Cigar end' symptoms on banana fruits.

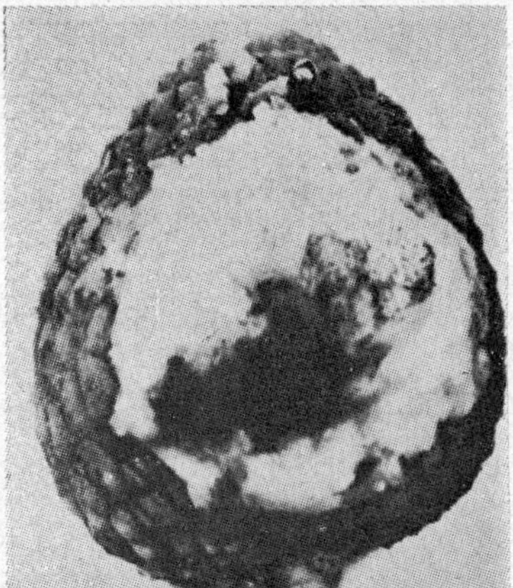

Fig. 17.4. *Botryodiplodia* soft rot of *Litchi* fruit.

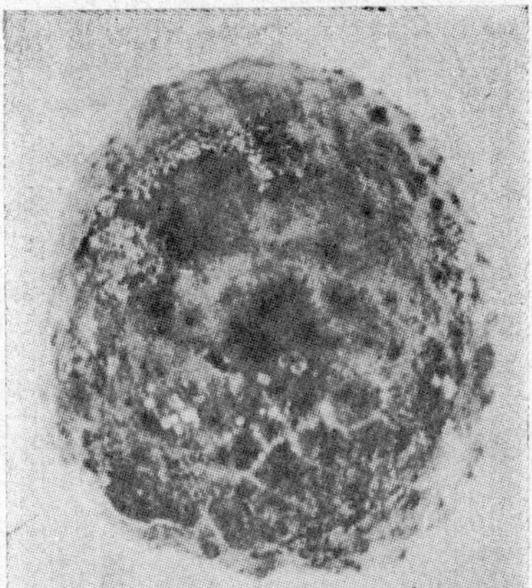

Fig. 17.5. *Pestolotia* soft rot of *Litchi* fruit. (*Courtesy* R. S. Bilgrami)

Fig. 17.6. *Pestalotia* soft rot and leaf spot on sapota.

(A)

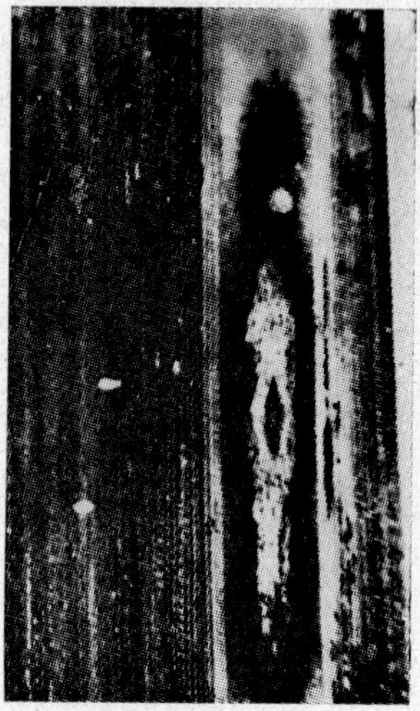

(B)

Fig. 17.7. (*A*) Longitudinal section of sugarcane, showing 'red rot' symptoms (*Courtesy* P.P. Mehta), (*B*) Acervuli on leaf spots.

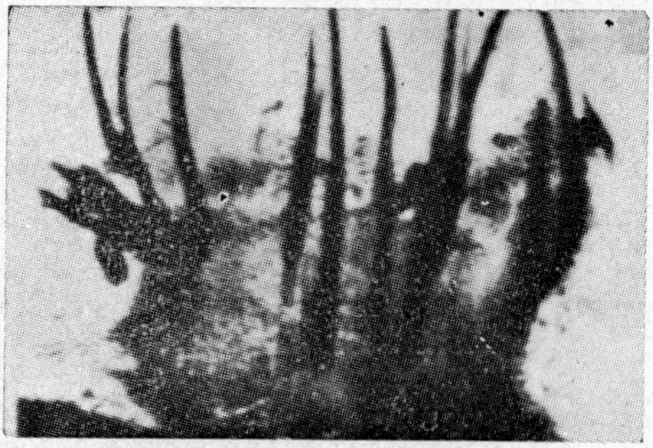

Fig. 17.8. Acervulus of *Colletotrichum falcatum* showing setae, conidiophores and conidia.

Fig. 17.9. Collapsing of maize plant due to stalk rot.

Fig. 17.10. Shows symptoms of Tondu disease, viz. twisting of stalk, curling of leaves and rot of ears.

Fruit Rot of Cucurbits

During summers a large number of cucurbitaceous crops are cultivated in India. Due to trailing habit of these plants, most of the fruits are in constant touch with the soil surface. Some of the soil inhabitants like *Pythium*, *Phytophthora*, *Fusarium* and *Rhizoctonia* attack the fruits of different cucurbits like bottle gourd (*Lagenaria bulgaria*), sponge gourd (*Luffa aegyptiaca*), parwal (*Trichosanthes dioica*) and several others. Slight injury to the host surface facilitates infection.

Pythium aphanidermatum (Eds.) Fitz. is the most common fungus which is associated with such diseases. In presence of moisture, there is profuse growth of cottony mycelium on the surface of the fruits. Prior to development of cottony growth the fruits develop water soaked areas which clearly exhibit the symptoms of soft rot.

Any type of injury to the fruits or contact with diseased fruits should be avoided.

Soft Rot of Fruits

Most of the fruits suffer with soft rot diseases which are more common during the post-harvest period. The frequency of these diseases is, therefore, very high during storage, transit and marketing. According to Smith *et al* (1964), "There are more than 250 known parasitic diseases of fruits and vegetables that cause decay and blemishes during the marketing process. The damage and losses incurred vary with the crop, growing conditions in the field, handling during harvest, transit and the storage conditions." The post-harvest diseases of fruits have received scant attention in India. Tandon and his students (1967) have made detailed investigations on the post-harvest and storage diseases of mango, papaya, banana, guava, sapota, litchi and pomegranate. On the basis of extensive surveys, they concluded that inoculum for most of the storage diseases was provided by the orchards and the diseased leaves used for packing the fruits. Injury to the fruits during transit were important factors for high incidence of soft rot diseases.

Soft Rot of Mango (*Mangifera indica* L.)

Botryodiplodia theobromae, *Colletotrichum gloeosporioides*, *Pestalotia mangiferae*, *Aspergillus niger* are some of the common fungi responsible for soft rot of mango.

The symptoms produced by different pathogens vary considerably. In case of *Botryodiplodia* rot, water soaked areas appear around the stalk which increase and cover about 1/3rd area of the fruit. The central portion of the infected zone acquires dark brown colour while the peripheral region remains light brown. In case of severe infection, minute pycnidia also develop in the infected regions.

Infection by *Colletotrichum gloeosporiodes* is characterized by formation of circular dark brown specks during the initial stages which subsequently enlarge and coalesce. After about a week of infection acervuli may also appear on the infected regions.

In *Pestalotia* rot, the patches are irregular, dark brown to black. Sometimes, black shining drops containing numerous conidia ooze on the host surface.

Soft rot of mango caused by *Aspergillus niger* originates as light brown circular patches around the fruit stalk. The lesions grow and form large circular spots encircling the stalk. Black conidial heads of the fungus become distinctly visible in the diseased regions.

Dipping the mango fruits in 6—8 per cent borax solution and 300-400 mycostatin solution after harvesting prevents the soft rot by *Botryodiplodia theobromae*.

Soft Rot of Papaya (*Carica papaya* L.)

Papaya is extensively cultivated in different parts of India. A number of fungi like *Botryodiplodia theobromae*, *Gloeosporium payayae*, *Colletotrichum papayae* and *Rhizopus nigricans* are consistently associated with soft rot of Papaya fruits. *Ascochyta caricae*, *Fusarium roseum*, *Alternaria tenuis* and *Trichothecium roseum* may also cause soft rot.

Botryodiplodia rot manifests itself in the form of water-soaked lesions. The lesions subsequently become dark brown and the tissues shrink. Under moist conditions the lesions get covered with light grey fungal growth of the causal organism. At a relatively late stage pycnidial bodies may also develop in the infected lesions.

Rhizopus nigricans causes watery rot of Papaya fruits. It is characterized by production of irregular water-soaked lesions. The lesions get enlarged and covered by sporangiophores which are white when young but subsequently turn black. Tandon and coworkers (1967) inoculated some of the soft rot fungi on papaya fruits and recorded the post infection changes in sugar, amino acids and ascorbic acid contents. They observed rapid hydrolysis of the sucrose in fruits infected with *Colletotrichum papayae* and *Gloeosporium papayae*. A gradual decrease in glucose, fructose and ascorbic acid contents also takes place. The concentration of certain amino acids like leucine and isoleucine increased in the diseased fruits.

Several control measures like storage at low temperature (15°C±1), hot water dip, antibiotic and borax solution treatment have been tried but none of them was effective in checking the disease.

Soft Rot of Banana (*Musa* spp.) The 'Cigar end' disease

Banana soft rot (Fig. 17.2) is very common and it is recorded from various parts of India. *Gloeosporium musarum* and *Botryodiplodia theobromae* are the main pathogens. In addition, the association of certain saprophytic fungi like *Rhizopus nigricans*, species of *Aspergillus* and *Nigrospora oryzae* is quite consistent. *Gloeosporium musarum* and *Curvularia fallax* have also been obtained from fruits expressing "Cigar end" symptoms (Fig. 17.3). Cigar end disease was first recorded by Dade (1926) on Cavendish bananas. The disease proceeds backwards from the tip of finger The diseased skin becomes greyish and tissues shrink. Subsequently, the infected regions get covered with salmon to brown coloured fruiting bodies of *Gloeosporium musarum*. The diseased surface appears corrugated showing close resmemblance to burnt cigar tips.

Infection by *Gloeosporium musarum* results in rapid loss of sugars (sucrose, glucose, fructose) and starch. The concentration of some amino acids like serine and glycine increases in the infected fruits.

Storage at low temperature, hot water dip and treatment with mycostatin and borax solutions have been tried but all proved ineffective.

Soft Rot of Guava (*Psidium guajava*, L.)

Gloeosporium psidii, *Pestalotia psidii*, *Botryodiplodia theobromae*, *Phoma psidii*, and *Aspergillus niger* are some of the common fungi associated with the soft rot disease of guava fruits.

Development of water soaked areas on the fruit surface is a consistent feature of all the rots. Rot caused by *Aspergillus niger* is characterized by formation of black mouldy growth. Infection by *Gloeosporium psidii* leads to production of salmon coloured spore masses at the centre of the lesions. In *Botryodiplodia* rot, the lesions are initiated from the region of fruit stalk and develop in irregular manner. At advanced stage, numerous pycnidia of *B. theobromae* are produced in the infected region.

In *Pestalotia* rot brownish spots are produced which enlarge in size. After about a week the central tissues form a depression. Subsequently, black minute acervuli with viscid spore masses appear on the infected tissue. The infection is caused by *Pestalotia psidii* Pat.

Tandon and co-workers (1967) analysed the sugar and amino acid contents of guava fruits under pathogenesis and concluded that *Pestalotia psidii* and *Gloeosporium psidii* were capable of hydrolysing entire sucrose contents of the fruit and that glucose was utilized in preference to fructose. There was a sharp decrease in the amino acid content of the diseased fruits. They also recorded a fall in the vitamin C content of the infected fruits.

Storing the fruits at 15°C minimizes the soft rot of guava.

Soft Rot of Litchi (*Litchi chinensis* Sonners)

More than a dozen fungi have been reported to be associated with soft rot of Litchi. Rots caused by *Botryodiplodia theobromae* (Fig. 17.4) and *Pestalotia pauciseta* (Fig. 17.5) are, however, quite serious during storage. Diseased leaves provide the inoculum during storage and transit. *Botryodiplodia* rot is usually initiated through the stalk region. In *Pestalotia* rot (Fig. 17.6) profuse development of conidia takes place in the affected regions. Prasad (1963) carried out extensive pathological investigations on four varieties of Litchi common at Muzaffarpur. Bilgrami (1970) recorded the changes in sugar and vitamin contents of the fruits under pathogenesis.

Proper care at the time of packing and transport and storing the fruits in cold storage at 15°C considerably diminishes the fruit rot of Litchi.

Soft Rot of Sapota (*Achras Sapota* L.)

Soft rot of Sapota is recorded from Tamil Nadu, Andhra Pradesh, Maharashtra, Uttar Pradesh, West Bengal and Rajasthan. The disease is caused by *Botryodiplodia theobromae* Pat., and *Pestalotia sapotae* P. Henn.

Botryodiplodia rot starts as irregular water soaked areas; within 8-10 days the fruits get completely rotten. At this stage pycnidia become evident on the fruit surface. The disease is characterized by soft watery rot.

Pestalotia rot (Fig. 17.6) is characterized by appearance of light brown spots on the surface of the fruits. Gradually they are covered with white fluffy mycelial growth. At a later stage, acervuli, oozing dark green viscid masses of spores, appear on such lesions. The rot is moderately soft. Diseased leaves provide inoculum during transit and storage.

A number of organic acids viz., oxalic acid, citric acid, malic acid, succinic acid and fumaric acid have been recorded in healthy fruits of Sapota. *Botryodiplodia theobromae* consumes all the organic acids quite rapidly. The concentration of the amino acids also decreases in the diseased fruits.

Storing the fruits at low temperatures (10 to 15°C) reduces the incidence of disease.

RED ROT OF SUGARCANE

The disease is quite severe in India, specially in Eastern Uttar Pradesh and Bihar. The disease is caused by *Colletotrichum falcatum* Wint., which is the conidial stage of *Glomerella tucumanensis* (Speg.) Arx and Mueller or *Physalospora tucumanensis* Speg.

Practically all the above ground parts are affected. Stems are the main targets of attack. Plant growth is adversely affected after infection. Subsequently the canes get shrivelled; the rind shrinks and longitudinal wrinkles are formed. The affected canes show a considerable loss in weight and become lighter. If split longitudinally, the affected stems show red streaks (Fig. 17.7). The red colour is interrupted by white patches. Vascular bundles show maximum reddening which may extend upto pith. At advanced stage of infection, the juice may emit unfavourable odour due to conversion of sucrose to glucose and alcohol. Red longitudinal bands are also seen on and along the midrib of leaf. Black acervuli of the fungus are profusely developed in these regions. White and cream coloured hyphae are profusely formed in the diseased stems. Chlamydospore formation is quite common. Chlamydospores, being comparatively more resistant, help in perpetuation of infection. *C. falcatum* is known to grow and produce acervuli (Fig. 17.8) in soil (Chona and Nariani, 1952). This helps in perpetuation of disease and initiation of infection to the new crop. Secondary infection and disease propagation is accomplished by conidia which are abundantly produced in the mid-rib lesions.

Red rot is spread through setts. Use of healthy setts is therefore, essential to reduce the incidence of disease. Rotation of crops, after a lapse of 2 to 3 years, is also beneficial. Use of resistant varieties, accompanied by healthy cultural practices, is the most effective method of controlling the disease. Most of the resistant varieties, which have been recommended, are, however, of regional value.

Stalk Rot of Maize

Stalk rot of maize was first reported from Pusa in Bihar by Prasad (1930). Now it is known from all the maize growing regions of the country as well as from USA, Israel, South Africa, Greece and Mexico.

A number of bacterial pathogens like *Pseudomonas lapsa* (Ark.) Starr and Burk., *Erwinia dissolvens* (Rosen) Burk, and *Erwinia carotvora* var. *zeae* Sabet have been reported to cause stalk rot of maize in different parts of the world. The disease induced by *E. carotovora* var. *zeae* is very important.

Sometimes the disease is so serious that within a few days the affected plants are killed. In beginning the upper leaves show signs of wilting and the basal internodes become soft and discoloured. Lower leaf sheaths and leaves covering these internodes may also express discolouration. The rind loses its green colour and becomes pale yellow. With the advancement of disease, the pith gets destroyed and the vascular bundles get disorganized. The stalk becomes so soft that with slightest application of mechanical force it collapses and affected plants topple down (Fig. 17.9). Sometimes, a disagreeable fermenting odour is emitted from the affected plants.

The organism is soil borne (Anil Kumar and Chakravarti, 1970) and makes its entry through wounds and injuries on the host surface. Borer insects play a significant role in initiation of the disease. High humidity, coupled with high temperature (30°C and above), are conducive for disease development. Avoidance of water logging and dense plantation, reduce the disease incidence. Use of agricultural antibiotics like streptocycline and agrimycin-100 as well as bleaching powder or clorox (Thompson, 1965; Rangarajan and Chakravarti, 1970) is recommended for control of the disease.

ROT OF WHEAT EARS (= Ear-cockle or Tundu disease)

Yellow ear-rot of wheat (Fig 17.10), popularly called as *Tundu*, is known in India since long. The disease is also reported from United Arab Republic, Australia and China. It is incited by the nematode *Anguina tritici* in association with the bacterium, *Corynebacterium tritici*. The disease is also called as Ear-cockle of wheat.

During the initial phase, the lower and central leaves show wrinkling. This is followed by exudation of bright yellow slime or gum on abortive ears and the leaves which are in contact with such ears. The affected ears are incapable of producing any grains. In humid weather, slime production is profuse and the slime may trickle down, while in dry season it becomes hard, brittle and brown in colour.

Ear cockle nematode galls, carried in soil as contaminant of wheat seeds, serve as primary source of inoculum (Vasudeva and Hingorani, 1952). Bacterium is carried to the plants by nematode larvae as surface contaminant (Gupta and Swarup, 1968). In soil the nematode galls, mixed with the seeds become moist and release larvae. The bacteria-laden larvae find their way to the growing point of germinating young wheat seedling. They are mechanically carried to the developing ear heads alongwith the lengthening of the culms. Atmospheric conditions determine the further stages. If a moist weather is available, the bacterium multiplies and produces slimy mass. A complete Tundu ear is developed under such conditions. However, if conditions do not entirely favour the bacterium partial ear-cockle and partial tundu symptoms can be produced. In the latter case, however, the nematode larvae continue to carry a fresh inoculum of the bacterium into the galls which perpetuate the disease.

Control measures are generally directed against the nematode which is responsible for carrying the tundu bacterium to the host plants. Galls are eliminated by floating the seeds in 2 per cent brine solution. Uprooting and burning of affected plants and maintaining a good drainage in fields is recommended. Some chemicals like Diazinon (0.15%) and Zinofos are known to minimize the incidence of infection

SOFT ROT IN VEGETABLES

A number of vegetables like tomato, brinjal, cabbage, cauliflower and various cucurbits express soft-rot symptoms. Initially, small water soaked areas appear on the host surface which enlarge rapidly and the affected tissues turn soft and pulpy. *Erwinia carotovora*, *Bacillus subtilitis*, *B. megatherium* and *Pseudomanas marginales* have been found to be associated with the soft rots of vegetables. Graham (1964) suggested that all soft-rot causing *Erwinia* species should be placed under a separate genus *i.e. Pectobacterium*. Soft rot bacteria oversummer/overwinter in soil, plant debris and fleshy plant organs in storage. Usually such organisms enter through wounds. The dissemination takes place through insect maggots like *Hylemyia cilicura* and *H. brassicae*. Several pectolytic and cellulolytic enzymes are known for soft rot bacteria.

Proper sanitary conditions in the godowns and warehouses are absolutely essential to control the soft rot diseases. Debris should be regularly removed from these places and walls should be disinfected with formaldehyde (1 lb. in 10 gallons of water) or copper sulphate (1 lb in 5 gallons of water). Wounds and injury to the plant parts should be avoided. Low temperature and humidity in the warehouses minimize the incidence of disease. Spray with antibiotics like streptomycin (0.1%) also reduces the incidence of disease.

REFERENCES

ANIL KUMAR, T.B. and CHAKRAVARTI, B.P., 1970, "The nutrition of two strains of *Erwinia carotovora* causal organism of stalk rot of maize in India," *Zbi. Bkt.*, 125: 707-712.

ASHOUR, W.E , 1954, "Pectinase production by *Botrytis cinerea* and *Pythium debaryanum*," *Brit. Mycol. Soc. Trans.*, 37: 343-352.

BILGRAMI, R.S., 1970, "Physiological and pathological studies on some fungi causing diseases of *Litchi*," Ph.D. Thesis, Bihar University.

BRADENBURG, E., 1950, "Uber die Bildung Von Toxinen in der enttung *Pythium* und ihre Wirkung anf die Pflanzen," *Nachrbl. dent. Pflanzenschutzdienst.* (Braunschw) 2: 69-70.

BROWN, W., 1955, "On physiology of parasitism in plants," *Ann. Appl. Biol*, 43: 325-341.

CHONA, B.L. and NARIANI, T.K., 1952, "Investigation on survival of sugar-cane red rot fungus in soil," *Indian Phytopath.*, 5: 152-157.

DADE, H.A , 1926, "Annual report (1925-26)" *Report. Agric. Dept. Gold Coast Govt.*, 33-36.

DAMLE, V.P., 1952, "Enzymic studies of certain parasitic fungi," *J. Indian. Botan. Soc.*, 31: 13-55.

FUSHTEY, S.G., 1957, "Studies on physiology of parasitism XXIV. Further experiments on killing of plant cells by fungal and bacterial extracts," *Ann. Botany* (London) 21. 273-286.

GRAHAM, D.C., 1964, "Taxonomy of the soft rot coliform bacteria," *Annu. Rev. Phytopath.*, 2: 113-42.

GUPTA, S.C., 1956, "Studies on the Physiology of parasitism XXII. The production of pectolytic enzymes by *Pythium debaryanum* Hesse," *Ann. Botany* (London), 20: 179-190.

GUPTA, P. and SWARUP, G., 1968, "On the earcockle and yellow ear rot disease of wheat. I Symptoms and histopathology," *Indian. Phytopath.*, 21: 318-322.

HIGGINS, B.B., 1927, "Physiology and parasitism of *Sclerotium rolfsii* Sacc," *Phytopathology*, 17: 417-448.

HUSAIN, A., and RICH, S, 1958. "Extra-cellular pectic and cellulolytic enzymes of *Cladosporium cucumerinum*," *Phytopathology*, 48: 316-320.

———, and KELMAN, A., 1958, "Role of pectic and cellulolytic enzymes in pathogenesis of *Pseudomonas solanacearum*," *Phytopathology*, 48: 377-385.

KOHLMEYER, J., 1956 "Uber den cellulose Abban durch einige phytopathogene Pilze," *Phytopathol*. Z, 27: 147-182.

MIDDLETON, J.T., 1943 "The taxonomy, host range, and geographic distribution of the genus," *Pythium. Mem. Torrey. Bot. Club*, 20.

OVRELL, B.T., 1952, "A toxin in culture filtrates of *Sclerotinia cierotiorum*," *Australian J. Sc.*, 14: 197-198.

PIERSON, C.F. and WALKER, J.C., 1954, "Relation of *Cladosporium cucumerinum* to susceptible and resistant cucumber tissue," *Phytopathology*, 44: 459-465.

PRASAD, H.H., 1930, "A bacterial stalk rot of maize," *Indian J. Agric. Sci.*, 25: 72.

PRASAD, S.S., 1963, "Pathological and physiological studies on some fungi causing leaf spot diseases," Ph.D. thesis, Bihar University.

RANGARAJAN, M. and CHAKRAVARTI, B.P., 1970, "Studies on the control of Bacterial stalk rot of maize with antibiotics," *Hindustan Antib. Bull*, 13: 16-19.

SMITH, M.A., RAMSBY, G.B. and GREEN, R.J., 1964, "Market diseases of fruits and vegetables," U.S. *Dept. Agric. Ext. Civ.*, 523: 19 pp.

TANDON, R.N., 1967, "Post Harvest diseases of tropical and sub-tropical fruits," *Tech. Rept.* P. L. 480 Project Pg. In. 133. Alld. University.

THOMPSON, D.L., 1965, "Control of bacterial stalk rot of corn by Chlorination of water in sprinkler irrigation," *Crop. Sci.* 5: 369-370

TRIBE, H.T., 1955, "Studies on the physiology of parasitism XIX. On the killing of plant cells by enzymes from *Botrytis cinerea* and *Bacterium aroideae*," *Ann. Botany* (London) 19: 351-368.

VASUDEVA, R.S. and HINGORANI, M.K., 1952, "Tannan or tundu disease of wheat," *Indian Farm* 2: 14-19.

VENKATARAM, C.S., 1956, "Studies on celluloytic activity of *Fusaria* with reference to bacterial and other cellulose substrates," *Proc. Natl Inst. Sci India*, 22: 104-211.

WOOD, R.K.S. and GUPTA, S.C., 1958, "Studies in the Physiology of parasitism XXV. Some properties of pectic enzymes secreted by *Pythium debaryanum*," *Ann. Botany* (London), 22: 309-320.

18

GALLS AND ABNORMAL GROWTH

INFECTION often leads to abnormal growth and reproduction of the host. Elongation of internodes and proliferation of floral parts are some of the common symptoms observed in many cereals, grasses and other infected plants. Several fungi, bacteria, viruses, nematodes and insects are capable of stimulating excessive localized outgrowths in plant cells. Usually, these areas of stimulated growth have richer pools of metabolites from which the organisms draw materials for their growth and development. Galls are developed practically on all the plant parts. Formation of galls is accomplished by cell enlargement as well as by rapid cell division which, in turn, are dependent upon specific substances that may be synthesized by such cells. Root nodules induced in leguminous plants by bacteria (*Rhizobium spp*) are highly specialized types of galls which are beneficial to the host. Certain metabolites elaborated by plant pathogens are quite similar in their physiological action to the growth regulating hormones of plants. Such substances alter the normal growth pattern with the result that plants show abnormal growth. Sometimes, the chemical stimuli produced by the pathogen are quite different from any of the growth promoting substances so far recognized in green plants. Being different, they divert the normal growth pattern of the plant into new and unusual directions. Increased auxin (IAA) levels have been demonstrated in many diseased plants. The production and role of auxin in plant disease have been studied more extensively in bacterial diseases. Morphological abnormalities due to viral infections are quite frequent and manifestation of such symptoms are usually similar to those produced due to imbalance of growth substances.

CLUB ROOT DISEASE OF CABBAGE
(*Brassica oleracea* var. *capitula* L.)

Pathogen. ***Plasmodiophora brassicae***

Nearly all the cultivated varieties of Crucifers are susceptible to it. The disease is common practically in all the parts of the world. It has also been referred as 'Finger and Toe' disease.

Symptoms

Symptoms produced on above ground parts include flagging of seedlings and chlorosis of leaves. Young plants are killed within few days of infection. Infected plants fail to produce marketable heads. Roots swell by 10-12 times and develop numerous spindle shaped, knobby or clubshaped galls. This results in failure of translocation of nutrients and water to aerial parts and leads to starvation and death. Several bacteria and fungi may colonize the infected roots and cause their decay. Numerous uninucleate haploid spores are produced in the infected roots, which are left in the soil after decomposition of the roots. They are highly resistant and may remain viable in soil upto eight years. According to Kunkel (1918), infection by penetration may be divided into the following four parts:

(*i*) primary infection of the cortical tissue upto the cambium.
(*ii*) infection of the cambium in all directions from the point of original penetration.
(*iii*) passage of the plasmodia out of the cambium into the cortex and the xylem.
(*iv*) infection of the medullary rays.

The infected cells in a 'club' usually lie scattered in small groups. There is no evidence to show that the affected cells themselves get permanently altered or that these cells are capable of continued abnormal growth in absence of the pathogen. Several factors like temperature, moisture as well as soil type and reaction have pronounced effect on the disease. High moisture content and a temperature of about 25°C favour disease development. The disease has been recorded only in the neutral or slightly acidic soils. It is believed that the stimulus resulting in abnormal cellular growth travels in advance of the infection. It is a malignant disease and in most of the cases the plants succumb to infection.

Control

A careful selection of the seedlings as well as proper drainage and aeration of the soil reduces the incidence of disease. Liming of the soil to make it slightly alkaline and distinfestation with 10 oz. corrosive sublimate in 10 gallons of water is recommended.

WART DISEASE OF POTATO (*Solanum tuberosum* L.)

Pathogen. *Synchytrium endobioticum* (Schilb) Per.

This is an obligate parasite. The disease is serious in temperate climate, more common in Europe and North America. In India, Ganguly and Paul (1953) reported this disease for the first time from Darjeeling where it was endemic in the beginning. The host range of this fungus is limited to Solanaceous crops. Besides potato and tomato, it is known to infect *Solanum nigrum*, *S. dulcamara* and *Nicandra physaloides*.

Symptoms

Buds on stems, stolon and tubers are the main areas of abnormal growth. Lumps of warts are attached to all the affected parts; they are conspicuous on tubers (Fig. 18.1). Their size varies and they are spherical and delicate. Initially they are green, when above ground and cream coloured when underground. At advanced stage, they become black and are often invaded by secondary saprophytes. The wart consists of distorted, prolified, branched structures grown together into a mass of hyperplastic tissue. Only meristematic tissue is prone to infection. It is characteristic that the tissue adjoining the infected cells responds with the development of tumours.

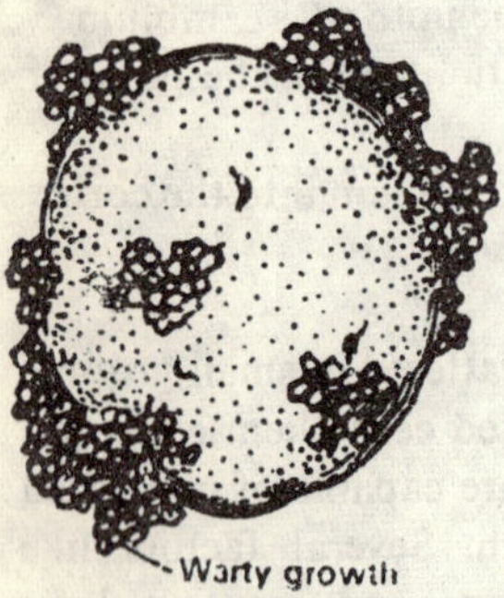

Fig. 18.1 Potato wart disease Symptom.

The disease is mainly soil borne. Resting sporangia survive in soil for more than ten years. Infection takes place through diseased tubers, infested soil, plant debris and contaminated manure. Zoospores in the soil are responsible for secondary infection. It is reported that 200 resting sporangia of *S. endobioticum* per gram of soil are necessary to cause infection and that the potato plant secretes certain substances into the soil which tend to break the dormancy of the resting spores. Enough soil moisture, a temperature of about 21°C, neutral or slightly acidic soil are most favourable for infection.

Several physiological races of *S. endobioticum* are known to exist.

Control

It is difficult to control the disease once it has been introduced in the field. The incidence of disease can be reduced by

(*i*) preventing the entry of diseased material into the healthy areas.

(*ii*) the soil treatment, steam sterilization, application of mercuric chloride, ammonia sulphocyanate, copper sulphate and 5% formaline.

(*iii*) application of lime to raise the pH of the soil is also recommendod.

(*iv*) use of resistant varieties.

STEM GALL OF CORIANDER

(*Coriandrum Sativum* L.)

Pathogen. ***Protomyces macrosporus*** **Unger.**

Symptoms

Only one species i.e., *P. macrosporus* occurs in India. The fungus causes tumour like swellings on the stem, leaf stalk, leaves, peduncle and fruits. Infection percentage is high in excessive moisture and shady conditions. Severe infection may prevent fruiting and even kill the plants. The organism confines itself to tumours only.

Disease Cycle

The fungal mycelium is branched and septate; it produces a large number of three walled elliptical, oval or rounded chalmydospores in the infected regions. Chlamydospore on germination produces a vesicle like structure in which a large number of uninucleate spores are developed. The vesicle ruptures and the spores are set free, and they multiply by budding in yeast like manner and cause infection of the host. Spores germinate and directly penetrate the cell wall of the host.

Control

Nene *et al* (1966) reported effective control by treating the seeds (0.25 kg per 100 kg of seed) and soil (8.00 kg per acre) with thiram.

CROWN GALL OF STONE FRUITS

Pathogen. *Agrobacterium tumefaciens*

It is common in Europe and the USA. Galls are usually round, with irregular surface (Fig. 18.2). In India, occurrence of the disease was reported by Singh (1943) but pathogenicity was established only recently by Durgapal (1971). The disease has so far been recorded on fruit trees like cherry, plum and peach

Fig. 18.2.

Symptoms

The disease first appears as small overgrowths on stems and roots, particularly near the soil line. Initially the galls are spherical, soft, white or cream coloured, which subsequently increase in dimension and become convoluted and dark brown. They, mostly, develop at the graft union, but may be produced on veins, and other aerial parts. This is primarily a disease of parenchyma. Infection results in rapid proliferation of cells.

Disease Cycle

Entry is affected through the wounds. After stimulation the tumour cells become entirely independent of the causal bacterium. Autonomous growth of tumour cells continues at the expense of the host. They may sometimes attain enormous dimensions and weight. Occasionally, complex tumours are formed from which roots and buds may arise. Such type of tumours are called 'teratomas', and are usually devoid of bacteria. Unlike other galls, the crown gall is an example of non-self-limiting tumour. Braun and White (1943) reported that the cells isolated from sterile (bacteria free) tumours were capable of self multiplication on suitable culture media. Such isolated sterile tissues are also capable of producing tumours on the host. Except the heavily lignified cells all other nucleated cells get transformed to crown gall cells. It is believed that the tumour inducing principle involves at least two auxins; one induces cell enlargement and the other promotes cell division. The behaviour and growth rate of these tumour cells can be controlled by supplementing the culture medium with various substances. According to Bräun (1951), alteration of normal cells to tumour cells is a gradual but progressive process. During transition from a normal cell to crown gall tumour cell, a series of distinct and well defined growth-substance-synthesizing systems become progressively activated. Tumour cells contain higher than normal amounts of IAA and cytokinins. The starvation of the host plant is caused by the neoplastic tissues which grow totally at the expense of normal tissues. Some bacteriologists have reported that tumour inducing property can be transferred to non-pathogens like *Agrobacterium raidobactor* and *Rhizobium leguminosarum* by cultivating them in synthetic media containing extracts of sterile tumour tissues or nucleic acid of virulent *A. tumefaciens*. Bräun (1962) has divided the process of tumour formation into two phases viz., (*i*) inception phase, the period between entry of bacterium into the wound and appearance of tumour cells, and (*ii*) developmental phase. The bacterium is a soil inhabitant, remaining viable for several years. It infects the host only through fresh wounds formed by cultural practices, grafting and insects etc.

Control

All sanitary precautions should be adopted. Clean and healthy stocks should be used Grafting knife should be frequently sterilized. Certain antibiotics like penicillin and vancomycin have been found to be effective in inhibiting the galls.

ROOT KNOT DISEASE

Several cultivated and wild plants suffer with root knot disease. This is caused by nematodes belonging to genus *Meloidogyne*. Infection results in reduced growth of the host, sudden wilting and formation of galls on the roots, which may result in their clubbing. The infection is initiated by larval penetration in large numbers, due to which, while growth along the pith is arrested, the cortical cells enlarge considerably to form giant cell Sometimes, cell walls of adjacent cortical cells may disintegrate. The mature female nematode is embedded in the plant tissues while the eggs are commonly found to be clinging to the surface of the roots. When the eggs hatch, young larvae may again penetrate the young roots just below the growing point. The enlarged cells of the host contain 6-20 nuclei, although in certain plants e.g. *Vitis gongylodes*, the giant cells may contain upto 500 nuclei. Often the galls may show severe rotting due to saprophytic organisms.

The females and the larvae survive in soil and on plant debris. Their concentration decreases with the depth of the soil. In light sandy soil they may be present upto 8 ft. deep. Various biotic and environmental factors influence their population and action. Root exudates also play a decisive role in attracting the larvae. Under normal conditions only females are produced and they reproduce parthenogenitically. When the food supply is scarce, and the concentration of the larvae in the cells is high, formation of male nematodes takes place.

Dropkin (1954) has reported a high positive correlation between gall area and number of larvae in the gall. He concluded that the size of the gall depends on the amount of stimulation provided by each larva in the outgrowth Some of the important root knots caused by the species of *Meloidogyne* are as follows:

(1) Root knot of potato and tomato	*Meloidogyne hapla*
(2) Root knot of wheat, barley, groundnut	*M. arenaria*
(3) Root knot of sugarcane, garden balsam	*M. javanica*
(4) Root knot of coffee	*M. exigna*

Larvae of *Meloidogyne* may act as vectors for certain soil fungi like *Fusarium* and *Pythium*.

The species of this nematode have a wide host range; therefore, rotation of crop is not expected to be of much advantage for controlling the disease. Application of soil fumigants, after destruction of root residues, eliminates infection to a considerable degree. Simple ploughing of the fields twice or thrice during summer months also helps in reducing the incidence of disease. In nature, there is possibility of some biological interference. Roots of *Tagetes* have been reported to have toxic effect on certain species of *Meleidogyne*. More than fifty predators and parasites of *Meloidogyne* that attack pine apples in Hawaii have been counted. Majority of them were fungi and other nematodes.

REFERENCES

BRAUN, A.C., 1951, "Recovery of crown-gall tumor cells," *Cancer Research*, 11: 839-844.

———, 1962, "Tumor inception and development in crown-gall disease," *Annu. Rev. Plant Physiol.*, 13: 533-558.

———, and WHITE, P.R., 1943, "Bacteriological sterility of tissues derived from secondary crown gall tumors," *Phytopathology*, 33: 85-100.

DROPKIN, V.H., 1954, "Infertility and gall size in tomato and cucumber seedlings infected with *Meloidogyne incognita* var. *aerita* (root-knot nematode)," *Phytopathology*, 44: 44-49.

DURGAPAL, J.C., 1971, "A preliminary note on bacterial diseases of temperate plants in India," *Indian Phytopath.*, 24: 379-382.

GANGULY, A. and PAUL, D.K., 1943, "Wart disease of Potatoes in India," *Sci. and Cult.*, 18: 605-606.

KUNKEL, L.O., 1918, "Tissue invasion by *Plasmodiphora brassicae*," *J. Agric. Research*, 14: 543-572.

NENE, Y.L., SIDDIQUI, I.A., and KHARBANDA, P.D., 1966, "Control of stem gall of Coriander by fungicide," *Mycopath. et. Mycol. Appl.*, 29: 142-144.

SINGH, U.B., 1943, *Misc. Bull. Indian Council Agric. Res.*, 51: 5-11.

WORONIN, M., 1934, "*Plasmodiophora brassicae*, the cause of cabbage bernia," *Phytopath. Classics*, No. 4.

19

DOWNY MILDEWS

THE term "Downy mildew" refers to both a group of fungi and the diseases they cause. These fungi are obligate parasites belonging to family Peronosporaceae (Class—Oomycetes). The diseases are usually characterized by white or grey bloom, usually called "downy growth", which is due to profuse production of sporangiophores on the lower surface of infected leaves. The downy mildews have been a major deterrent to cultivation of some of the most important crops. Nonetheless, these diseases have not received attention they deserve. A century back, downy mildew of grapes threatened the wine industry in France and the story of the disease is similar to the Irish famine due to *Phytophthora infestans*. The fungicide 'Bordeaux mixture' was developed by Millardet to control this disease. Downy mildew of hops caused by *Pseudoperonospora humili* reached catastrophic proportions in several European countries in 1924-26 and many hop gardens had to be abandoned. These diseases aroused great concern and had stimulating effect in attracting the attention of plant pathologists.

Symptoms

The symptoms usually comprise the "downy growth" on the host leaves. In some cases *e.g.* crazy top of maize, green ear disease of Bajra, the effect is very much pronounced and drastic. In addition to the downy growth, chlorotic spots or streaks on the leaves, tillering, stunting, phyllody (in ears) are the other common symptoms.

The pathogens

These fungi grow intercellularly, forming haustoria of different shapes. The hyphae fill the intercellular spaces and thus assume irregular shape. In humid environment and at low temperature, which occurs mostly during dewy nights, the hyphae emerge through stomata and form sporangiophores and sporangia (or conidia). The sporangia germinate either directly by forming germ tubes or form zoospores. Conidia always germinate directly and thus show a constant behaviour. The sporangia or conidia can be distinguished even without observing their germination. Sporangia always possess a lens shaped operculum which lacks in conidia. Environment determines the mode of germination of the sporangia. The sporangia of *Sclerospora graminicola* form germ tubes at higher temperatures and zoospores at low temperatures. The genera of downy mildew fungi are identified on the basis of the structure of sporangiophore, pattern of their branching and the behaviour of the sporangia. The pathogens produce conidia and zoospores in profuse numbers which serve as inoculum for secondary infection. Sexual reproduction occurs in the late phase of the disease, resulting in the formation of oospores which abound in necrotic tissues. The oospores which lie in the plant remains, or in the soil, perenate the fungus and cause primary infections on arrival of the host. Oospore morphology, specially the markings on their exospore, may be used in the taxonomy of these fungi (Safeeulla, 1970). The four species of *Sclerophthora* can be identified on the basis of such markings on the oospore wall. The oospores undergo a period of rest (sometimes referred as weathering) before germination occurs. The physiology of oospore germination is only little known. Oospores which germinate readily in soil or in contact with host seeds, fail to do so in laboratory even under most favourable conditions for germination.

The Genera

There are twelve genera which are identified on the basis of sporangiophore characteristics (Fig. 19.1). Some of the more common ones are as follows: In *Plasmopara* the branches of sporangiophores arise at right angles and are irregularly spaced. In *Sclerospora* the sporangiophore comprise a swollen hypha bearing several upright branches at the apex. The genus has two subgenera *Eusclerospora* and *Peronosclerospora* which are differentiated by the formation of sporangia in the former and conidia in the latter. Shaw (1970) has proposed that these should be raised to generic level. The genus *Sclerophthora*, as is evident from the name, combines characters of *Sclerospora* and *Phytophthora*. The sporangiophores are branched like *Sclerospora* but their indeterminate growth and comparatively bigger sporangia resemble *Phytophthora*. In *Peronospora* the dichotomously branched sporangiophores taper to fine points bearing conidia. If similar branchings bear sporangia, the fungus is designated as *Pseudoperonospora. Bremia*, too, has similar branches but the branches terminate in disc like structures

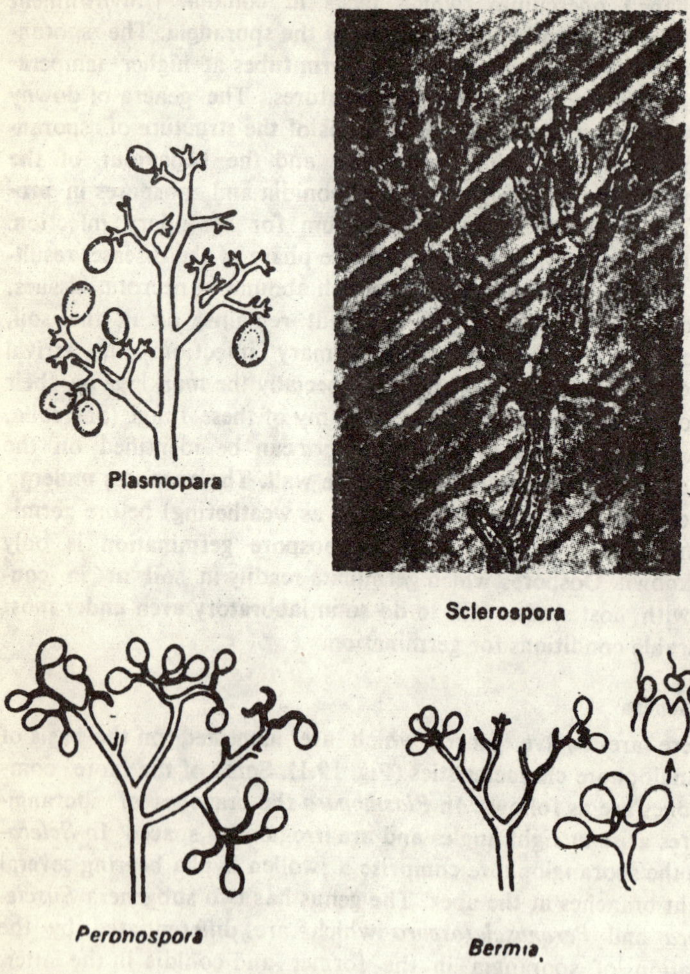

Fig. 19.1. Shows Conidophore branching of some principal downy mildew genera

which bear sporangia on sterigmata. The genus *Basidiospora* bears club-shaped sporangiophore bearing sporangia on sterigmata. The structure appears basidium like. Shaw (1970) has given the following key for the identification of Graminicolus downy mildews.

A—Sporangia produced
B—Sporangiophores indeterminate — *Sclerophthora*
BB—Sporangiophores determinate
C—Sporangiophores unbranched — *Basidiospora*
CC—Sporangiophores branched —
d—Penultimate branches ending in disc-like swellings — *Bremia*
dd—Penultimate branches not enlarged — *Sclerospora*
AA—Conidia produced—*Peronosclerospora*

The Disease Cycle

The downy mildews, inspite of the diverse symptoms, show a remarkable similarity of disease cycle, the mode of perenation, infection and spread. The chlorotic spots on leaves are the site of the sporangial (or conidial) formation. Emergence of sporangiophores through stomata (more abundantly from the lower surface) form the 'downy growth'. High humidity combined with low temperature is the *sine qua non* for the formation of the downy growth. This takes place in dewy nights following cooler days. Nights following warmer days fail to induce the downy growth. These structures wither by 10 A.M. or so. There are experimental evidences for almost all downy mildews to suggest that a relative humidity of 90-100 per cent and the temperature range of 20-25°C is most conducive. "Materials" of downy growth can be collected in the late afternoon. The conidia, sporangia and the zoospores bring about secondary infections and propagate the disease. Their quick drying up (in 3-4 hours) prevents their long distance dispersal. Thus the spread of these diseases is limited both in time and space. During this "grand period" of disease the various symptoms develop. When the crop season approaches its end, sexual reproduction takes place resulting in the formation of oospores which abound in necrotic tissues. The oospores reach the soil through the debris where they lie quiscent till the arrival of the host, next season. Downy mildews are thus soil-borne. The oospores in most diseases cause the primary infection. They infect the underground parts of the young host. The fungus develops intercellularly resulting in systemic infection of the host. Seedlings upto 15 days have been found to be most susceptible. In some diseases, perenation through mycelium (lying in soil, debris or seeds) or collateral hosts also occurs. The extent of perenation in seeds is not known. If these diseases are proved to be internally seed borne, it may prove hazardous and seed movement to distant areas may have to be prevented through quarantines. Before imposing restrictions on trade, we should have sound and reliable information concerning the efficacy of seed transmission. Mycelium of *Peronospora destructor* survives in onion bulbs and plants raised from such infected bulbs are infected. This results in a continuous supply of the inoculum. Collateral hosts are other important "ground keepers" which supply inoculum for primary infection. This happens in *Sorghum* and sugarcane downy mildews.

CONTROL MEASURES

Cultural practises, fungicides and resistant varieties constitute the control measures.

Cultural Practices

Where oospores perenate the pathogen, their elimination or reduction in number is the best method of control. This can be done by (*i*) quick removal and burning of infected plants and their debris, and (*ii*) by abandoning the crop for some years depending on the viability of the oospores, which is usually 2-3 years. The use of a "trap crop", which could induce oospore germination at a time when host crop plants are not available greatly minimizes the inoculum.

In diseases, where recurrence is only through collateral host, destruction of the latter may eliminate the disease.

Fungicides

Every plant has a critical time when primary inoculum is effective. If this period is overcome with the help of fungicides, control will be achieved. This can be done with the help of good stickers. Copper compounds like Bordeaux mixture, which was used first to control a downy mildew, is still widely used. Zineb, Thiram, Maneb (dithiocarbamates) are being used in Australia, USA and some other countries. Dexon, a systemic fungicide, is effective against *Sclerophthora* but not against *Sclerospora*. Further details regarding fungicidal control of *Sclerospora* are lacking and unless the resistant varieties are available, there is no way out than to use fungicides. However, economic considerations always outweigh the efficacy of fungicides. This is more true for commercial crops.

Resistant varieties

Breeders are looking towards multiple disease resistance in many crops. Onion varieties resistant to *Peronospora destructor* are now available. For some other downy mildews also cultivars are available which bring down infection to very low percentage. The Ganga 5, 17 and other maize varieties, released by coordinated maize improvement scheme, have shown very good resistance against Brown stripe downy mildew of maize. Similarly downy mildew of *Sorghum* can be easily controlled by using GM_2-31, shallu and CS H-2 varieties.

SOME IMPORTANT DISEASES

Important crops like maize, sorghum, pearl millet grapes and opium poppy are worst hit by downy mildews.

Maize

Several downy mildew diseases of maize are known. Brown stripe downy mildew of maize is more common. Some important downy mildews of maize are discussed here.

(1) Brown Stripe Downy Mildew of Maize

Pathogen *Sclerophthora rayssiae* var. *zeae* Payak and Renfro.

The disease is restricted to India and severe outbreaks have occurred in several states. The disease was first observed in Pantnagar (UP) in 1962.

Symptoms

Only the leaves are affected Narrow, chlorotic spots (3-7 mm wide) with well defined margins extend in parallel fashion between the veins. The stripes later turn reddish to purple. If the development of the chlorotic stripes occurs prior to flowering, seed development fails and plants die prematurely. Collapse of the tissues has never been observed. Malformation or deformation is also not on record.

Disease Cycle

Oospores present in soil serve as source of primary infection. They form sporangia and the infection appears in form of minute flecks (characteristic of zoosporic infections) on the leaves, which grow to form the brown streaks. Secondary infection is brought about by sporangia formed on the lower surface of infected leaves. The downy growth is more persistent in this disease and can be collected until late afternoons. Singh *et al* (1970) reported that highest production of sporangia occurred from 12 A.M. to 4 P.M. In this respect this downy mildew is different from other downy mildews in which sporulation occurs during nights.

Control

(1) By destroying the infected plants at the earliest.

(2) Spraying $ZnSO_4$ (.5%) with lime, eliminates or reduces the disease to very mild form, Singh *et al* (1967).

(3) By using resistant varieties. Ten hybrid and seven composite varieties have been released by the coordinated maize improvement scheme. Ganga 5 shows maximum resistance.

(2) Sugarcane Downy Mildew of Maize

Pathogen *Sclerospora sacchari* Miyake.

The disease does great damage to maize in several tropical and subtropical countries. It is most severe in Taiwan and is a potential threat to India also. It is already causing great loss in many parts of India.

Miyake (1911) discovered this downy mildew in Taiwan first on sugarcane and later on maize growing adjacent to sugarcane fields. Maize is more susceptible than sugarcane. Thus, the major threat is to maize than to sugarcane. The disease is transmissible between the two crops. Singh *et al* (1967) found the disease to be prevalent in zinc deficient soils.

Symptoms

The infection starts as localized chlorotic spots which grow and form chlorotic streaks of varying lengths. Downy growth under the chlorotic leaves are characteristic of the disease. In advanced infections distortions and malformations are seen in ears, tassels and the stalks.

Disease Cycle

The oospores, present in soil, initiate the disease by infecting seedlings. One month old plants develop high resistance to infection. Singh *et al* (1967) have reported the possibility of their being internally seed borne also. In Taiwan, however, the primary inoculum for infection comes from diseased sugarcanes, which grow round-the-year and oospores have been found to play no role in the disease cycle (Sun, 1970). Secondary infections are brought about by conidia. Free water is essential for conidial germination. Both oospores and conidia germinate directly through germ tubes.

Control

(1) In Taiwan, where the disease is restricted to maize, growing adjacent to sugarcane, control can be achieved by isolating maize fields. The destruction of infected sugarcanes would minimize the inocula and, in turn, the disease.

(2) Maneb gives effective control but is economically feasible only on small experimental plots. Singh *et al* (1967) found that spraying $ZnSO_4$ (.5%) with lime, eliminates or reduces the disease to very mild form.

(3) High degree of resistance has been found in many inbred lines. Chang (1970) has shown that resistance is controlled by a single pair of genes. A resistant variety "Taiwan no. 8" has given good results.

Sorghum Downy Mildew of Maize

Pathogen. *Sclerospora sorghi* Weston and Uppal.

The disease is severe in Rajasthan and Mysore. The use of new high yielding varieties is responsible for its spread. The downy mildew is transmissible between maize and sorghum.

Fig 19.2. 'Downy growth' of *Sclerospora sorghi* on maize leaf.

Symptoms

Drying up of infected seedlings is a common symptom. In late infections, leaves become chlorotic from base upwards and plants get stunted. Whitish 'downy growth' appears on both sides of the infected leaves (Fig. 19.2). Oospores have not been observed, but for the solitary report of Patel (1949) on Kashmir sweet corn. More pronounced symptoms have been observed in USA, where the disease is more severe in Texas and Indiana and is fast spreading to other states. The symptoms also express considerable variation.

(1) Pith shows browning, found to be a physiological reaction.
(2) Excessive base root formation.
(3) Basal half of infected leaves narrows but upper portion remains normal—"half narrow half diseased symptom"
(4) Ears either do not form or develop few seeds, stalks twisted.
(5) Excessive phyllody and twisting of the entire plant. The phylloided tassels can not be differentiated from those caused by *Sclerophthora macrospora* in "Crazy top of maize", unless oospore locations can be determined. When both pathogens infect simultaneously the symptomatology gets more confused.

Disease Cycle

The recurrence of the disease has been reported by Dange *et al* (1974) to occur through a collateral host *Heteropogon contortus* on which the fungus perpetuates in absence of the host. The role of *Heteropogon* as a collateral host was proved by the cross inoculation tests and occurrence of the disease in fields adjacent to it. Both conidia and oospores, formed on the collateral host, serve as the primary inoculum. For *Heteropogon*, host oospores serve as the primary inoculum.

Early maize crops escape infection because they develop resistance by the time conidia are produced on the collateral host. Susceptibility occurs only upto about 15 days of age.

Control

(1) By destruction of or separation from the collateral host.

Fungicide application, in general, appears to enhance infections.

Bajra, "Pearl millet" (=*Pennisetum typhoides* Stapf.)

Downy Mildew of Bajra (Green ear disease of Bajra)

Pathogen. *Sclerospora graminicola* (Sacc.) Schroet.

Butler in 1917 reported this disease from India.

Symptoms

The most pronounced symptoms appear in the ears which turn green and leafy (Fig. 19.3). Bristles, glumes, stamens and the pistils become green and leafy. Sometimes only a portion of the ear is affected, the remaining portion remains compact. Chlorotic spots

Fig. 19.3. A. Green ear of bajra. B. The diseased ear collapsed.

on the leaves, twisting of infected leaves and tillering are the other symptoms. The tillers remain short and chlorotic.

Disease Cycle

Soil-borne oospores (viable upto 5 years) initiate the primary infection by infecting root hairs or coleoptile of host seedlings. The infection soon becomes systemic. 'Downy growth' appears on the lower surface of the chlorotic leaves and is formed mostly from 12 P.M. to 4 A.M. in dewy nights. Oospores formed late in the season are confined to interveinal areas of chlorotic-turned brown and necrotic spots. The oospores perenate in soil.

The fungus infects *Setaria italica* but its role in recurrence is not known.

Control

(1) By quick removal and burning of infected plants.
(2) By using resistant varieties like HBI.

Sorghum, "Jowar" (*Sorghum vulgare* Pers.)

Downy Mildew of Sorghum

Pathogen. *Sclerospora sorghi* (Kulk) Weston and Uppal.

The disease is often referred to as the "leaf shredding disease". It is one of the most important diseases of sorghum and is more prevalent in southern states of India.

Symptoms

Infected leaves become chlorotic, later turn brown and start shredding. Ears either fail to form or remain infertile and turn leafy.

Disease Cycle

The soil-borne oospores cause primary infection of seedlings which later becomes systemic. Secondary spread is through conidia, which germinate directly and are short-lived.

Control

(1) By removal of infected plants.

No resistant varieties are, unfortunately, available.

Puttarudrappa *et al* (1969) reported resistance to be controlled by more than two genes with complementary interaction.

REFERENCES

CHANG, S.C., 1970, "A review of studies on downy mildew of maize in Taiwan," *Indian Phytopath.*, 23: 270-274.

DANGE, S.R.S., JAIN, K.L., SIRADHANA, B.S. and RATHORE, R.S., 1974, "Perpetuation of sorghum downy mildew of maize (*Sclerospora sorghi*) of maize on *Heteropogon contortus* in Rajasthan, *India*," *Plant Dis. Reptr.*, 58: 285-286.

MIYAKE, T., 1911, "On a fungus disease of sugarcane by a new parasitic fungus, *Sclerospora sacchari*," *Rep. Sugar Expt. Sta. D.V. Path. Bull.* 1.

PATEL, M.K., 1949, "Production of oospores by *Sclerospora sorghi* in corn," *Indian Phytopath.*, 2: 52-57.

PUTTARUDRAPPA, A., PATIL, B.G., KULKARANI and KAJJARI, N.B, 1969, "Studies on inheritance of resistance to downy mildew *Sclerospora sorghi*," *Sorghum newsletter*, 12: 57-58.

SAFEEULLA, K.M., 1970, In "Downy mildew issue." *Indian Phytopath.* 23: Round Table Discussion, pp. 399-412

SHAW, C.G., 1970, "Morphology and physiology of downy mildews, significance in taxonomy and pathology," *Indian Phytopath.*, 23: 364-370.

SINGH, J P., RENFRO, B.L. and PAYAK, M.M., 1970, "Studies on the epidemiology and control of brown stripe downy mildew of maize (*Sclerophtkora rayssiae* var. *zeae*)," *Indian Phytopath.*, 23: 194-208.

SUN, M.H., 1970, "Sugarcane downy mildew of maize," *Indian Phytopath.*, 23: 262-269.

20

POWDERY MILDEWS

POWDERY mildews' are easily recognised by the white powdery coating on the plants, which consist of enormous conidia formed by the fungus (Fig. 20.1). Like downy mildews, morphologic signs of the pathogen, rather than the disease symptom, are more conspicuous. Powdery mildews were recognised and named, at least, as early as 1753 by Linnaeus (cf. Salmon, 1900). Both the pathogen and the disease, as also in downy mildews, are designated by the same name.

Fig 20.1. Shows typical powdery mildew symptom on *Dalbergia sisso* leaves and on different parts of Rose.

All powdery mildew fungi belong to the family Erysiphaceae of the order Erysiphales of class Ascomycetes. They are ectoparasites and the superficial hyphae are anchored to the epidermis by means of haustoria. These fungi, which are widely distributed, attack a wide variety of cultivated plants. However, only angiosperms are attacked. The damage caused by powdery mildews are less spectacular than those caused by downy mildews, rusts, viruses and root rot fungi. The powdery mildews usually cause chronic infections, which cause moderate loss every year. The lack of systemic infection, lack of root infection, relatively slow growth, infrequent death of the host and the availability of effective control measures, are the reasons for their being less harmful. Nonetheless, the loss due to powdery mildew of grapes, caused by *Uncinula necator*, in France in 1850, is unforgettable. Thanks to the fungicides, the yields rose rapidly.

Some of the marked characteristics of the powdery mildews are:

(1) The superficial mycelium,
(2) Haustoria in the epidermal cells of the host,
(3) Preference for dry weather,
(4) The high water content of their conidia,
(5) The diurnal periodicity for several characters,
(6) Host specialization, and
(7) Vulnerability to control measures.

The Powdery Mildew Fungi

There are seven commonly recognized genera, which are easily identified on the basis of the location of the mycelium, the type of appendages on the cleistothecium and the number of asci (Fig. 20.2). The genera can be identified with the help of the following key.

Key to Identification of Powdery Mildew Fungi

A. One ascus in cleistothecium
  Appendages myceloid—***Sphaerotheca***
  Appendages dichotomously branched—***Podosphaera***
AA. Several asci in cleistothecium
  Appendages myceloid
    mycelium superficial—***Erysiphe***
    mycelium internal—***Levillula***
  Appendages with bulbous base—***Phyllactinia***
  Appendages dichotomously branched at the tip—***Microsphaera***
  Appendages with coiled tips—***Uncinula***

In contrast to the simple identification up to generic level, there is disagreement over the characters for species identification.

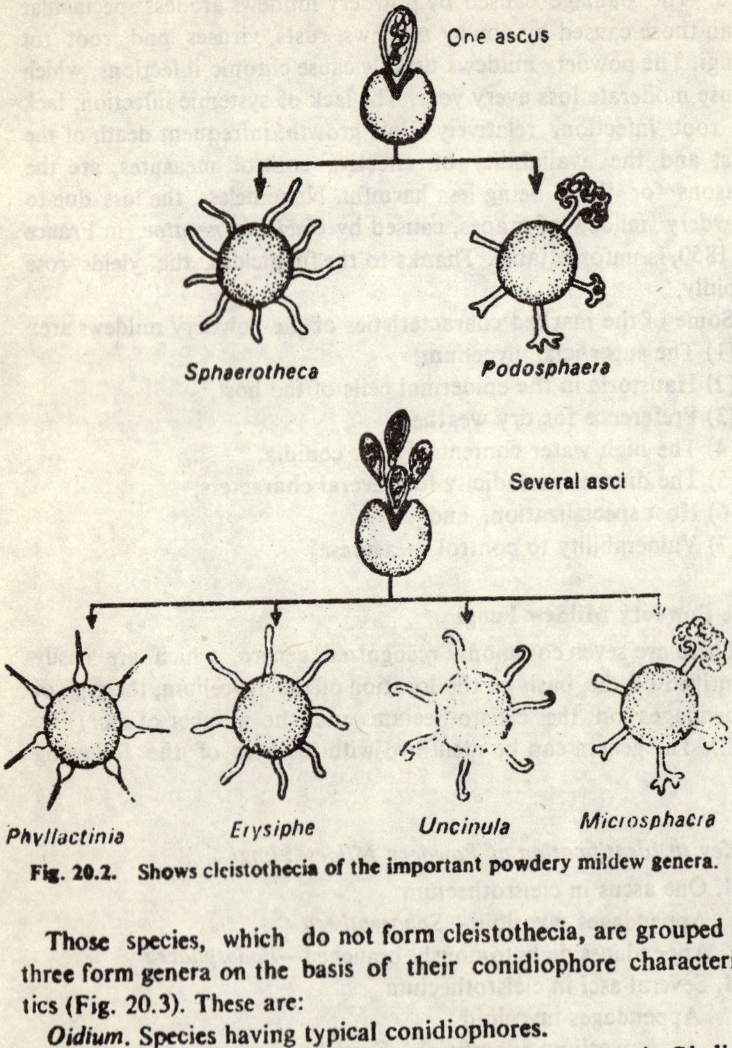

Fig. 20.2. Shows cleistothecia of the important powdery mildew genera.

Those species, which do not form cleistothecia, are grouped in three form genera on the basis of their conidiophore characteristics (Fig. 20.3). These are:

Oidium. Species having typical conidiophores.

Ovulariopsis. Species having clavate conidia, as occurs in *Phyllactinia*, and

Oidiopsis. Species having branched conidiophores that emerge through stomata, as in *Levillula*.

Sawada (1914) advocated that the powdery mildew fungi should be classified on the basis of their conidiophore and conidia and also gave a scheme of such a classification. This is a good proposition as most collections of powdery mildews either do not form cleistothecia, or the diagnostic cleistothecial characteristics are poorly formed.

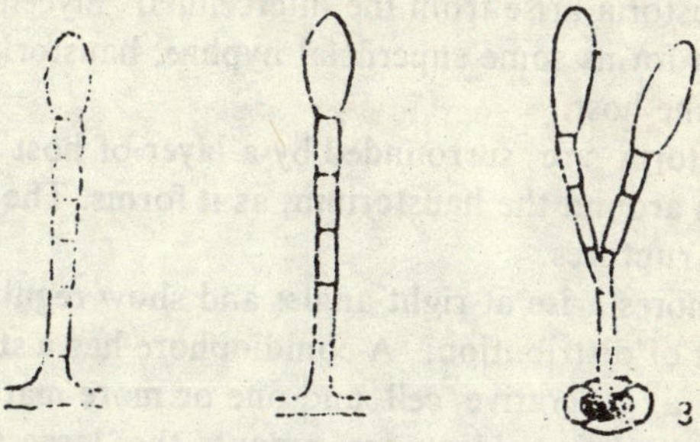

Fig. 20.3. Types of conidiophore: 1. *Oidium* type; 2. *Ovulariopsis* type, and 3. *Oidiopsis* type.

These fungi are obligate parasites and have not been cultured on synthetic media yet. Typically, the hyphae grow superficially, closely appressed on the host and only haustoria enter, that too only up to the epidermis. Exceptions to the superficial existence are found

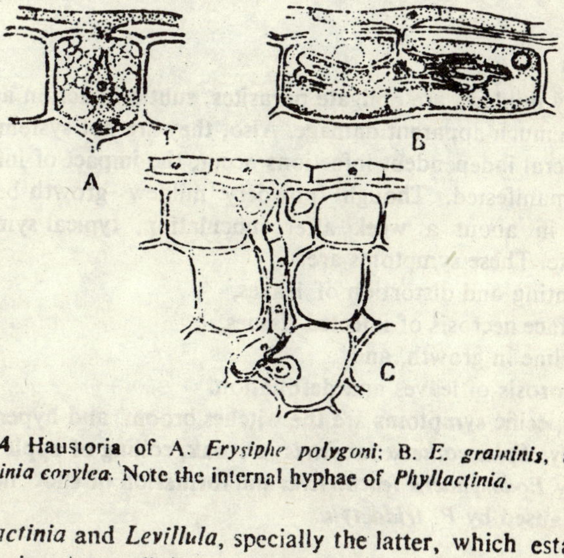

Fig. 20.4 Haustoria of A. *Erysiphe polygoni*; B. *E. graminis*, and C. *Phyllactinia corylea*. Note the internal hyphae of *Phyllactinia*.

in *Phyllactinia* and *Levillula*, specially the latter, which establishes an extensive intercellular mycelium (Fig. 20.4). The hyphae bear lateral swellings, the appressoria, on alternate cells which attach the hyphae to the host surface and also give rise to fine penetration tube which, in turn, form haustoria. The haustoria may be globular (e. g., *Uncinula necator, Erysiphe polygoni*) or may have finger like projections (e.g., *E. graminis*). In *Phyllactinia* and *Levillula* globular haustoria arise from the intercellular mycelium. *Phyllactinia*, though forms some superficial hyphae, haustoria are formed only inside the host.

The haustoria are surrounded by a layer of host wall material, which grows around the haustorium, as it forms. The plasma membrane never ruptures.

Conidiophores arise at right angles and show regular or irregular arrangement of distribution. A conidiophore has a stipe of one or more cells, a generative cell and one or more maturing conidia. Conidia of powdery mildews are among the largest single-celled spores in fungi: (20—50 × 10—24 μ). They are typically ovate, thin walled, colourless, smooth and vacuolate when young. They may also be cylindrical, clavate or barrel-shaped, with many intergrading forms. Only conidia with vacuoles germinate.

The superficial cleistothecia represent the perfect stage of the fungus and bear appendages that are many times longer than the diameter of the cleistothecium. Only in *Phyllactinia*, the appendages are short.

Symptoms

Powdery mildews are obligate parasites, subtle in action and do not make much apparent damage. Also, they are not systemic and unless several independent infections occur, the impact of infection is little manifested. Though 'powdery mildew' growth becomes apparent in about a week after inoculation, typical symptoms appear late. These symptoms are:

(1) Stunting and distortion of leaves,

(2) Surface necrosis of infected tissues,

(3) Decline in growth, and

(4) Chlorosis of leaves and defoliation.

Some specific symptoms are the witches broom and hypertrophy caused by *Sphaerotheca lanestris* on oak, rolling of apple leaves, caused by *Podosphaera leucotricha* and formation of shot holes in *Prunus*, caused by *P. tridactyla*.

Disease Cycle

Figure 20.5 depicts a generalized disease cycle. The disease is initiated by the conidia or ascospores, both of which germinate similarly. Germination is favoured by (*i*) absence of water, (*ii*) light, and (*iii*) host contact. A short germ tube emerges, which bears appresorium at its tip, from where the penetration tube originates and enters

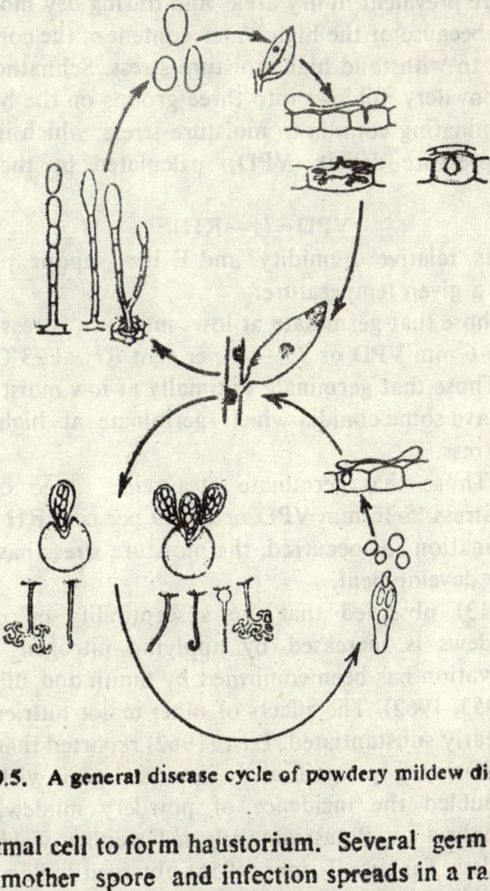

Fig 20.5. A general disease cycle of powdery mildew diseases.

host epidermal cell to form haustorium. Several germ tubes arise from the mother spore and infection spreads in a radial manner which is characteristic of powdery mildews. On all hyphae, appresoria are formed laterally. Conidial formation starts after four-five days of inoculation. The conidia are exclusively wind-disseminated and may be blown to comparatively long distances. On falling on a suitable host, the conidia repeat the asexual cycle and bring about a multiplication of the fungus and spread of the disease. At the end of the host season, cleistothecia are formed, though in some species they are regularly formed. The role of cleistothecia in providing the inoculum, for primary infection in the next season, is insignificant. The fungus may perpetuate itself as mycelium in host debris or on collateral host or volunteer plants.

Factors affecting Disease Development

Of all the factors moisture stress is most important. Powdery mildews are more prevalent in dry areas and during dry months of the year. This is because of the high water content of the conidia which enables them to withstand high moisture stress. Schnathorst (1965) has divided powdery mildews into three groups on the basis of response of germinating conidia to moisture stress, which is measured by vapour pressure deficit (VPD), calculated by the following formula:

$$VPD=(1-RH)E$$

where RH is relative humidity and E the vapour pressure at saturation at a given temperature.

Group I. Those that germinate at low moisture stress, between .5-6 mm VPD or 75—99 per cent RH at 23°C.

Group II. Those that germinate optimally at low moisture stress, have some conidia which germinate at high moisture stress.

Group III. Those that germinate at a wide range of moisture stress ·5-36 mm VPD or 0—99 per cent RH at 23°C.

Once germination has occurred, the moisture stress has no effect on the colony development.

Spinks (1913) observed that the susceptibility of cereals to powdery mildews is increased by applying nitrogen fertilizers. Spinks' observation has been confirmed by Smith and Blair (1950) and Last (1953, 1962). The effects of other major nutrients, P and K, are less clearly substantiated. Last (1962) reported that applying N and P to barley crop significantly increased grain yields. Nitrogen nearly doubled the incidence of powdery mildew, whereas phosphorus halved it. Potassium affected neither yield nor the powdery mildew fungus. Turner (1956) obtained higher incidence of powdery mildew during May-June than in winter months (which he observed by measuring the atmospheric spore concentration) to be the effect of temperature rather than relative humidity. The effect of light and temperature are less important, though the severe occurrence of some species in shade suggest the retarding effect of light and high temperature.

Control

The powdery mildews are controlled effectively by sulphur dust and certain other fungicides. These have been partially replaced by organic fungicides such as dinocap. Since the mycelium is external, the fungicides do not have to penetrate the plant tissues to be effective eradicants, and for this reason, most powdery mildew fungicides serve as eradicants also. Systemic fungicides have a promising career in disease control. For example, procaine hydrochloride, applied to roots, has been found by Dekker (1961) to be systemically effective against several powdery mildews but not other plant pathogens. It was inactive *in vitro*. Root applications of 6-azauracil, barbiturates, and the antibiotics griseofulvin and cycloheximide have shown systemic activity against powdery mildews.

Powdery mildews are unique among plant pathogens in that they succumb to sulphur and some other fungicides through most part of their life cycle. Only cleistothecia and mycelium in dormant buds are exceptions. Thus powdery mildews can be controlled by protective, eradicant and therapeutic applications. Sulphur dust kills ungerminated conidia already present (eradication); it will kill conidia which arrive subsequently (protection) and kills mycelium already established (therapy).

The other potent method of control is the use of resistant varieties. Resistant varieties of grapes, hop, wheat, bean, and many other crops are available. Resistance to powdery mildews may be dominant or recessive and controlled by single or multiple factors.

Powdery Mildew of Grapes (*Vitis vinifera* L.)

Pathogen. *Uncinula necator* (Schw.) Burr.

The disease caused unprecedented damage to vine orchards in France in 1850 and continues to occur freqtiently in epiphytotic proportions. In India, the disease occurs in Rajasthan, Maharashtra and other grape-cultivating areas.

Symptoms

All the plant parts, leaves, stems, flowers and fruits are infected. In addition to the powdery coating of conidia, the following symptoms are frequently seen.

(1) Malformation and discolouration of leaves.

(2) Dwarfing of infected plants.

(3) Berries become deformed, fail to ripen, develop cracks and, in early infections, are shed.

Morphology of the Pathogen

Conidia egg-shaped; cleistothecia are not formed in Indian conditions.

Disease Cycle

Figure 20.6 shows the disease cycle. In absence of the perfect stage, conidia are believed to provide the inoculum, spread the disease and also to perpetuate the fungus.

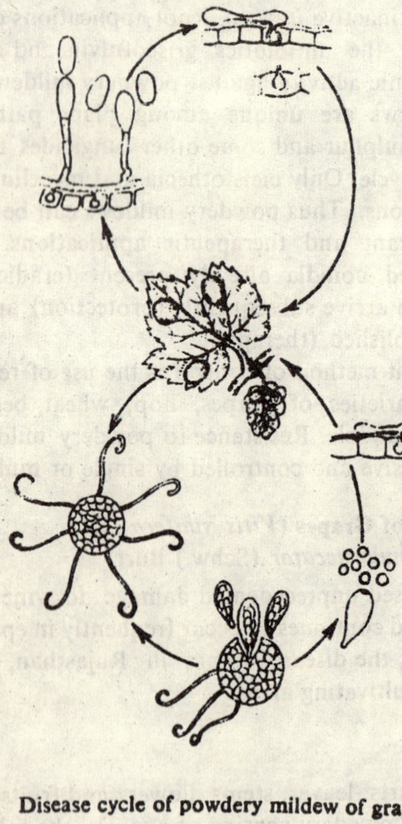

Fig. 20.6. Disease cycle of powdery mildew of grapes.

Control

The disease is kept in complete check by use of sulphur dust. First dusting is done very early when plants are 3—6 inches long and a second when the plants are in flowering stage. A third dusting after 2 months ensures that the fourth is not needed.

POWDERY MILDEW OF CUCURBITS

Powdery mildews are most common on various cucurbits (viz. pumpkin, cucumber, ash gourd, snake gourd, bitter gourd and ribbed gourd, etc.) and are worldwide in distribution. The disease, which, as a rule, occurs during dry season, is widespread in India.

Pathogens

(1) *Sphaerotheca fuliginea* (Schlecht.) Poll.

(2) *Erysiphe cichoracearum* DC.

Butler (1918) reported the occurrence of both the organisms on cucurbits. According to Kapoor (1967), only *S. fuliginea* attacks cucurbits and *E. cichoracearum* parasitises non-cucurbitaceous hosts, mainly members of compositae. The powdery mildew occurs in nature mostly in the conidial stage and cleistothecia are only rarely formed.

Symptoms

The infected leaves turn yellow and wither off. The fruits are reduced in size and number.

Disease Cycle

Since the fungus has a wide host range, it perpetuates on the collateral hosts in conidial form. Cleistothecia, that are seldom formed, are not known to be effective in disease recurrence.

Control

Dusting sulphur powder keeps the disease in check.

Powdery Mildew of Wheat (*Triticum vulgare*)

Pathogen. *Erysiphe graminis tritici* (DC) Marchal.

The disease is reported from several parts of the world. In Ohio, USA, the disease is regarded more destructive than the rusts. In India, the disease being confined to hills, is not much important, as 95 per cent of wheat cultivation is done on the plains. Mehta (1930) and Arya and Ghemawat (1953) have reported only sporadic appearance of the disease on the plains in Rajasthan. The cleistothecial stage, though not common, is reported to be formed late in the season.

Symptoms

Superficial white powdery patches appear on the leaves (upper surface), stem and sheath, and with passage of time, turn brown. Occasionally, late in the season, cleistothecia appear as brown black dots. In severe infections the leaves get deformed by twisting.

Disease Cycle

Conidia, blown down from the hills, where they survive the hot months, cause the primary infection. Arya and Ghemawat (1953) suggested that Mount Abu, in Rajasthan, might be playing an important part in the annual recurrence of the disease in the neighbourhood of Jodhpur. The role of cleistothecia is not known.

Morphology of the Pathogen

Haustoria form characteristic finger like projections. Conidia hyaline, elliptical and single-celled. Cleistothecia bear 9-30 asci but ascospores are not formed (Arya, 1964).

Control

(1) The disease is effectively controlled by use of sulphur dust. The use of 1 per cent sodium thiosulphate, which gives equally good control and is cheaper, is preferable.

(2) In USA and Canada, resistant varieties are in use. Arya (1962) has reported the prevalence of three physiological races (3, 4 and x) in Jodhpur area. He also reported that improved varieties (E 750, C 591, RS 31-1, N.P. 718, N.P. 710 and K 53), were resistant to the mixture of these races at the adult stage.

SOME OTHER IMPORTANT POWDERY MILDEWS

The following are the other important plants infected by powdery mildews:

Powdery Mildews *Host/Pathogen*	*Remarks*
1. Apple (*Malus* sp.) (*Podosphaera leucotricha*)	The disease attacks all plant parts; fruits get deformed.
2. Barley (*Hordeum vulgare*) (*Erysiphe graminis hordei* (DC) Marchal.)	Cleistothecia not recorded in India.

Powdery Mildews Host/Pathogen	*Remarks*
3. *Dalbergia sisso* (Shisham) (Fig. 20.1) (*Phyllactinia corylea*)	The pathogen has a wide host range and attacks also birch, chestnut, oak and barberry.
4. Pea (*Pisum sativum* L.) (*Erysiphe polygoni*)	Haustoria globose, powdery masses appear on the leaves and pods. Ascocarps are frequent and, along with dormant mycelium in seeds, may cause primary infection.
5. Rose (*Rosa*) (Fig. 20.1) (*Sphaerotheca pannosa* var. *rosae*)	The disease is very commonly encountered on all parts of the plant.

REFERENCES

ARYA, H.C., 1962, "Studies on the physiological specialization and varietal reaction of wheat to powdery mildew in India," *Ind. Phytopath.*, 15: 127-132.

———, 1964, "On the physiology of ascospore formation in powdery mildew of wheat," *Ind. Phytopath.*, 17: 27-34.

———, and GHEMAWAT, M.S., 1953, "Occurrence of powdery mildew of wheat in the neighbourhood of Jodhpur," *Ind. Phytopath.*, 6: 123-130.

BUTLER, E J., 1918, *Fungi and Diseases in Plants*, Thacker, Spink and Co Calcutta, 314-315.

DEKKER, J., 1961, "Systemic activity of procaine hydrochloride on powdery mildew," *Tllijdschr. Plziekt.*, 67: 25.

KAPOOR, J N., 1967, "*Sphaerotheca fuliginea, Erysiphe cichoracearum*:" C.M.I. description of pathogenic fungi and bacteria, Nos. 152-159.

LAST, F.T., 1953, "Some effects of temperature and nitrogen supply on wheat powdery mildew," *Ann. Appl. Biol.*, 40: 312-322.

———, 1962, "Effect of nutrition on the incidence of barley powdery mildew," *Pl. Path.*, 11 (3): 133-135.

MEHTA, K.C., 1930, "Studies on the annual recurrence of 'Powdery Mildews' of wheat and barley in India," *Agric. J. India.*, 25: 283-286.

SALMON, E.S. 1900, "A monograph of the Erysiphaceae," *Torrey Bot. Club, Mem.*, 9. 292 P.

SAWADA, K , 1914, "The classification of the Erysiphaceae based on the conidial stages," *Formosa Agr. Exp Stat. Spec. Rep.*, 9: 1-102 (in Japanese).

SCHNATHORST, W.C., 1965, "Environmental relationships in the powdery mildews" *Annu. Rev Phytopathol.*, 3: 346-366.

SMITH, H.C. and BLAIR, I.D , 1950, "Wheat powdery mildew investigations," *Ann. Appl. Biol.*, 37: 570-583.

SPINKS, G.T., 1913, "Factors affecting susceptibility to disease in plants, *J. Agric. Sci* 5: 231-247.

TURNER, D.M., 1956, "Studies on cereal mildew in Britain," *Trans. Brit. Mycol. Soc.*, 39: 495-506.

21

THE SMUTS AND BUNTS

DISEASES caused by fungi of the order Ustilaginales (Class—Basidiomycetes) are popularly known as smuts because of the sooty black powdery masses of spores they produce on their hosts. The term smut is generally used to designate both the disease as well as the pathogens but the reference to the latter as smut fungi is considered to be more accurate. Like rust fungi, they also parasitise a large number of cereals and grasses. The smut fungi cause more severe damage to their hosts than the rust fungi. Since they have been cultured on media in laboratory, they are not obligate parasites; nonetheless in nature they are not usually found as saprobes. Smut fungi are facultative saprophytes. The maize smut fungus, *Ustilago maydis*, however, is a vigorous saprophyte growing on manure heaps. These are specialized parasites showing host specificity upto variety and strain level.

Hyphae of smut fungi (except *Ustilago maydis*) grow intercellularly deriving nutrition through haustoria. The nuclear behaviour in the life cycle of smut fungi involve: (*i*) the haplophase, (*ii*) the dikaryophase, and (*iii*) the diplophase. The primary mycelium, formed by the germination of sporidia, is of short duration and represents the haplophase. Dikaryotization soon establishes the dikaryotic secondary mycelium which forms the prolonged somatic phase of these fungi. Due to the extensive dikaryotic phase, as occurs also in the rust fungi, the fusion of the paired nuclei is delayed. The true diploid phase, represented by the diploid nucleus prior to meiosis, is restricted only to a brief period in the mature teleutospore. Raper (1954) has classified smut fungi on the basis of dikaryotic cycle. .

In most of the smut fungi there is a unique relation between the nuclear phase and the mode of nutrition. The haploid sporidia are usually saprophytic but the dikaryotic mycelium is obligately parasitic. However, Tirone and Metzger (1962) succeeded in growing the dikaryotic mycelium of *Tilletia caries* in culture and obtained teleutospores which germinated normally to produce the sporidia. Tirone (1964) was able to cultivate the dikaryotic mycelium of *T. controversa*, the wheat dwarf bunt fungus and obtained teleutospores. However, the sporidia produced by the teleutospores grown in special media failed to infect the host.

At the time of spore formation, the hyphae aggregate in certain localized areas of the host and become profusely septate, and transform into smut spores. The extent of this transformation varies in different genera. In *Ustilago* all the mycelium is utilized in spore production whereas in other genera like *Sphacelotheca* the outer layer of the mycelium does not form spores but adheres to form a "false membrane" which acts like an effective peridium. Since the spores are formed by rounding off of the hyphal cells, much in the manner resembling chlamydospore formation, these spores are often designated as chlamydospores, though wrongly. True chlamydospores are asexual in nature whereas the smut spores are binucleate in the beginning; become uninucleate and diploid, which during germination undergoes a reduction division. The intercalary origin of teleutospores is another point of resemblance with chlamydospores but several species are known to produce only terminal spores. The smut spores are analogous to teleutospores of rust fungi in their mode of formation, function and germination. Fischer and Holton (1957) suggest the use of the term 'teliospores' to designate smut spores, which was originally used by Arthur and followed by Bessey.

The outer wall "epispore" of teleutospores is thick, pigmented, smooth or variously ornamented; the inner layer "endospore" is thin and hyaline. In some genera the spores aggregate to form "spore balls" which may consist of all fertile cells (e.g. *Tolyposporium*, *Thecaphora*) or may have sterile cells also e.g. *Urocystis*. Usually the site of sporulation is the ovary of flowers but there are several exceptions to this. The location of sporulation is characteristic of a particular genus or species. *U. violacea* sporulates only in the anthers of the members of family Caryophyllaceae (e.g. *Dianthus*, *Melandrium*); entire flower is damaged in sugarcane smut, and the whole flower axis is smutted in loose smut of wheat. *Entyloma oryzae* and *Urocystis brassicae* are the other exceptions which produce sori on the leaves of rice and galls on the roots of *Brassica* sp respectively.

The intervening time between infection and sporulation generally varies from about 10 days in corn smut (*U. maydis*) to 2 to 3 years in common stem smut of grasses, *U. spegazzinii.* Except in local infections e.g. corn smut, the site of infection and sporulation are generally widely separated.

CLASSIFICATION OF SMUT FUNGI

The teleutospores germinate by forming a germtube of determinate growth called promycelium.[1] This gives rise to hyaline, fragile spores called sporidia. On the basis of the structure of the promycelium and pattern of formation of sporidia, the order Ustilaginales is divided into 2 families—Ustilaginaceac and Tilletiaceae. In Ustilaginaceae the promycelium is comparatively longer, septate bearing lateral and terminal sporidia. In members under Tilletiaceae the promycelium is non-septate with terminal sporidia. Some mycologists have recommended the inclusion of the members of the family Graphiolaceae (often called false smuts) within the Ustilaginaceae. Certain smut fungi (e.g. *U. tritici* and *U. nuda*) produce infection hyphae instead of sporidia while some others do not form any promycelium and directly form the mycelium. Also, some genera placed in Tilletiaceae form septate promycelium with lateral sporidia. These anomalies led to the suggestion to merge the two families into one family Ustilaginaceae. This is supported by Fischer and Holton (1957). There is another practical problem in this scheme of identification based on spore germination. The use of spore germination restricts its utility to viable spores only. Except in few cases where teliospores retain their viability upto 15 years, the majority of smut fungi remain viable for two to three years only. Identification of old herbarium specimens can not be made. Fischer and Holton (1957) have given an "artificial key" to the genera of Ustilaginales without taking spore germination into consideration, thus ignoring their family affinities. This necessitates the relegation of the two families into one.

The localization and variety of sori and spore structures are the characters used in the identification of genera and species. The spores may be single, united in pairs or spore balls or multicellular structures.

[1]The terminology of promycelium and sporidia now in general use for the products of the germinating teliospores was given by Tulasne (1847) and followed by De Bary (1887).

Some Important Genera of Smut Fungi

Ustilago. Sori formed in various parts of hosts such as leaves, stems and fruits; are black and dusty at maturity. The sorus is covered by host tissue. The spore germinates in various ways but in general conform to the type characteristic of Ustilaginaceae.

Sphacelotheca. The genus is differentiated from *Ustilago* by the possession of an outer false membrane covering the sorus (formed by fungus hyphae hence false), and a central columella formed by host tissue. In many cases the false membrane is evanescent and can be confused with *Ustilago*.

Tolyposporium. Sori formed in various parts of the host, contain spore balls. The spore balls may consist of same types of spores or differentiated into an outer layer of coloured spores and an inner mass of hyaline spores. The characteristic feature is the development of outgrowths on the outer surface of the spores which intercolate with those of neighbouring spores forming the spore balls.

Tilletia. Sori mostly formed in ovaries, occasionally occur in other parts also. Spores are associated with sterile cells. Spores germinate forming septate or unseptate promycelium bearing 8 terminal sporidia.

Neovossia. Bear numerous sporidia which usually do not conjugate *in situ* and show distinct apiculus, a small hyphoid appendage.

Entyloma. Sori formed mostly in leaves. Spores intercellular, hyaline or pale yellow, germinating by forming unseptate promycelium. Some species form secondary conidia which are extruded out through stomata.

Urocystis. Sori formed in various parts including roots. Spore balls consist of 1 to 4 dark fertile teliospores surrounded by a layer of sterile cells. The spores germinate *in situ*.

MODES OF INFECTION

The smut fungi infect their hosts in the following three ways.

Embryo infection (=blossom infection, flower infection, floral infection and intraseminal infection).

This mode of infection occurs only in the loose smuts of wheat and barley. The seeds contain the dormant mycelium in their embryo which on germination causes systemic infection in the plant. The symptoms appear during heading of ears. The loose smut teleutospores present on the grain or mixed with the soil will not cause infection. If the teliospores are put on the ovary about the time the pollen mature, the grain will produce the disease the following year. In nature the embryo becomes invaded via the blossom. Wind-blown teleutospores infect the ovary and establish dikaryotic mycelium in the embryo, the seed looking perfectly normal externally. It is curious that out of hundreds of smut fungi studied, this mode of infection occurs only in two smuts.

Seedling Infection

In such type of infections the smut pathogen enters the host by penetrating the young seedlings. This is accomplished by spores or mycelium present on the seed or in the soil. Since the seedlings are vulnerable to infection only during the earliest stages, the timing of spore and seed germination is crucial. Occasionally, as happens in covered smut of barley (*U. hordei*) or loose smut of Oats (*U. avenae*), the mycelium of the pathogen lies dormant in the pericarp of the seed or lodged within the seed hulls.

The infection becomes systemic and subsequent development is similar to that of embryo infection.

Shoot Infection

In this mode of infection the teleutospores present in the soil germinate to form sporidia in abundance which, blown by wind, fall on young buds and cause infection. The infection is systemic.

Shoot infection occurs in the smut of sugarcane (*U. scitaminea*) "long smut" of Sorghum and several others. Prasad (1945) showed that in long smut of Sorghum, infection results from air-borne sporidia infecting buds and initiating a systemic infection which later expresses itself in the heads.

Smut of maize (*U. maydis*) and smut of rice (*Tilletia barclayana*) too are examples of shoot infection but with a difference. The above three infections become systemic while in these the infections are localized; each sorus representing a separate infection. In systemic infections the sporulation occurs in parts remote from the point of infection, in local infections the mycelium and subsequent sporulation is confined to the region of infection.

SEXUAL REPRODUCTION AND HETEROTHALLISM

The occurrence of regular alternation of karyogamy and meiosis in the life cycle of all smut fungi, proves the existence of sexual reproduction. However as Fischer points out "sex in smuts is relative". There are no criteria to differentiate male and female elements. The occurrence of more than two groups of haplonts (compatibility of mating groups) derived from the same promycelium can not be adequately referred by the term "sex". In the absence of sex organs the nuclei of opposite mating types are brought together during sporidial fusions which fuse and undergo meiosis at the time of promycelium formation. Heterothallism is thus of paramount importance in smut fungi. Three conditions of sexuality are recognized in smuts. Homothallism, bipolar heterothallism and tetrapolar heterothallism.

Majority of smut fungi are heterothallic except few species which exhibit homothallism. e.g. *Tilletia cerebrina* and *T. elymi.* Solopathogens, behaving as true homothallic forms are also known. Solopathogenicity is a phenomenon which closely resembles true homothallism. Single sporidia can cause infection and form teliospores. They are designated as "true diploids" and are comparable to "illegitimate diploids" of yeasts. Failure of normal meiosis results in the formation of diploid sporidia and thus of the solopathogens. Such diploid strains combine the nutritional abilities of the two phases i.e. they grow *in vitro* saprophytically like the haploid sporidia and also *in vivo* like the dikaryotic mycelium. Solopathogenic strains have been reported mostly in *Ustilago maydis*, *Sphacelotheca* species and *Tilletia caries.*

Heterothallism is an outbreeding or outcrossing mechanism to attain heterozygocity in fungi. It provides a greater store of heritable variation for future adaptive changes in the physiology or morphology of the pathogens.

It is generally agreed that most of the smut fungi are "bipolar *Ustilago cynodontis*, *U. turocamina*, *Tilletia caries* and *Urocystis cepule* are examples in which bipolar heterothallism has been clearly established. The claim of Whitehouse (1951) that all species of Ustilaginales are bipolar is questioned by several workers. *Sphacelotheca sorghi*, *S. reliana*, *Ustilago maydis* and *U. striiformis* display multiple mating or compatibility groups and are described as "tetrapolar". The tetrapolar heterothallism in *U. maydis* is now generally accepted, but doubts have been expressed about others. The question of bipolarity versus tetrapolarity in *Ustilago maydis* has been worked out in great detail. It has been found that two loci *a* and *b* are operative in the corn smut fungus. While the *a* locus has only two alleles a^1 and a^2 the *b* locus is multiallelic—b^1, b^2, b^3, . . . b^n. The *a* locus governs the process of sporidial fusion, since a^1 mated only with a^2 whereas the *b* locus controlled the stability, vigour and pathogenicity of the dikaryotic mycelium. If only *a* locus is considered, then *U. maydis* is clearly "bipolar". On the other hand, if both *a* and *b* loci are considered, then it is tetrapolar. Since both loci function to produce the zygote—teleutospore in the life cycle of *U. maydis*, the species is truly tetrapolar. Without the *b* locus teleutospores will not be formed. Rowell observed that sporidia with dissimilar *a* and similar *b* readily fused but were not pathogenic. So heterozygocity at both loci is a precondition for sporulation. The designation of *U. maydis* as "bipolar" or "tetrapolar" has become a matter of interpretation.

The tetrapolarity in this case (*U. maydis*) shows marked differences if compared to tetrapolar higher basidiomycetes (Hymenomycetes and Gasteromycetes).

(1) The presence of a single pair of alleles at one locus and several alleles at a second locus is peculiar to *U. maydis* not known in any other organism. In higher basidiomycetes both loci are multiallelic.

(2) The two loci *a* and *b* control compatibility and pathogenicity respectively; in tetrapolar basidiomycetes the two loci control the formation of clamp connections and nuclear migration respectively.

A notable contribution to smut genetics is the demonstration of parasexual cycle in *U. maydis* by Holliday in 1961. Several methods of determining the compatibility of monosporidial lines in the smut fungi have been outlined by Fischer and Holton (1957). In species forming gametic sporidia "Bauch test" is used to find out the compatibility groups. Sporidia from same or different teliospores of a single smut fungus are plated on water agar, in all possible combinations. Pattern of anastomosis indicates the compatibility groups. If the sporidia fall into two mating groups, the species is termed 'bipolar'. Tetrapolar heterothallism is recognized by the presence of many different mating types. Holton (1953) devised a more efficient method of determining compatibility groups by observing the mating pattern of secondary sporidia in *Tilletia caries*.

Physiological Specialization

Physiological specialization has been defined as the occurrence of entities within morphologic species that differ from each other in one or more physiologic characters including pathogenicity, host morphology, biochemical properties, cultural variability, spore germination and ecological relationships.

Pathogenic specialization is the most important criterion used in differentiating physiological races of pathogens. The existence of widespread pathogenic races has been established in several species of smut fungi.

Inheritance of virulence and avirulence has been shown to be of the Mendelian type in *Ustilago avenae.* Virulence is found to be monogenic and recessive to avirulence. Specific genes for virulence have been identified in some species. Pathogenic behaviour also depends on the transmission of several other factors controlling virulence. When these factors are present in heterozygous condition meiotic reassortment during spore germination may greatly affect the pathogenic capability.

Flor's gene for gene hypothesis has been invoked to explain the host-parasite relationship in *Tritricum Tilletia, Triticum/Ustilago,* and *Avena/Ustilago* systems. Genes for resistance and avirulence have been identified but their interaction in biochemical terms has not been indicated.

Mutation and hybridization bring about new characters and thus the evolution of new morphologic and physiologic characters. Due to the intervention of meiosis during each spore generation there is a constant re-assortment of genes at all loci controlling virulence. This brings about great pathogenic variability which enables the pathogen to adapt to the changing environment. Studies on induced hybridization between species and races of smut fungi produce intergrading types. The frequent occurrence of such intergrading types in the fields clearly indicates the occurrence of hybridization in nature, resulting in the formation of new varieties of smut fungi.

CONTROLLING THE SMUTS

The control measures depend on the type of infection. For the internally seed-borne embryo infections e.g. loose smut of wheat and barley, a differential killing of the mycelium without damaging the embryo is achieved by hot water treatment invented by Jensen (1888). Several modifications of this method are in practise. A systemic fungicide vitavax (2, 3 dihydro 5 carboxanilido 6 methyl 14 oxanthinn) has been reported by Jones and Barnett (1967) to control loose smut diseases of wheat and barley. Seed treatment with fungicides is of no avail.

Externally seed-borne smut diseases are successfully controlled by treating seeds with fungicides. The discovery of seed treatment as a method of control is interesting and dates back to the time when the parasitic origin of diseases was not known. Wheat seeds salvaged from a wrecked ship near England when sown gave a clean crop, free from the bunt disease, which was a menace to wheat crop raised from local seeds. With this, the brine treatment of seeds became a practise with the farmers. The use of copper sulfate as a fungicide was discovered by Prevost in 1807 which stayed as an effective seed-disinfestant until formaldehyde was discovered in 1895. Formaldehyde too being a wet treatment, left much to be desired and the need for a chemical involving dry treatment was felt. Copper carbonate as a dust provided the answer and replaced formaldehyde. Though it controlled wheat smut successfully, it was ineffective against barley and oat smuts for which formaldehyde continued to be preferred. Organic mercury dusts, having several advantages, gradually replaced copper carbonate dusts by 1930, but the enthusiasm with dust fungicide did not last long. While the wet treatment, if not dried quickly and completely, was injurious to the seeds; the dusts when inhaled proved poisonous to the workers. The use of fungicide in the form of a slurry provided a middle course which did not wet the seeds.

Control becomes difficult and economically unfeasible if the spores persist in the soil. In such situations, the resistant varieties provide the only remedy. However, treatment of the soil with chemicals like formaline prevented some diseases, especially onion smut. Several genetic factors for resistance, pooled in the resistant varieties, control various physiological races of the smut fungi.

EFFECTS ON THE HOST

In addition to the characteristic smut sori, these fungi bring about several less conspicuous secondary effects on their hosts, some of which are of great economic value while others have attracted attention due to certain basic values in the biological processes.

The dwarfing of the infected plants to varying degree is very common in infections by several genera, more particularly with bunts. In fact, wheat plants infected with stinking smut (bunt) are first recognized by their reduced height. Of the two species *Tilletia foetida* causes less dwarfing as compared to *T. caries*. Thus, the infections by *T. foetida* and *T. caries* came to be called "high bunt" and "low bunt" respectively. *T. controversa* causes extreme dwarfing and the disease is thus called "dwarf bunt". However, the degree of dwarfing is variable depending on host variety and physiologic race of the fungus.

Stimulation of tillering (shoot development) and inhibition of root development are other vegetative modifications associated with smut diseases. It must, however, be noted that all records of smut diseases do not show dwarfing or tillering.

One of the most interesting cases of floral modification is found in the infections on the members of the family Caryophyllaceae by *Ustilago violacea.* The fungus attacks only the anthers and replaces the pollen grains with spores. When the fungus infects female plants of *Melandrium album* or *M. diocum*, the stamens, which are present normally only in rudimentary state in the female flowers, grow to full size, only to get filled up with smut spores later. A similar induced hermaphroditism in pistilate flowers of *Knantia arvensis* infected with *U. scabiosae* has been reported. Thus, the parasite changes the sex of the female plants. The nature of the morphogenetic stimulus which overcomes the arrested development of stamens in female flowers is still unknown. The role of a fungus induced (or produced) hormone, though envisaged, is not experimentally substantiated.

Tilletia tritici changes the growth pattern of wheat heads from club to elongate or vulgare type. Induced phyllody of reproductive organs is very pronounced in several cases in the form of bizarre of leafy structures. An unusual effect in corn infected by *U. maydis* is the transformation of kernels into long tubular structures which terminate into equally long threads. This accompanies general phyllody of all reproductive organs. Similar phyllody is reported to be induced also in infections by *Sphacelotheca reiliana* on corn (head smut) and sorghum.

It is amusing that some of these secondary modifications have misled botanists into creating new varieties based on the new morphological variants of the hosts. Linnaeus, thus, described the hermaphrodite forms of *Melandrium album* induced by *U. violacea* as a new variety.

PHYSIOLOGY OF SMUT FUNGI

Some interesting observations on the physiology of smuts have been made during the last two decades. Presence of glycolytic enzymes has been reported. Some major and minor pathways for glucose catabolism in smuts are as follows.

Smut	*Major Pathway*	*Minor Participation*
Tilletia caries (mycelium)	EM	HMP
T. caries (spores)	ED	—
T. controversa (spores)	HMP	EM
Ustilago maydis (mycelium)	HMP	EM
U. maydis (spores)	EM	

EM—Embden-Meyerhof pathway; ED—Entner-Doudoroff pathway, HMP—Hexose monophosphate pathway.

Profound metabolic changes take place when spores germinate into vegetative cells. Enzymes of the HMP and TCA cycles are lacking in the spores of *Ustilago maydis* until 12 hours after germination (Gottleib and Caltrider, 1962). *Tilletia caries* spores apparently use the ED pathway as the sole glycolytic pathway while mycelium of *T. caries* does not catabolize glucose by this pathway. A complete cytochrome system has been reported in *Ustilago zeae* (Grimm and Allen, 1954).

Loose Smut of Wheat

Pathogen. *Ustilago tritci* (Pers.) Rostr. (=*U. nuda*)=(Jens.) Rostr.

Symptoms

Entire 'ear' except rachis and awns smutted (Fig. 21.1); plants systemically infected ; look normal till the ears appear. After the rupture of the covering, the loose mass of teleutospores is extensively dispersed by wind.

Disease Cycle

The internally seed-borne pathogen becomes active during seed germination (Fig. 21.2); grows along with the plant and becomes systemic. The dikaryotic mycelium, at the time of ear formation, transforms into teleutospores, which infect the embryo of new host plants. Teleutospores fall and germinate on the stigma of wheat

Fig. 21.1. Loose-smutted 'ear' of wheat.

Fig. 21.3. 'Bunt' of wheat ear.

Fig. 21.6. Whip-like smutted shoot of sugarcane.

Fig. **21.9.** (*A*) Smut of bajra; (*B*) Smut of oats, and (*C*) Smut of jowar.

flowers and form 'infection threads' on a septate promycelium. The dikaryotization establishes secondary mycelium which grows and

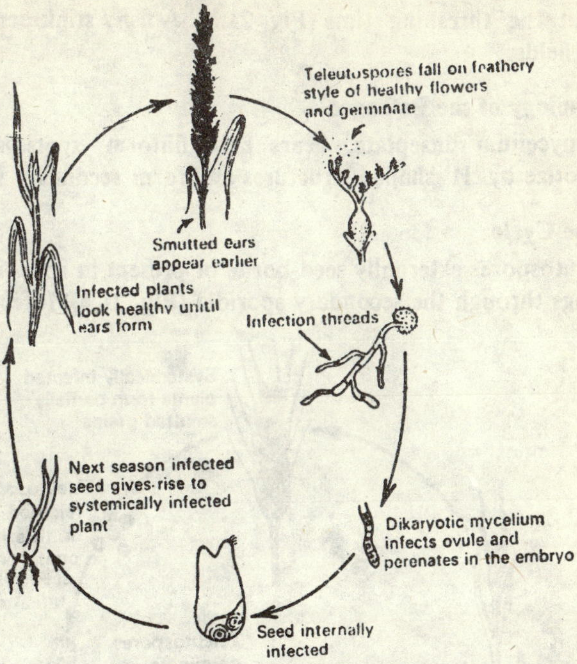

Fig. 21.2. Shows disease cycle of loose smut of wheat.

establishes in the embryo (Embryo infection). After infection, the mycelium becomes dormant in the seed until next sowing season.

Control

(1) Hot water treatment followed by 4-5 hours of drying in sun effectively disinfects the seeds.

(2) Use of resistant varieties. Mundkur and Pal (1941) and Pal and Mundkur (1945) found several immune or resistant varieties. Since then, several resistant varieties have been evolved at I.A.R.I., New Delhi.

Bunt of Wheat (=*Stinking smut*=*Hill bunt*)

Pathogen. *Tilletia caries* (DC.) Tul (Rough spores
T. foetida (Wallr) Liro (Smooth spores)

Symptoms

Infected plants are shorter and bear smutted ears. Generally all the ears are smutted. The thick covering allows dispersal of spores only at the threshing time (Fig. 21.3). A fishy stink pervades infected fields.

Morphology of the Pathogen

Promycelium unseptate, bears 8-16 filiform sporidia, which dikaryotize by 'H' shaped structures and form secondary sporidia.

Disease Cycle

Teleutospores, externally seed-borne or present in soil, infect the seedlings through the secondary sporidia (Fig. 21.4). Infected seed-

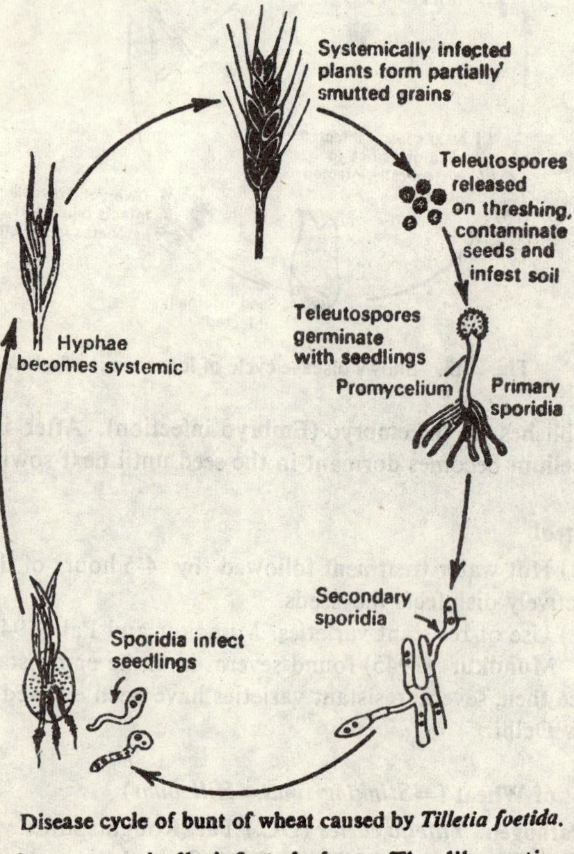

Fig. 21.4. Disease cycle of bunt of wheat caused by *Tilletia foetida*.

lings grow into systemically infected plants. The dikaryotic mycelium, starting right with the infection, ends up with formation of

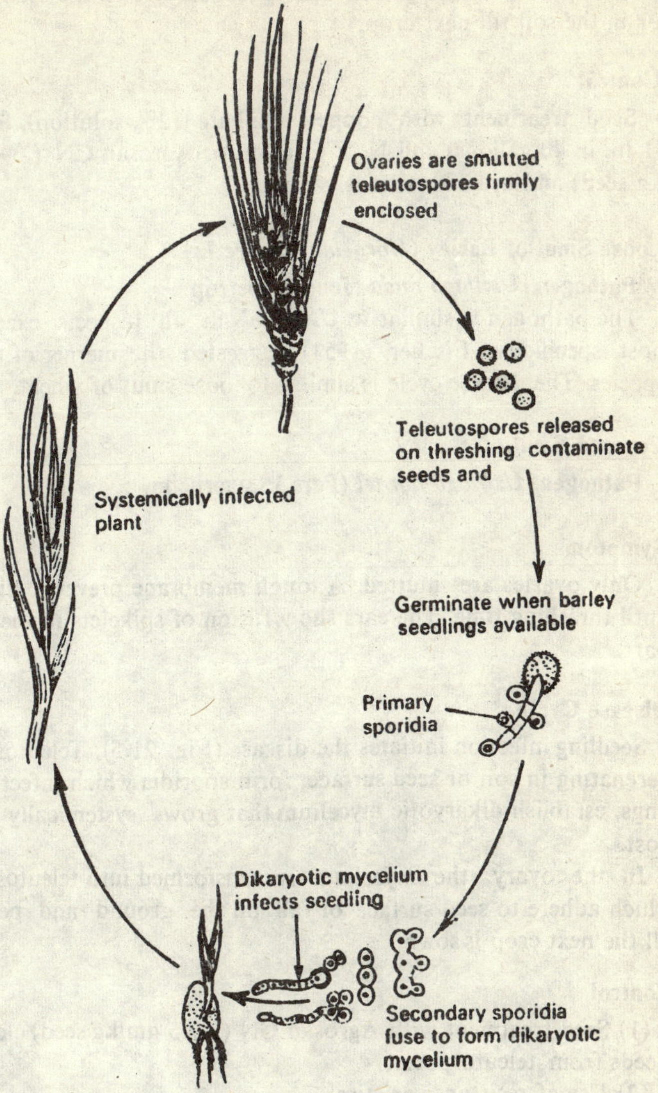

Fig. 21.5. Disease cycle of covered smut of barley, caused by *Ustilago hordei*.

teleutospores in the grains. The spores survive on the seed surface or in the soil till next crop.

Control

Seed treatment with copper sulphate (2% solution), formalin (1 lb. in 40 gallons) and N. I. Ceresan or Agrosan CN (2—2.5 kg/kg seed) and use of resistant varieties.

Loose Smut of Barley (*Hordeum vulgare* L.)

Pathogen. *Ustilago nuda* (Jens.) Rostrup.

The pathogen is similar to *U. tritici* in all respects except the host specificity. Fischer (1953) suggested the merger of the two species. The disease cycle is similar to loose smut of wheat.

Covered Smut of Barley

Pathogen. *Ustilago hordei* (Pers.) Lagerheim

Symptoms

Only ovaries are smutted. A tough membrane prevents dispersal until threshing time. The ears show fusion of spikelets in the lower part.

Disease Cycle

Seedling infection initiates the disease (Fig. 21.5). Teleutospores, perenating in soil or seed surface, form sporidia which infect seedlings, establish dikaryotic mycelium that grows systemically in the host.

In the ovary, the mycelium is transformed into teleutospores, which adhere to seed surface or fall on the ground and perenate till the next crop is sown.

Control

(1) Seed treatment with Agrosan CN (2-2.5 gm/kg seed) clear the seeds from teleutospores.

(2) Use of resistant varieties.

Smut of Sugarcane (*Saccharam officianarum* L.)

Pathogen. *U. scitaminea* var. *sacchari-officinarum* Mundkur.

Symptoms

Shoots smutted; appear like a long whip (Fig. 21.6). A membrane covers the teleutospores, which later slough off. The pathogen is systemic.

Disease Cycle

The disease is initiated by internally infected 'seed'; or by teleuto spores hibernating in soil (shoot infection, Fig. 21.7). Stem pieces,

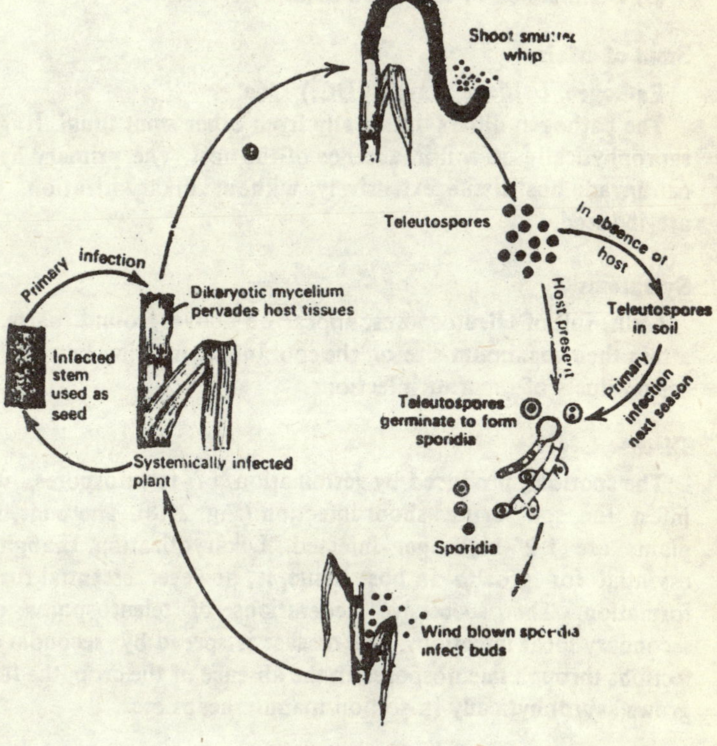

Fig. 21.7. Disease cycle of smut of sugarcane, caused by *Ustilago scitaminea.*

used as seeds, if derived from infected plants, produce systemically infected plants, which produce smutted whip like shoots. The teleutospores, blown to healthy plants, infect axiliary buds through sporidia.

The sporidia, after infection, form primary hyphae, which by dikaryotization form the secondary mycelium. The smutting occurs only in shoots. Teleutospores perennate in soil and cause infection when crop is available.

Control

(1) Use of healthy stem pieces, obtained from disease-free fields. for sowing.

(2) Burning of infected plants.

(3) Disinfection of suspected seeds by formalin.

Smut of Maize

Pathogen. *Ustilago maydis* (DC.) Cda.

The pathogen differs drastically from other smut fungi. It grows saprophytically in soil in absence of the host. The primary hyphae can invade host tissue extensively without dikaryotization. Galls are induced.

Symptoms

Galls, full of teleutospores, appear on above ground parts, but attain their maximum size on the cob. Infection being local, all galls are products of separate infections.

Disease Cycle

The sporidia, produced by germination of teleutospores, which infest the soil, cause shoot infection (Fig. 21.8). The buds, when plants are 1-2" high, get infected. Dikaryotization, though not essential for growth in host tissue, is, however, essential for gall formation. The successive generations of teleutospores cause secondary local infections. The disease is spread by secondary infections through teleutospores. In the absence of the crop, the fungus grows saprophytically in soil on manure heaps etc.

Control

Soil sanitation and use of resistant varieties.

Some other important smut diseases are described in Table 21.1.

Table 21.1. Some Smut Diseases

Mode of Infection	Disease cycle	Control

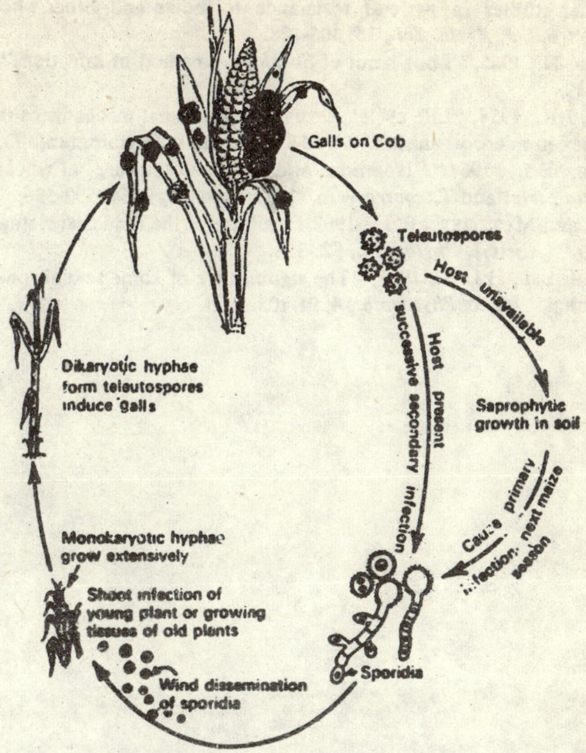

Fig. 21.8. Disease cycle of maize smut, caused by *U. maydis.*

REFERENCES

FISCHER, G.W. and HOLTON, C.S., 1957, *Biology and Control of the Smut Fungi*, The Ronald Press Co., New York. 622 p.

GOTTLEIB, D. and CLATRIDER, P.G., 1962, " Synthesis during the germination of Uredospores of rust fungi," *Phytopathology*, 52: 11. (Abstract).

GRIMM, P.W. and Allen, P.J., 1954, *Plant Physiol.*, 29: 369-377.

HOLTON, C.S., 1953, "Physiologic specialization and the genetics of smut fungi," II *Bot. Rev.*, 19: 187-208.

JONES, J.P. and BARNETT, R.D., 1967, "A promising new control for wheat and barley loose smut," *Arkans. Fm Res.*, 16, 2. (*Rev. appl. Mycol.* 47: 133).

MUNDKUR, B.B. and PAL, B.P., 1941, "Studies in Indian cereal smuts, II. Varietal resistance of Indian and other wheats to loose smut," *Ind. J. Agric. Sci.* 11: 675.

PAL, B.P. and MUNDKUR, B.B., 1945, "Studies in Indian cereal smuts. VII.

Further studies in varietal resistance to Indian and other wheats to loose smut," *Ind. J. Agric. Sci.*, **15**: 106-108.

PRASAD, N., 1945, "Long smut of *Sorghum*— method of infection," *Curr. Sci.*, **14**: 239.

RAPER, J.R., 1954, "Life cycles, sexuality and sexual mechanisms in the fungi" in "Sex in microorganisms." *AAAS Monograph*, Washington, D.C., 362 p.

TIRONE, E.J., 1964, "Isolation and *in vitro* culture of wheat bunt fungi *Tilletia caries* and *T. controversa*," *Phytopathology*, **54**: 592-596.

———, and METZGER, R.J., 1962, "Culture of the binucleate stage of *Tilletia caries*," (Abstr.). *Phytopath.*, **52**: 366.

WHITEHOUSE, H.L.K., 1951, "The significance of some sexual phenomena in the fungi," *Indian Phytopath.*, **4**: 91-105.

22

RUSTS

Rust is the greatest enemy of cereals. ***Pliny***

RUST diseases have attracted maximum attention from the plant pathologists because of their destructiveness and vexed biology. Butler in 1903 said that "there are few diseases of plants so obscure as these and the difficulties instead of diminishing appear to increase with each succeeding advance of knowledge." Their 'peaceful coexistence' with the 'special' hosts makes them elite among plant pathogens. Damage to host is subtle and prolonged. Their obligate parasitism and extreme host specialization baffles the mycologists and the breeders alike.

All the rust fungi are put together under the order Uredinales of class Basidiomycetes. They are identified on the basis of their teliospore morphology. Rust diseases occur on wide variety of economically important crops as well as on ornamental plants. Some of the important genera (Fig. 22.1) and the diseases they cause are given below:

(1) *Puccinia*—Teliospores typically two-celled.	
P. graminis	Causes cereal rusts, comprises several sub species. Some of the important ones are as follows:
P. graminis tritici	Wheat rust
P. graminis hordei	Barley rust
P. graminis avenae	Oat rust
P. malvacearum	Hollyhock rust
P. coronata	Crown rust of oats
P. antirrhini	Snapdragon rust
(2) *Gymnosporangium*	Two-celled teliospores embedded in tongue like gelatinous masses. lack Uredospores.
G. juniperi—virginianae	Juniper and Apple rust.

(3) *Phragmidium* — Teliospores many celled, long stalked, surrounded by gelatinous sheath.

P. mucronatum (rose rust)

(4) *Uromyces* — Unicellular teliospores

U. pisi (pea rust)

U. appendiculatus bean rust

U. fabae (gram rust)

(5) *Ravenelia* — Fused teliospores (stalks free) form a plate like structure.

R. emblicae attacks *Phylanthus emblica* (amla).

(6) *Hemileia* — Turnip shaped teliospores are borne on swollen tips of stalks and share a common sorus with uredospores. Uredospores look like an orange segment.

H. vastratix Coffee leaf-rust

(7) *Melampsora* — Sessile teliospores unite laterally to form a palisade layer below the epidermis.

M. lini flax rust.

Spore forms. Five types of spores are encountered in the life cycle of some (not all) rust fungi. These spore types are designated in Roman numerals 0—IV as spore stages.

The spore forms are discussed here in the sequence in which they occur in the life cycle. These are as follows:

Stage 0	Spermogonium
Stage I	Aecium
Stage II	Uredium
Stage III	Telium
Stage IV	Basidium

Stage 0. Spermogonium (plu. spermogonia, syn. pycnidium, pycnium)

Spermogonia are produced from haploid primary mycelium developed by germination of basidiospores on the host. This stage was represented as 0 stage as the function of these structures was not known. Now the function of spermogonia is known; these bear the sex organs of rust fungi.

Spermogonia are flask shaped (though globose or flattened types are also known) ostiolate structures usually borne on the upper surface of the infected leaves. From the base of the spermogonia arise elongated cells called spermatiophores. These cut off uninucleate, unicellular spermatia which act as the sperm cells. Spermatia are extruded out through the ostiole in a sugary exudate. A few long flexuous hyphae, called the "receptive hyphae", arise from the wall of the spermogonium and project

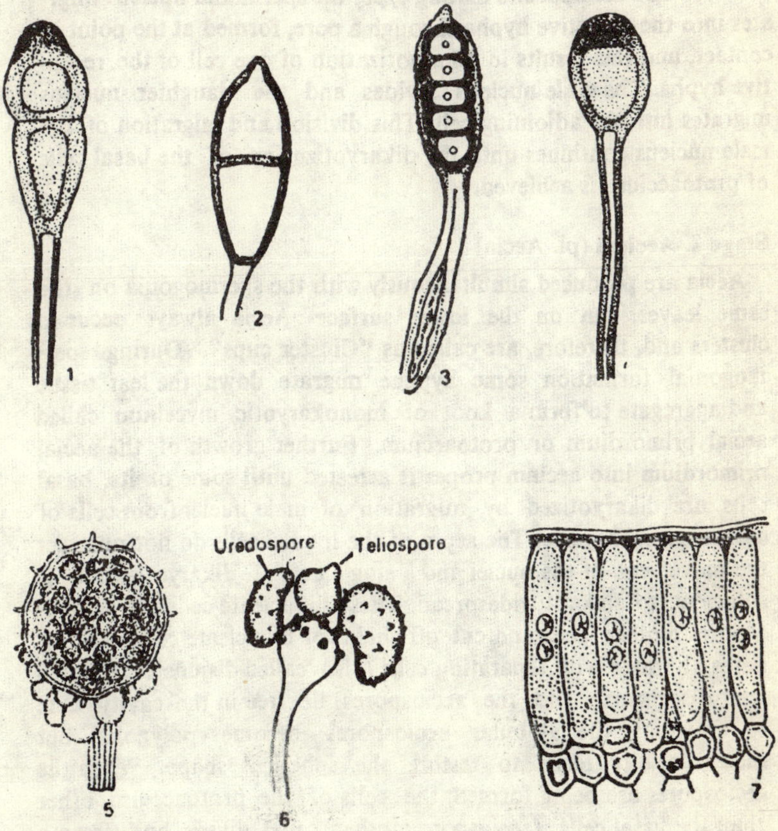

Fig. 22.1. Teliospores of some important rust genera: 1. *Puccinia*; 2. *Gymnosporangium*; 3. *Phragmidium*; 4. *Uromyces*; 5. *Ravanelia*; 6. *Hemileia* and 7. *Melampsora*

through the ostiole. These are the female organs which receive the male nucleus from spermatia. Stiff hair like structures called Periphyses (sing.=periphysis) arise from the uppermost part of the spermogonium which support the spore bearing exudate drop Rusts are mostly heterothallic and the morphologically similar spermogonia belong either to plus or to minus strain. Thus, spermatia and receptive hyphae, produced by the same mycelium, are incompatible. The exudates of adjacent spermogonia usually fuse and spermatia find access to receptive hyphae of opposite strains. Sometimes, insects carry the exudate containing spermatia to other spermogonia. When a spermatium comes in contact with the receptive hypha of the opposite mating type, the spermatial nucleus migrates into the receptive hypha through a pore, formed at the point of contact, and this results in dikaryotization of one cell of the receptive hypha. The male nucleus divides and the daughter nucleus migrates into the adjoining cell. This division and migration of the male nucleus continues until the dikaryotization of the basal cells of protoaecium is achieved.

Stage I. Aecium (pl. Aecia)

Aecia are produced simultaneously with the spermogonia on the same leaves, but on the lower surface. Aecia always occur in clusters and, therefore, are called as "Cluster cups". During spermogonial formation some hyphae migrate down the leaf tissue and aggregate to form a knot of monokaryotic mycelium called aecial primordium or protoaecium. Further growth of the aecial primordium into aecium proper is arrested until some of its basal cells are dikaryotized by migration of male nuclei from cells of spermogonial hyphae. The septa of the hyphal cells do not obstruct the movement of the nuclei and a single act of dikaryotization is sufficient to make it widespread. These binucleate cells act as aeciospore mother cells and cut off chains of binucleate cells. Aeciospores alternate with separating cells (also called disjunctors), which later disorganize and the aeciospores lie free in the cavity. Due to crowding, the globular aeciospores become polygonal but their turgidity tends to restore the spherical shape. When the aeciospores are being formed, the cells of the protoaecium, other than the basal cells (aeciospore mother cells), divide and form a covering around the spores. This is called periderm. The pressure of the aeciospores pushes the periderm which bulges out and ultimately breaks. This pressure breaks the overlying lower epidermis also and the periderm usually protrudes out of the leaf tissue. A mature aecium looks like an inverted saucer or cup. Chains of aeciospores hang from the base of the cup.

The aeciospores are the first binucleate spores in the life cycle of rust fungi and are discharged with force. They have a thick wall and contain oil as food reserve. The aeciospores when land on the proper host, germinate and enter through the stomata to form a dikaryotic mycelium which spreads intercellularly, drawing nutrition from host cells through haustoria. The dikaryotic mycelium aggregates and forms characteristic rust pustules bearing first crop of uredospores.

Five types of aecia are known. These are designated as "form genera" until the teliospore stage is traced to identify them. The aecia are distinguished on the basis of extent of peridium development. Though deceptive, yet it is good for tentative placing when the search for the teliospore stage is on. The five form genera are *Aecidium*, *Peridermium Roestelia*, *Caeoma* and *Uraecium*.

In the form genus *Aecidium* the peridium is short (*Puccinia* and *Uromyces*). It is strongly developed in *Peridermium* (*Cronartium*, *Milesia* and *Coelosporium*); horn like protruding and shreded in *Roestelia* (*Gymnosporangium*) and absent in *Caeoma* (*Phragmidium*, *Melampsora*) and *Uraecium*. In addition, the aecium of *Caeoma* has characteristic paraphyses. The aecia of *Uraecium* (e.g. *Uropyxis*, *Ravenelia*) look like uredia but can easily be distinguished by the presence of spermogia, and the 'cluster cup' that the aecia form by aggregation.

Stage II. Uredium (pl. Uredia, syn.=Uredosorus)

Uredia, which give the characteristic rusted appearance to the leaves, are produced by dikaryotic mycelium formed initially by germination of aeciospores and further by uredospores. Uredospores, because of several successive crops of spores, carry the infection to large number of plants over a wide area. This stage is also known as the repeating stage of the fungus.

Uredia are subepidermal, flat pustules, bearing stalked uredospores. The epidermis bulges out and ultimately breaks exposing the uredospores to wind for dissemination. The uredia, with the approach of end of the host crop season, gradually change to telia by producing lesser number of uredospores and more teliospores, ultimately producing only teliospores.

Uredospores vary in shape and may be single (sessile or stalked) or in chains. The spore wall may be echinulate, verrucose, reticulate or smooth. Uredia of *Pucciniopsis* have a well developed periderm, while paraphyses are present in *Uromyces*.

Stage III. Telium (pl. telia; syn=Teleutosorus)

Telia constitute the most important stage as the identification of these fungi is based on the morphology of teliospores. Being the site of karyogamy and meiosis, these spores represent the perfect stage of rust fungi. Teliospores can be distinguished from other spores by the fact that on germination they produce a small hollow tube like structure called the basidium.* Teliospores are called 'resting' or 'winter' spores because of their ability to stay viable after exposure to unfavourable weather and in the absence of the host. About 50% of the rust fungi have teliospores that can germinate without a rest period.

Telia may be erumpent (*Puccinia*); extruded out of host tissue as columns (*Cronartium*); formed within the epidermal cells (*Milesia*); scattered in mesophyll cells (*Urediniopsis*) or grouped in horizontal columns (*Melampsora*). In *Masseeella*, though telia are subepidermal, the teliospores extrude out in a gelatinous mass to form much branched, hair like structures. Teliospores may be sessile or stalked, unicellular or multicellular with varying pattern of arrangement of the cells.

Stage IV. Basidium*

On germination, the teliospores give rise to a tubular pro-mycelium or basidium. The diploid nucleus from the teliospore migrates into the tube and undergoes meiosis forming four haploid nuclei. Segregation of sex takes place during the reduction division and out of the four nuclei formed, two are of plus strain and the other two of minus strain. Transverse septa divide the basidium into four cells each containing a haploid nucleus. Sterigmata (sing. sterigma) arise from these cells which terminate into a fine point and bear basidiospores. Basidiospores are usually discharged with force. The basidiospores, when reach the proper host, germinate and restart the haploid phase by forming plus or minus mycelium, which later form the spermogonia.

The haploid phase, initiated during basidiospore formation, continues in the spermogonial and a part of the aecial stage. Comparatively, the haploid phase is of shorter duration and is limited by the incapability of its haploid spores (spermatia) to propagate the

*If the teliospore is considered as hypobasidium then the tube called basidium above should be designated as epibasidium and the two together as the basidium. This is mycologically correct but long usage can allow the retention of the terms basidium and telium instead of epibasidium and hypobasidium respectively.

haploid phase. The dikaryophase, once initiated at the site of receptive hyphae, is propagated by aeciospores and numerous successive generations of the uredospores. The disappearance of the host winds up the dikaryophase and teliospores are formed which terminate the dikaryophase by allowing the karyogamy of long restrained nuclei of opposite strains. The metabolic versatility, enjoyed by the dikaryotic mycelium, comes to an end when the two nuclei get separated in the two types of basidiospores.

THE LIFE CYCLES OF RUST FUNGI

All the spore forms may not be formed in the life cycle of a particular rust fungus. The following three types of life cycles represent the variations.

(*A*) **Macrocyclic Life Cycle.** Rust fungi forming all the five types of spores (or lacking in spermogonia only) are called macrocyclic or long-cycled. These macro-cyclic rust fungi may form all the spore stages on a single host (autoecious) or on two alternating hosts (heteroecious) *Puccinia helianthi* (Sunflower rust) *Uromyces straitus* (Alfa-alfa rust) and *Phragmidium mucronatum* (Rose rust) are examples of autoecious macrocyclic rusts. Heteroecious, macrocyclic forms are represented by the well known *Puccinia graminis tritici* wheat (black stem rust) and *Cronartium ribicola* (white pine blister rust).

(*B*) **Demicyclic** life cycles are those which have spermogonia (sometimes missing), aecia and telia; uredospores are not produced. Demicyclic rust fungi too may be heteroecious e.g. *Gymnosporangium juniperi-virginianae* (cedar-apple rust) or autoecious e.g. *Gymnoconia peckiana* (orange rust of *Rubus*).

(*C*) **Microcyclic** life cycles are those which have only telia as the binucleate spores. Spermogonia may or may not be present. By necessity, all microcyclic rust fungi are autoecious.

In some microcyclic rust fungi telia look like aecia because of their 'cluster cup' appearance, but contain teliospores. This telial condition is found in *Puccinia malvacearum*. But in some cases, not only the telia look like aecia but teliospores too resemble aeciospores. In such situations only germination can help in the identification; if the spore is a teliospore, a basidium will be formed. This type of microcyclic rust, in which the telia simulate the habit of aecia, is called 'endotype'. Depending on the shape of the telia, the species are identified as *Endophyllum* (telia aecidioid) or *Kunkelia* (telia caemoid). Telia of other aecial shapes are not known.

It is generally agreed that the macrocyclic rust fungi represent the existing primitive forms and that the microcyclic or demicyclic rust fungi are descendants of the macrocyclic forms.

EFFECT OF RUSTS ON THE PHYSIOLOGY OF THE HOST

Studies on rust physiology have been carried out with special reference to the following aspects:

(*i*) metabolism of spores;

(*ii*) altered translocation of infected host;

(*iii*) respiratory pattern and photosynthetic activity of infected tissues;

(*iv*) nucleic acid metabolism of the host-parasite complex,

(*v*) metabolism of growth substances in the infected host.

Metabolism of Spores

Mostly work has been carried out with Uredospores. The study of the spore metabolism is rendered difficult by the limited availability of material and due to their failure to grow in the absence of the host. It is interesting because their incapacity to grow in absence of host is not due to any metabolic deficiency. Radioisotopic studies have established that the uredospores possess the usual metabolic pathways and give no clue to the reason of their obligate parasitism. Their turn-over rates are like any other fungus.

However, rust spores are characterized by surprisingly low level of metabolism, low respiratory rate and low rate of protein synthesis. Exogenously supplied substrates are very slowly utilized. Fatty acids, especially those of high molecular weight, however, are consumed abundantly and quickly. This is not so with sugars, organic acids or amino acids, which are very poorly utilized. Thus fatty acids have a special status in uredospore metabolism because they considerably stimulate oxygen uptake and spore germination even in the presence of endogenous inhibitors. Their quick consumption and quick penetration is related to the high fat content of these spores. Fatty acids decrease during the spore germination.

The nature of endogenous substrate utilization is less understood. Conditions which stimulate oxygen uptake and spore germination (e.g. treatment with pelarogonaldehyde) considerably enhance endogenous substrate utilization also. No such effect is seen on utilization of exogenously supplied substrates.

The operation of both tricarboxylic acid (TCA) and glyoxylate* cycles in uredospores have been demonstrated by employing specially labelled fatty acids C_1-C_6 in feeding experiments. A substantial part of the label can be detected in the intermediate steps of the TCA cycle, and amino acids associated with TCA and glyoxylate cycles. Thus the utilization of the fatty acids takes place via TCA and the glyoxylate cycles. The labelling data indicate that the fatty acids undergo β-oxidation and enter these cycles as acetyl units. Most of the enzymes of the two cycles have been detected in cell-free extracts of the uredospores. However, their relative contribution to the overall metabolism is not clear. When pelarogonate is fed to the uredospores, the glyoxylate cycle seems to be more pronounced whereas with methyl labelled acetate or other fatty acids, with one corresponding carbon only, the TCA cycle is operative. The low C_6/C_1 ratio indicates that sugars are broken down in the spore via the HMP pathway. Thus, whereas the utilization of fatty acids takes place via TCA and glyoxylate cycles, sugars are broken down via HMP pathway.

The uredospores are unique in that during spore germination, though protein turnover does occur, there is no net gain of proteins. This indicates that the protein synthesizing system is present and active. Uredospores can be induced to produce variety of enzymes and the turnover levels compare well with higher plants and animal tissues. Polygalacturonase, mucilaginase, hemicellulase, cellulase and several other enzymes have been shown to be formed inductively during spore germination. Since proteins do not accumulate, the synthesis of these enzymes must be at the cost of pre-existing protein molecules. This is an important aspect which explains as to why the germinating uredospores cease growing. Amino acid incorporation into proteins of uredospores is dependent on oxidative metabolism and shows sensitivity to inhibitors of the protein synthesis (e.g. flurophenylalanine, chloromycetin and puromycin).

Thus, the uredospores, though equipped with the normal metabolic pathways, operate at low ebb. This sluggishness could be the effect

*Glyoxlate cycle is a shunt of the TCA cycle which bypasses the α-ketoglutarate and thus two oxidative decarboxylations. Splitting of isocitrate gives rise to succinate and glyoxylate. The latter on condensation with Acetyl-CoA results in the formation of malate.

of some metabolic block (blocks) which retards the normal speed of metabolism. The nature of this block is not yet known.

ALTERED TRANSLOCATION OF THE INFECTED HOST

The infected tissues show accumulation of substances in the vicinity of the infected tissue at the cost of neighbouring and also remote tissues. Gottleib and Garner (1946), using labelled substances, detected accumulation of radioactive P^{32} around rust pustules. Shaw *et al* (1954) used xylose 1-C^{14} in the feeding experiments and detected accumulation of the radioactive carbon at the loci. The infected tissues show accumulation of all elements in quantities several hundred times of that present in the healthy tissues.

Thus the infection centre acts as a metabolic sink, which competes with other growth centres like the meristematic regions. The normal distribution of substances is thus disturbed in the whole plant and directed towards the infected tissues. It is interesting that even non-metabolizable substances accumulate in the infected tissues. Maximum incorporation of labelled substances is seen in the mycelium and the uredospores. However, tritiated thymidine was shown to be incorporated in the host nuclei and not in the fungal nuclei.

In addition to the influx of solutes, the accumulation is also due to enhanced synthetic activity in the infected host tissue. The inhibition of accumulation by dinitrophenol (DNP) and similar inhibitors of aerobic respiration, suggests that the phenomenon of accumulation is dependent upon aerobic metabolism. The significance of accumulation, apart from an obvious advantage to the fungus, is not clear. Thus, the 'peaceful coexistence', that the fungus establishes with the host, is a later phase. Initiation is by develop ment of a field of dominance.

RESPIRATORY PATTERN AND PHOTOSYNTHETIC ACTIVITY OF INFECTED TISSUES

An increased rate of respiration, which is known to be associated with several host-pathogen systems, occurs in rust infected tissues also. The increased respiration is first recorded when rust flecks are visible and attains maximum rate when sporulation is vigorous. A shift from EMP to HMP pathway is indicated by low

C_6/C_1 ratio in the infected tissues (Shaw and Samborski 1957 and Daly *et al* 1957, 1961).

'Pasteur effect', which is abolished during infection, is made good by the increased rate of respiration. Cytochrome oxidase acts as the terminal oxidase. The rate of photosynthesis, which shows an increase in the initial stages of infection, is transitory and goes down, showing maximum decrease at the time of sporulation of the fungus. The photosynthates accumulate in the developing spores which affects the whole plant.

NUCLEIC ACID METABOLISM OF THE HOST-PARASITE COMPLEX

Rust-infected tissues show marked increase in rate and level of RNA synthesis. It is accompained by enlargement of nucleolus. This was clearly proved by increased incorporation of ^{32}P into RNA (Heitefuss, 1961). Autoradiography has indicated increased synthesis of organic P-compounds (glucose phosphates, nucleotides). In studies with cytidine H^3, Nielsen and Rohringer (1963) observed less labelling in infected cells than in distant healthy cells. Also, no label was detected in the fungal hyphae. These results do not explain the increased RNA synthesis in host cells, in addition to the high label in the fungus. The labelled cytidine, perhaps, failed to reach the DNA precursor pool due to the breakdown of the transport system of the infected plant.

An increased DNA synthesis, too, as is evident by microautoradiographic studies,occurs in the host nuclei (Staples and Ledbetter, 1960) but cell division failed to show up. The number of labelled nuclei was particularly high near the infected centre.

METABOLISM OF GROWTH SUBSTANCES

Rust-infected tissues, as a rule, show an increased level of growth substances of the indole acetic acid type and sometimes exhibit growth abnormalities also. In susceptible leaves. the IAA concentration initially declines but later increases with the progress of infection. In resistant tissues, the decline continues steadily. The increased level of IAA, particularly at the sporulation stage, corresponds with a decreased level of IAA-oxidase activity. However, the correlation between IAA and IAA oxidase concentration is far from conclusive. The rate of degradation may not be the only factor determining the pool size of IAA. The disrupted transloca-tion, release of bound IAA or different rate of synthesis could be the cause of difference in IAA content of healthy and diseased tissues. Increased concentration of tryptophan (a precursor in IAA synthesis) in diseased tissues suggests an increased rate of synthe-sis. Kim and Rohringer (1969), using radioactive tryptophan, found increased rate of IAA synthesis in rust (*Puccinia*)-infected susceptible wheat leaves; no such increase occurred in rust-resistant tissues.

CULTURING THE RUST FUNGI

When we know how the parasite breathes, feeds, excretes, and uses within its host and at its host expense, whatever biochemical proce-sses link it to that host, we may learn to control it. Anonymous
Editorial, *Nature* 1951

Growing fungi in culture is important and growing obligate parasites is, in addition, challenging. Our knowledge of the physio logical activities of the rusts is limited due to the failure of these fungi to grow in the absence of the host. In this situation, the relative contribution of the pathogen and the host to the deranged physiology of infected plants, can not be sorted out.

The problem of growing rust fungi saprophytically is pesky. Many workers had to leave the problem half way. Some felt that the equilibrium between host and pathogen was too precarious to be within the range of experimental analysis. Arthur (1928) said that "repeated attempts to grow the rust on artificial media or even to accelerate the germination of the spores by using nutritive solu-tions have met with uniform failure." The frustration is apparent in Duggar's statement (1909): "it would be useless to try to culti-vate these fungi in nonliving substrate". Chester, as late as 1946, wrote of many investigators seeking in vain "the philospher's stone underlying the problem". Nonetheless, the efforts were continued and the progress that has been made since 1951 has belied the overt hopelessness underlying the above statements.

Early work (Ezekiel 1930; Stock, 1931) showed that uredospores germinated in distilled water but the growth of the germ tubes ceased within 24 hours. Greater spore density in the germination drop increased the percentage of germination. Mains (1917) beli-eved that the secret of obligate parasitism lay in some transient metabolites available in the host plant. Hit and miss method of knowing the essential nutrient was pursued by several workers but with no result. It was thought that the infection structures, formed by the fungus (appresorium, infection peg, substomatal vesicle etc.), were necessary for growth in culture. Brown (1936) suggested that once these structures were formed, much of the nutritional specificity will disappear. Thus emphasis was laid on knowing the factors that determined the formation of these infection-structures. Much information has been gathered on the subject. Appresorium formation in artificial media was found to be dependent on various stimuli, like contact (Dickinson, 1949; Rowell and Olien, 1957), chemicals (Allen, 1957), light and temperature (Emge, 1957). Thus Maheshwari *et al* (1967) reported hundred per cent formation of infection structures by some rusts on collodion membrane impregnated with mineral oil or hydrocarbons isolated from surface wax of leaves. Unidentified volatile compounds isolated from uredospores of *Puccinia graminis tritici*, inhibited appresorium formation. Another feature of interest was the relation observed between nuclear division and differentiation of germ tubes into infection structures. Maheshwari *et al* (1967b) noted active nuclear divisions in germ tubes of *Uromyces phaseoli* uredospores, differentiating into the infection structures. The non-differentiated germ tubes lacked such divisions. Formation of the infection structures was accompanied by increased RNA (6 to 10 times) and DNA synthesis (Ramakrishnan and Staples 1968). Fuchs and Gaertner (1958) made the significant observation that addition of certain amino acids (cysteine, leucine or glutathione) to the basal medium (consisting of glucose, egg yolk, Knop's mineral solution, pyridine and iron saccharate solidified in silica gel) increased the formation of the infection structures from 10 to 40 per cent. However, the growth did not continue for long and ceased after 10 to 14 days.

The first claim of successful culture of a rust, *Gymnosporangium juniperi virginianae*, in axenic culture, was made in 1951 by Hotson and Cutter. Later in 1960, Cutter reported successful axenic culture of *Puccinia malvacearum* (a microcyclic rust infecting Hollyhock) on callus of *Althaea* and observed teliospore formation on three months old callus tissues. However, his experiment could not be successfully repeated. In 1966 Williams, Scott and Kuhl successfully obtained the first saprophytic culture of an Australian isolate of *Puccinia graminis tritici*. They incubated uredospores on an agar medium containing Czapek's minerals, ·1% yeast extract and 3% sucrose. Germination, followed by germ tube elongation, occurred after 48 hours and branching commenced after 96 hours. The vegetative growth was found to be enhanced by addition of Evan's peptone in the medium. Some cultures showed formation of stromata, bearing uredo and teliospores after 2 to 4 weeks. Replacement of sucrose by glucose and addition of citrate to the yeast extract-peptone medium resulted in more consistent growth. Uredospores grew vigorously on more chemically-defined medium containing Czapek's minerals, glucose, aspartate, citrate and either cysteine or glutathione (Scott and Maclean, 1969). However, the growth stopped after 3-4 weeks due to staling. Based on above experiments, axenic cultures of other rust fungi have since been obtained.

In the above saprophytic growth, it was found that: (*i*) the uredospores never behaved consistently in any two experiment under identical conditions, and (*ii*) that a high spore density was favourable for colony formation. In nature the fungus shows a very orderly development and the different stages of growth occur in a definite sequence. In saprophytic growth the behaviour of uredospores was unpredictable and erratic. This could be due to two reasons. Firstly, the difference in behaviour of uredospores could be due to several variable factors associated with sample collection. Secondly, the difference in behaviour could be due to the difference in their mode of growth. The infection structures, which are the *sine qua non* for pathogenicity, seldom appeared in saprophytic cultures. The two phases of growth, according to Scott and Maclean (1969), are controlled by different genomes. The expression of the saprophytic genome is under influence of numerous variable factors of the medium, which modify their behaviour frequently.

In parasitic growth, the extreme host specialization ensures a stable environment and expression of the parasitic genome. The other observation, viz. the impact of spore density on colony formation, is associated with leakage of amino acids and other essential metabolites by the leaky uredospores. The leakage has been demonstrated by several studies using radioactive labelled substances. In high spore density environment, some spores, which would have otherwise ceased growth, use the metabolites leaked out into the medium.

The next important thing observed was that the saprophytically grown cultures retained the pathogenicity and caused infection, forming the 'green islands', the characteristic symptom of rust infection. But there was a deviation in behaviour. In the presence of epidermis, infection failed. However, when uredospores were put directly on mesophyll cells, soft appeared by rupturing the epidermis on the other side. On leaves with intact epidermis, spore germination was high but appressoria, which were scantly formed, failed to give rise to infection pegs. Low potential of independent growth, rather than loss of pathogenicity, is ascribed as the reason of non-infectivity in presence of the epidermis. The uredospores formed by reinfection, using inoculum from saprophytic growth, again, could grow saprophytically and formed uredia.

Impact of Rust Culture on the Concept of Obligate Parasitism

De Bary in 1887 defined obligate parasites as species which require a parasitic existence for the attainment of their full development. He divided obligate parasites into two groups: "Facultative saprophytes (which can also live a part of their life as saprophytes) and strict obligate parasite" (which live only as parasites). De Bary based this on his observations of fungi as they exist in nature. Because of repeated failures in growing rust fungi on nonliving media, the obligate parasites came to be recognised as organisms which could not be cultured axenically (Yarwood, 1956). Brian (1967) grouped the rust fungi into (*i*) physiologically obligate parasites and (*ii*) ecologically obligate parasites. The rusts which failed to grow in even most complex media were put in the first group while those which failed in nature to grow saprophytically (due to lack of competitive saprophytic ability) but could be grown axenically were included in the second category. Thus the rusts which have been cultured axenically fall in Brian's second group of ecologically obligate parasites. The physiologically obligate parasites can never grow anywhere other than on the host.

RUST OF WHEAT

The rust of wheat has been studied more than any other disease in the world. In India also, K.C. Mehta at Agra and his associates studied this disease with exemplary zeal.

Three rusts of wheat are known and all the three occur in India.

(1) Black rust (stem rust) *Puccinia graminis tritici*

(2) Orange (brown) rust (leaf rust): *P. recondita* (Syn= *P. triticina*)

(3) Yellow rust (stripe rust): *P. striiformis*

According to Pal (1936), brown and yellow rusts cause principal damage because the black rust appears too late in the season to cause much harm.

Black Stem Rust

Pathogen. *Puccinia graminis tritici* Erikss & Henn.

Symptoms

On the plains the disease appears late in the month of March, when brown pustules of uredosori appear on the leaves (lower surface), the leaf sheaths and the stem. These pustules grow and fuse to form bigger lesions of dark brown colour. The overlying epidermis bursts. In the later stages, teliospores are produced in the pustules, which turn black. The change of spores from uredo to teliospores is gradual and hence, the pustules of different shades of brown to black appear on the leaves. Telia form oblong to linear lesions and are more frequent on leaf sheaths than on culms.

The other symptoms include the retarded plant height, poor tillering and poor shrivelled grains, which in extreme cases fail to form.

Morphology of the Pathogen

The dikaryotic mycelium grows intercellularly, drawing nutrition from host cells through haustoria. Uredospores are stalked, brown, oval, thick-walled with minute spines on the wall and have 4 equatorial germpores. They germinate and reinfect wheat. Teliospores are bi-celled, spindle shaped, dark-brown and have a thick smooth wall with 2 germpores—one at the apex, the other below the septum. The apex is round, rarely pointed. After a rest period, they germinate to form 4 sporidia on a septate promycelium. The sporidia germinate on barberry to form monokaryotic mycelium, which form flask shaped ostiolate spermogonia, bearing spermatiophores and periphyses. Aecia appear on the lower surface; are saucer shaped, and have well developed peridium, which protrudes out of the leaf. Chains of globular aeciospores are cut terminally by aeciosporophores, lining the base of the aecial cup. The aeciospores get pressed by crowding and become polygonal in shape. The dikaryotic aeciospores infect wheat.

Disease Cycle

This is heteroecious full-cycle rust; the other host being *Berberis walchiana* and *Mahonia*. On the plains of India, the source of

K. C. Mehta

primary infection are the uredospores formed on wheat on hills, where the crop is grown throughout the year (Fig. 22.2) The uredospores, blown by wind, reach the plains late in the season. This is supported by the fact that rust infections appear late in the season when winds blow from hills downwards. Uredospores initiate the infection and produce uredosori, which produce several successive generations of uredospores. This results in spread of

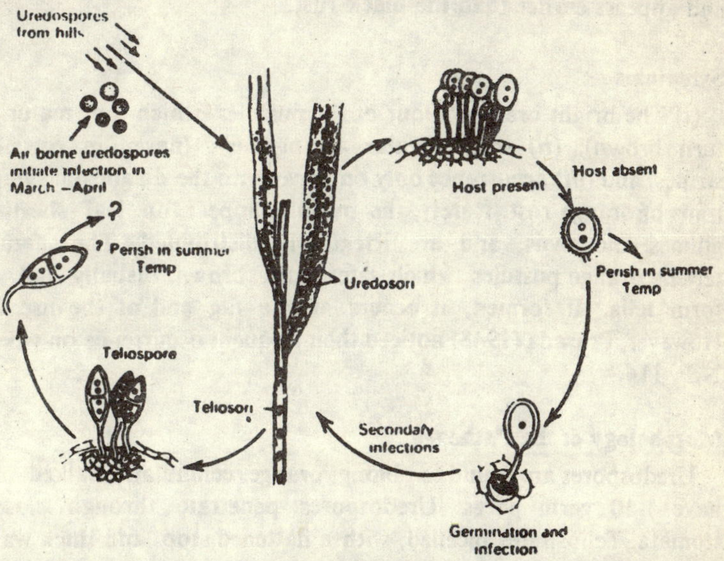

Fig. 22.2. Disease cycle of black rust of wheat on plains of India.

disease to new plants, and whenever most conducive factors are available, epiphytotics result. In the last few generations of spore formation, teliospores appear which die in the high heat of the intervening summer months. Next year, the winds again bring the inoculum (uredospores) from the hills. The aeciospores do not initiate the disease cycle as they do in America. Since *Berberis spp.* do not carry natural infections of this rust, the barberry eradication programme, which was so successful in USA, is ineffective in India.

Control

(1) Rust resistant varieties are available. In India 14 physiologic races and 6 biotypes of the fungus are reported. It is difficult to pool resistance against all races in one wheat variety. Thus such varieties, which are resistant to most virulent locally existing races, should be used in a particular area.

2. Chemical control of black wheat rust too is available but is unprofitable. Some systemic fungicides are giving promising results.

Orange (Brown) Rust of Wheat (=Leaf Rust)

Pathogen. *P. recondita* Roxb. ex. besm.

This is the most common rust in Northern and Eastern India and appears earlier than the black rust.

Symptoms

(*i*) The bright orange colour of the pustules (which on maturity turn brown); (*ii*) the irregular arrangement (never in rows or stripes) and (*iii*) occurrence only on leaves, are the diagnostic symptoms of orange rust. Rarely, the pustules appear on leaf sheaths, glumes and awns, and are irregularly distributed. The characteristic orange pustules, which later turn brown, usually fail to form telia. If formed, it occurs at the fag end of the disease. However, Prasada (1948) noticed their frequent occurrence on wheat N.P. 114.

Morphology of the Pathogen

Uredospores are round to oblong, orange, echinulate, stalked and have 1-10 germ pores. Uredospores penetrate through closed stomata. Teliospores bicelled, with a flattened top and thick wall, are always covered by the epidermis. The spermogonial and aecial stages are not reported in India.

In some other countries, the rust forms spermogonia and aecia on species of *Thalictrum* and *Isopyron fumarioides*. However, presence of such structures has not been reported from India. In fact *Thalictrum* has not been found to be the natural alternate host in any country.

Disease Cycle

The wind blown uredospores, which survive on volunteer wheat crops in hills, serve as the primary inoculum. The uredospores spread the disease and end up with the production of teliospores. Both spores fail to retain their viability in the hot intervening summer months.

Control

Like the black rust, the disease is controlled by using resistant varieties.

Yellow Rust (=Stripe Rust)

Pathogen. *P. striiformis* West. (=*P. glumarum* Schmidt Erikss & Henn.)

The rust does considerable damage particularly in Punjab, Haryana and Uttar Pradesh. It is rare in southern and western India.

Symptoms

The symptoms appear by January and are characterized by the bright yellow colour of pinhead like uredo-pustules, which are arranged in long streaks (mostly on leaves)—hence the name yellow or stripe rust. In severe infections, all parts of the plant are infected and the stripes, because of crowding of spots, disappear and from a distance the leaves appear dried. The abundance of uredospores results in the soil, underneath, being turned yellow. Telia too, unlike orange rust, are frequently formed and appear on the undersurface as black streaks. The pustules are always covered by epidermis.

Morphology of the Pathogen

The uredospores are nearly round, echinulate with a hyaline epispore and have 6-10 scattered germpores. Telia have a surrounding layer of brown paraphyses. Teliospores are oblong to cuniform and are more pointed at the apex than in *P. graminis*. The other spore forms and the alternate host, are unknown.

Disease Cycle

The uredospores, as in other two rusts, growing on hills on volunteer wheats, are wafted down to the plains to initiate the infection and later spread the disease. Some 90 weeds (e.g. *Agropyron semicostatum*, *Bromus catharticus*, *B. japonicus* and *Hordeum murinum* etc.) are known to serve as collateral hosts and thus may be the source of uredospores for primary infection.

Control

Use of resistant varieties, several of which are available, is the best way of overcoming the rust.

TABLE 22.1. COMPARISON OF THE THREE RUSTS

Black rust/*P. graminis*	Orange rust/*P. recondita* (=*P. triticina*)	Yellow rust/*P. striiformis* (=*P. glumarum*)
On plains appears first in March-April	January	January
Attacks stalks most severely	Attacks leaves almost exclusively, rarely the leaf sheaths and seldom the stalk	Attacks leaves most severely followed by leaf sheaths, stalks and ears.
Uredopustules long, elongated, burst early, brown, turn darker with the production of teliospores, formed on all green parts; uredospores oval, have 4 equatorial germ pores.	Uredopustules small, oval or round, never in rows, scattered, burst early, bright orange when fresh, later brown, found mainly on upper surface of leaves. Uredospores orange, round and have 7-10 scattered germ pores.	Uredo-pustules small like pin dots, arranged in long rows, burst late, lemon yellow, found on all green parts.
Telio-pustules resemble unredo-pustules but black; found on all parts, mostly on stems and least on leaf-blades, paraphyses absent.	Telio-pustules rarely formed, do not burst. Found chiefly on the under surface of leaves; paraphyses surround the sori.	Telio-pustules frequently formed, found chiefly on the under surface of leaves; paraphyses divide the sori into compartments.
Spermogonia and aecia formed on *Berberis* and *Mahonia*	Spermogonia and aecia formed on *Thalictrum* and *Isopyron*, but not in India.	Spermogonial and aecial stages not known.
240 races of the rust are known out of which 14 occur in India	Of over 100 known physiologic races, 14 are reported from India.	10 of the 14 known races occur in India.
The annual recurrence is by uredospores from hills.	By uredospores from hills.	By uredospores from hills or local collateral hosts.

PHYSIOLOGIC RACES

Erikson in 1894 first demonstrated that within the rust fungus *Puccinia graminis*, there existed several strains that differ in pathogenicity. He distinguished the several *formae speciales* (sub-species) which attacked one particular crop. The sub-species were designated by a third name to the binomial *Puccinia graminis* e.g. *tritici; avenae, secalis* and *hordei* etc. This is called biological (or physiologic) specialization and is known for some other pathogenic fungi also like *Albugo candida* and powdery mildews.

Stakman and Levine at Minnesota further observed that each biological sub-species comprises several physiologic races, which differ in their virulence on various agronomic varieties called differential varieties. The identification of the physiologic races is done by specialists. The procedure is as follows. For delineating the races of *Puccinia graminis tritici*, the uredospores, obtained from different areas, are inoculated on 'differential varieties' (which are 12 in number for wheat rust fungus). The infections produced, on the basis of number and size of pustules formed, are graded on a standard chart as follows:

0. Immune—no pustules produced, no sign of infection.
1. Very resistant—only faint chlorotic spots surrounded by necrotic tissues appear.
2. Moderately resistant—minute scattered pustules appear.
3. Moderately susceptible—medium size pustules and chlorotic spots appear.
4. Very susceptible—medium to big size confluent pustules formed.

X. Heterogeneous—size of pustules varying but mainly a mixture of type 1 and 4.

Small variations in the grade are indicated by one or two plus or minus signs e.g. 2^{-}, 2^{--}, 2^{+}, 2^{++} etc. The identification of physiologic races is a difficult task; the fluctuating effect of the environmental factors adds to the complexity. Thus, certain races produce the X-reaction under some conditions but not under others.

Fungi are a treacherous lot, and their physiologic races can be further demarcated into "biotypes" by adding another variety to the standard differentials. Uppal and Gokhale (1947), Gokhale (1950) and Prasada and Sreekantiah (1956) discovered the existence of two biotypes of race 42; one of race 15 and one of race 21 of *P. graminis tritici*, respectively.

A knowledge of all the existing physiologic races is essential before work on breeding for resistance is undertaken. Mehta in 1931 started the work on examining the physiologic races in India which was continued mainly by Vasudeva, Prasada, Gokhale and Uppal.

New races continue coming up in the nature's 'field laboratory' by the genetic recombinations and mutations. Once they arrive, they multiply unchecked, as other races are ineffective on the hitherto resistant host. The resistant host, thus, selects the new race and brings about, what is aptly called, the 'man-guided evolution'. Thus, the plant breeders have to be on the alert.

RUST OF FLAX (LINSEED, *Linum usitatissimum* L.)

Pathogen. *Melampsora lini* (Pers.) Lev.

The disease causes considerable damage to flax all over the world and occurs frequently in epiphytotic proportions in North India. The disease is of great importance also because it was with this host pathogen system that Flor (1956) gave the gene-for-gene concept of disease resistance.

Symptoms

Small, yellow spots, bearing spermogonia, appear very early on the upper surface of young leaves; orange-yellow aecia appear on the lower surface. The disease becomes conspicuous in late December or January when bright orange spots of uredosori, surrounded by necrotic zones, appear on leaves and stems. The leaves die and fall prematurely, hence telia, which appear late, are formed mostly on stems. The brown to black telia appear as crusts.

Morphological Characters of the Pathogen

It is an autoecious rust. Spermogonia are usually flask shaped but occasionally diffuse spermatiophores, without a definite structure, appear. Aecia lack peridium or paraphyses (*Caeoma* type, characteristic of *Melampsora*). Aeciospores are polygonal with verrucose epispore.

Uredia are circular on leaves but elongated on stems. Uredospores are ovate, stalked and intermingled with capitate paraphyses. Telia consist of a palisade layer of brown-black, sessile and cylindrical teliospores. The teliospores germinate only when subjected to freezing temperatures.

Disease Cycle

The disease cycle is depicted in Fig. 22.3.

Teliospores fail to survive the high summer temperatures and thus have no role in the recurrence of the disease. It is believed that uredospores (also teliospores) blown down from hills cause the primary infection on plains of India. The role of the collateral

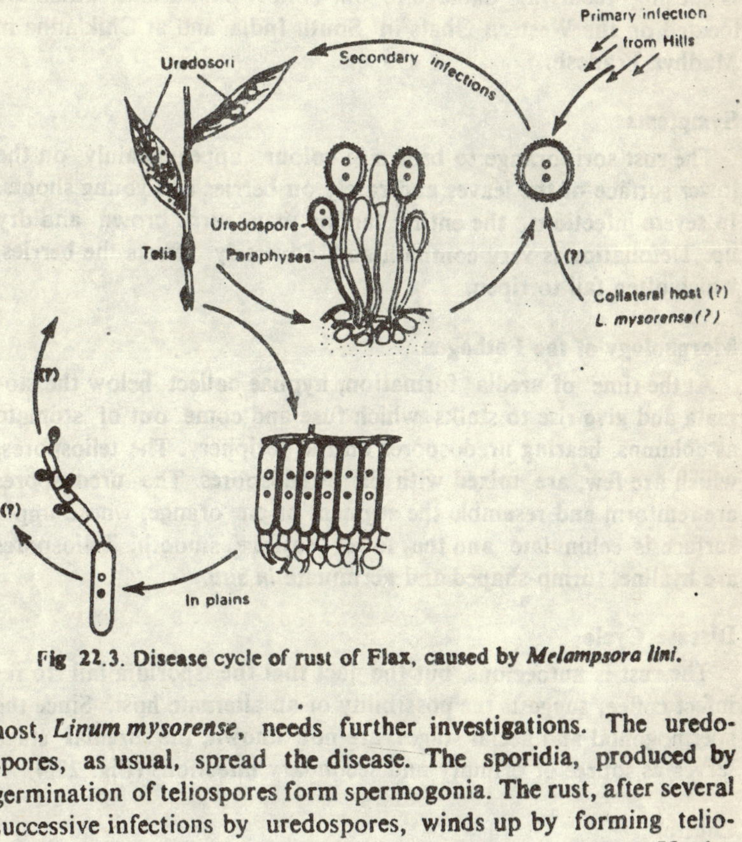

Fig 22.3. Disease cycle of rust of Flax, caused by *Melampsora lini.*

host, *Linum mysorense*, needs further investigations. The uredospores, as usual, spread the disease. The sporidia, produced by germination of teliospores form spermogonia. The rust, after several successive infections by uredospores, winds up by forming teliospores, which on the plains get lost to the hot summer. Uredospores either perpetuate on some collateral host, which is yet not known, or die. Next crop, again, gets the first infections from uredospores or sporidia, blown down from the hills.

Control

By use of resistant varieties; American and Australian varieties are immune to the rust in Indian conditions.

LEAF RUST OF COFFEE (*Coffea arabica* L.)

Pathogen. *Hemileia vastatrix* Berk. and Br.

The disease was first noticed in Ceylon in 1868 where it caused damage on such a big scale that the crop had to be abandoned and replaced by tea. It was reported from India in 1870 and since then is causing recurring damage to our coffee plantations, which are located on the Western Ghats in South India and at Chikladha in Madhya Pradesh.

Symptoms

The rust sori, orange to brown in colour, appear mainly on the lower surface of the leaves and rarely on berries and young shoots. In severe infections, the entire lamina may turn brown and dry up. Defoliation is very common and adversely affects the berries, which often fail to ripen.

Morphology of the Pathogen

At the time of uredia formation, hyphae collect below the stomata and give rise to stalks which fuse and come out of stomata as columns, bearing uredospores on the periphery. The teliospores, which are few, are mixed with the uredospores. The uredospores are reniform and resemble the segment of an orange, whose upper surface is echinulate and the lower surface smooth. Teliospores are hyaline, turnip-shaped and germinate *in situ*.

Disease Cycle

The rust is autoecious, but the fact that the sporidia fail to reinfect coffee, suggests the possibility of an alternate host. Since the spermogonial and aecial stages are not known, the uredial stage serves as source of primary and secondary infections (Fig. 22.4).

Control

1. By spraying Bordeaux mixture (5 per cent) or other copper fungicides.
2. The 'Coorg' and 'Kents' varieties are resistant.
3. Cultivation under shade prevents secondary infections.

NUCLEAR CYCLE IN *H. vastatrix*

Rajendren (1967) reported a new type of nuclear cycle in *H. vastatrix* (Fig. 22.5). He observed that the parasitic hyphae are dikaryotic in the initial stages. The dikaryons fuse to form diploid monokaryotic hyphae in the stromatal hyphae consisting of the

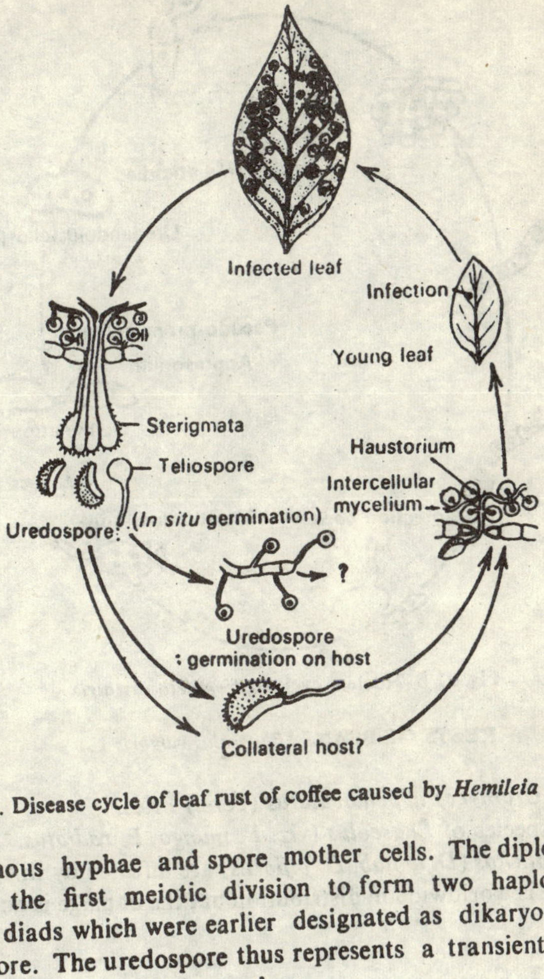

Fig. 22.4. Disease cycle of leaf rust of coffee caused by *Hemileia vastatrix.*

sporogenous hyphae and spore mother cells. The diploid nuclei undergo the first meiotic division to form two haploid nuclei (meiotic diads which were earlier designated as dikaryons) in the uredospore. The uredospore thus represents a transient unseptate haplophase and not a dikaryophase.

The fusion of nuclei and meiosis in rust fungi occur in teliospores. The abnormal meiotic diad may, for convenience, be designated as 'uredenoid teliospore', which on germination, forms a "pseudopromycelium", bearing an inflated tip, called appresosium in which the second meiotic division occurs. Four nuclei are formed. The 'appresorium' forms two peg like projections in each of which a pair of nuclei migrate. The bifid 'pegs' infect the host.

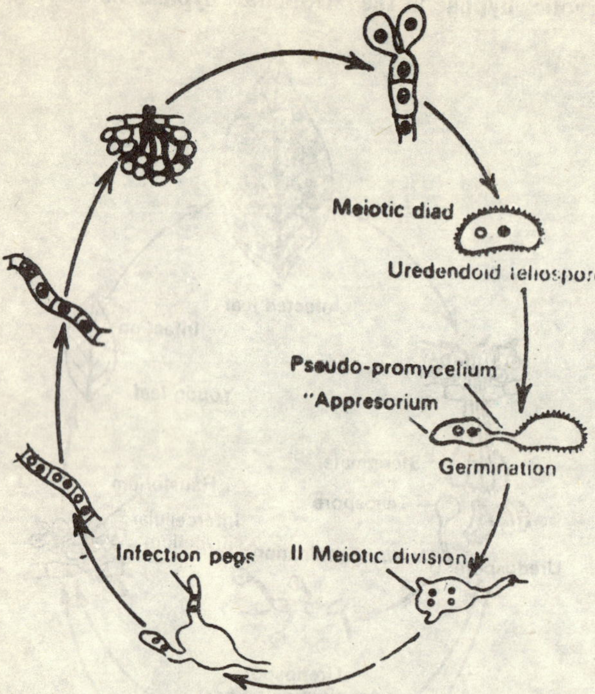

Fig. 22.5. Nuclear cycle in *Hemileia vastatrix*.

RUSTS OF BEANS (*Phaseolus vulgaris* L.)

Pathogen. *Uromyces appendiculatus* (Pers.) Fries.

Several species of *Phaseolus* (viz. *P. mungo*, *P. radiatus*, *P. lunatus*) and *Dolichos* (*D. lablab*, *D. biflorus*) are infected by this fungus. The disease is worldwide in distribution but the damage is not much.

Symptoms

Rust sori appear usually on the leaves and pods. The sori, which are more frequent on the lower surface of the leaves, are small, round, brown and frequently fuse. Telia are linear, dark brown and appear late in the season.

Morphology of the Pathogen

Uredospores are spherical or oval, echinulate, light brown and have two equatorial germpores. Teliospores thickwalled, subglobose or oval, one celled and have a broad apical hyaline papilla.

Disease Cycle

This is an autoecious macrocyclic rust but the spermogonial and aecial stages are rare.

Soil-borne teliospores initiate the primary infection (Fig. 22.6). They germinate to form sporidia which infect young leaves. A dikaryotic mycelium is established and uredosori appear in a week's time. The uredospores also can oversummer and perpetuate the fungus in addition to their major role of causing successive secondary infections.

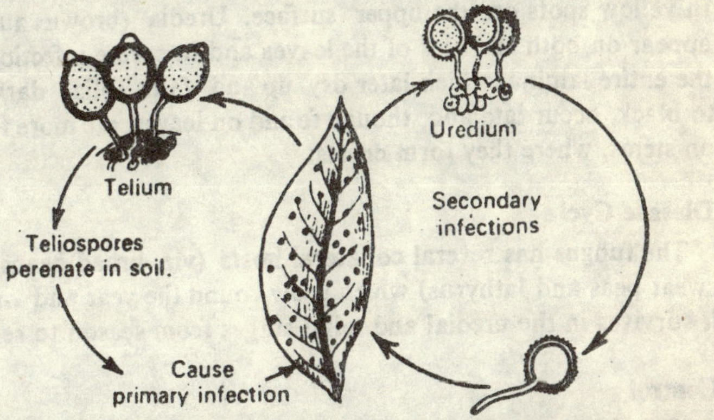

Fig. 22.6. Disease cycle of bean rust caused by *Phaseolus vulgaris*.

Symptoms appear in seven days from the time of inoculation. The rare occurrence of spermogonia and aecia suggests that dikaryotization of primary hyphae, formed by sporidia, may be occurring by anastomoses.

Control

1. Dusting powdered sulphur or spraying Bordeaux mixture before symptom appearance gives good control.
2. Resistant varieties are available but the physiological races of *Uromyces appendiculatus*, which are several, make them susceptible.

RUST OF PEA (*Pisum sativum* L.)

Pathogen. *Uromyces fabae* (Pers.) De Bary.

The disease occurs in several states of India. In North India, symptoms appear first in the month of January. It is a macrocyclic autoecious rust.

Morphological Characters of the Pathogen

Teliospores are subglobose with distinctly thickened apex (in *U. appendiculatus*, it is hyaline at the apex). The stalk persists on the fallen spore. Aeciospores are elliptical, yellowish brown and echinulate.

Symptoms

The aecial cups (with short peridium) appear first on the lower surface of leaves; accompanied by spermogonia which are formed in yellow spots on the upper surface. Uredia (brown and oval), appear on both surfaces of the leaves and in severe infections cover the entire lamina, which later dry up and fall. Telia, dark-brown to black, occur late and, though found on leaves, are more frequent on stems, where they form crusts.

Disease Cycle

The fungus has several collateral hosts (viz. broad beans, lentil, sweat peas and lathyrus) which grow round the year and on which it survives in the uredial and telial stages from season to season.

Control

Use of resistant varieties is urgently needed.

RUST OF GRAM (*Cicer arietinum* L.)

Pathogen. *Uromyces ciceris* (Gron.) Jacz.

The disease is common in North India, as well as, in parts of Maharashtra and Tamil Nadu.

Symptoms

Numerous small, oval, brown uredia appear on both surfaces of leaves; occasionally on stems, floral parts and pods. Telia appear late in the season.

Disease Cycle

The role of teliospores is not known and the annual recurrence may be through uredospores growing on collateral hosts. It has been shown by Saksena and Prasada (1955) and later confirmed by Payak (1962), that the rust perpetuates itself on *Trigonella polycerata*, a leguminous weed. This plant is widely prevalent in the Himalayas upto an altitude of 1,820 metres. Payak (1962) has

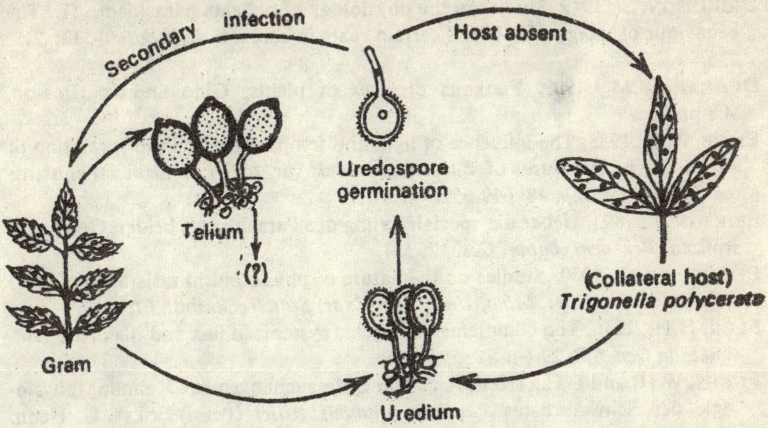

Fig. 22.7. Disease cycle of rust of gram caused by *Uromyces ciceris*.

observed that 'Methi' (*T. foenumgraecum*), which belongs to the same genus, is not susceptible to uredospores of the gram rust or *T. polycerata*.

REFERENCES

ALLEN, P.J., 1957, *Plant Physiology*, 32: 385-389.

ANONYMOUS, 1951, Future of parasitology. *Nature*, 168: 527-529.

ARTHUR, J.C., 1928, Progress of rust studies. *Phytopathology*, 18: 659-674.

BRIAN, P.W., 1967, Obligate parasitism in fungi. *Proc. Roy. Soc.* (London). Ser. B. 168: 101-118.

BROWN, W., 1936, The Physiology of host parasite relations. *Botan. Rev.* 2: 236-281.

BUTLER, E.J., 1903, The Indian wheat rust problem. Deptt. of Agri. India Bull. Nat. Part 1.

CHESTER, K.S., 1946, The nature and prevention of the cereal rusts as exemplified in the leaf rust of wheat. Chronica Botanica Co., Walthum, Mass. p. 269.

CUTTER, V.M., (Jr.) 1960, An axenic culture of *Puccinia malvacearum*. *Bull. Assoc.* Southern Biologists. 2: 26 (Abstract)

DALY, J.M., SAYRE, R.M. and PAZUR, J.H., 1957, The hexose monophosphate shunt as the major respiratory pathway during sporulation of rust of safflower, *Plant Physiol.*, 32: 44-48.

———, and JENSEN, S.G., 1961, Evidence for functional cytochrome oxidase in tissues infected by obligate parasites. *Phytopathology*, **51**: 759-765.

De Bary, A. 1887, Comparative morphology and biology of the fungi, Mycetozoa and Bacteria (Eng. Trans. by Garnsey, H.E.F.; Rev. Balfour, J.B.), Clarendon Press, Oxford.

Dickinson, S., 1949, Studies in the physiology of obligate parasitism. II. The behaviour of the germ tubes of certain rusts in contact. *Ann. Botany*. **13**: 219-236.

Duggar, B.M., 1909, Fungous diseases of plants. Ginn and Co., Boston, 508 pp.

Emge, R.G., 1958, The influence of light and temperature on the formation of infection type structures of *Puccinia graminis* var. *tritici* on artificial substrates *Phytopathology*, **48**: 649-652.

Erikson J., 1894, Ueber die specialisierung des Parasitismus beiden Geteidcrostpilzen. *Ber dent. botan. Ges.* **12**: 331.

Ezekiel, W.N., 1930, Studies on the nature of physiological resistance of *Puccinia graminis tritici. Minn. Univ. Agr. Expt. Stat. Tech. Bull.* **67**: 1-62.

Flor, H.H., 1956, The complementary genic systems in flax and flax rust. *Advances in Genet.* **8**: 29-54.

Fuchs, W.H. and Gaertner, A., 1958, Untersuchungen zur Keimungsphysiologie des Schwarzrostes *Puccinia graminis tritici* (Pers.) Erikss. U. Henn. *Arch. Mikrobiol.* **28**: 303-309.

Gokhale, V.P., 1950, A new biotype of race 15 of *Puccinia graminis tritici. Curr. Sci.* **19**: 214-215.

Gottleib, D. and Garner, J.M., 1946, Rust and phosphorus distribution in wheat leaves. *Phytopathology*, **36**: 557-64.

Heitefuss, R., 1961, Untersuchungen uber den phosphatstoffwechsel von weizenkeimpflanzen unterschiedlicher Resistenz gegen Schwarzrost mit hilfe von P^{32}. *Ber. Deut. Botan. Ges.* **74**: 359-369.

———, and Fuchs, W.H., 1963, Phosphatstoffwechsel and Sauerstoffaufnahme in Weizenkeimpflanzen nach Infektion mit *Puccinia graminis tritici. Phytopath. Z.* **46**: 174-198.

Hotson, H.H. and Cutter, V.M., 1951, The isolation and culture of *Gymnosporangium juniperi virginianae* Schev. *Proc. Natl. Acad. Sci* U.S. **37**: 403-407.

Kein, W.K. and Rohringer, R., 1969, Metabolism of aromatic compounds in healthy and rust infected primary leaves of wheat. III. Studies on the metabolism of Tryptophan. *Can. J. Bot.* **47**: 1425-1433.

Maheshwari, R., Allen, P.J. and Hildebrandt, A.C., 1967, Physical and chemical factors controlling the development of infection structures from uredospore germ tubes of rust fungi *Phytopathology*. **57**: 855-862.

———. Hildebrandt, A.C. and Allen, P.J., 1967 (b), The cytology of infection-structure development in uredospore germ tubes of *Uromyces phaseoli* var. *typica* (Pers.) Wint. *Can. J. Bot.* **45**: 447-450.

Mains, E.B., 1917, The relation of some rusts to the phylogeny of their hosts. *Am. J. Bot.* **4**: 179-220.

Nielsen, J. and Rohringer, R., 1963, *Can. J. Bot.* **41**: 1501.

Pal, B.P., 1936, Effect of brown rust attack on wheat. *Ind. J. Agric. Sci.* **6**: 127-128.

Payak, M.M., 1962, Natural occurrence of gram rust in uredial stage on *Trigonella polycerata* L. in Simla Hills. *Curr. Sci.* **31**: 433-444.

PRASADA, R., 1948, Studies on the formation and germination of teleutospores of rusts in India. *Indian Phytopath.*, **5**: 128-129.

———, and SREEKANTIAH, K.R., 1956, Identification of biotypes within physiological races of wheat stem rust. *Ind. Phytopath.* **9**: 209-211.

RAJENDREN, R.B., 1967, New type of nuclear life cycle in *Hemileia vastatrix*. *Nature*. **213**: 105-106.

RAMAKRISHNAN, L. and Staples. R.C., 1968, Template activity and RNA synthesis from germinating bean rust uredospores. *Phytopathology*. **58**: 1064 (Abstract).

ROHRINGER, R. and HEITEFUSS, R., 1961, Incorporation of P^{32} into Ribonucleic acid of rusted wheat leaves *Can. J. Bot.* **39**: 263-267.

ROWELL, J.B and OLIEN, C.R., 1957, *Phytopathology*. **47**: 650-655.

SAKSENA. H K and PRASADA, R , 1955; *Indian Phytopath.* **8**: 94.

SCOTT, K.J. and MACLEAN, D.J., 1969, Culturing of rust fungi. *Annu. Rev. Phytopath.* **7**: 123-146.

SHAW, M., BROWN, S A. and RUDD JONES, D., 1954, Uptake of radioactive carbon and phosphorus by parasitized leaves. *Nature*. **173**: 768.

———, and SAMBORSKI, D.J., 1957, The physiology of host-parasite relations. III. The pattern of respiration in rusted and mildewed cereal leaves. *Can. J Bot.* **35**: 389-407

STAKMAN, E.C. and LEVINE, M.N., 1922, The determination of biologic forms of *Puccinia graminis* on *triticum* spp. *Tech. Bull. Univ. Minn.* **20**.

STAPLES, R C. and LEDBETTER, M.C., 1960, Incorporation of *triticum* labelled thymidine into nuclei of rusted bean leaves. *Contrib. Boyce Thompson Inst.* **20**: 349.

STOCK, F, 1931, Untersuchungen uber Keimung and Keimschlauchwachstum der Uredosporen einiger Getreideroste. *Phytopath. Z.* **3**: 231-280.

UPPAL, B.N. and GOKHALE, V.P., 1947, A new race of *Puccinia graminis tritici* and two biotypes of race 42. *Curr. Sci* **16**: 61.

WILLIAMS, P.G., SCOTT, K.J. and KUHL. J.L., 1966, Vegetative growth of *Puccinia graminis* f. sp. *tritici in vitro* *Phytopathology*. **56**: 1418-1419.

———, SCOTT, K.J., KUHL, J.L. and MACLEAN, D.J., 1967, Sporulation and pathogenicity of *Puccinia graminis* f. sp. *tritici* grown on an artificial medium. *Phytopathology*. **57**:326-327.

YARWOOD, C.E., 1956, Obligate parasitism. *Annu. Rev. Plant Physiol.* **7**: 115-142.

23

WILTS

THE wilts comprise an important group of plant disease in which the plants suffer from dehydration and show symptoms of drought. The pathogens colonise the non-living xylem vessels; their propagules and metabolites are transported up in the transpiration stream. It is only when the plant becomes moribund that the pathogens invade the xylem parenchyma, medullary rays, phloem and cortex. The bacterial wilt pathogens, however, escape from the vessels earlier and form bacterial 'pockets' in the cortex. The conidia are transported easily in the vessels of angiosperms but fail to do so in the gymnosperms because of the discontinuity and resistance offered by the short and narrow tracheids. The penetration through pits has to be repeated many a times and this is the reason for the absence of wilt diseases in gymnosperms. Host tissues, meant for transport of water, become occupied by the pathogens and the transpiration stream gets polluted with their metabolites. Growth regulators, toxins, polysaccharides, glycopeptides, hydrolytic enzymes etc., begin to flow in the vessels, and become detrimental to the plant.

The Pathogens

The causal agents of vascular wilt diseases may be bacteria, fungi and rarely viruses. Organisms that bring about wilting symptoms by root damage i.e. the root rot pathogens and the nematodes are being excluded from this discussion. There are some examples of viruses causing wilt e.g. tobacco etch virus causing wilt of tobasco pepper, but as a group they do not cause wilt symptoms. Insects and angiosperms (witchweeds, broomrapes) are also excluded as they are not vascular pathogens and do not produce any of the wilt symptoms. Thus, bacteria and fungi are the two important groups of vascular wilt pathogens. *Stereum purpureum*, causing plum silver leaf disease, because of its capacity to invade ray parenchyma and cortex cells, can find better treatment as a "wood rotting fungus".

Among the bacterial wilt pathogens, ***Pseudomonas solanacearum*** causes wilt of important plants like potato, tomato, egg plant, groundnut and banana. The virulent and avirulent strains of ***P. solanacearum*** can be distinguished in culture by their colony characters. Buddenhagen and Kelman (1964) delimited this species into three races. Race 1 has a wide host range while race 2 and 3 are restricted to few plants. ***Xanthomonas***, ***Corynebacterium***, and ***Erwinia*** are the other bacterial genera that cause wilts.

The genera of fungi, though few, are very destructive wilt-pathogens. These are ***Ceratocystis***, ***Fusarium*** and ***Verticillium***. ***Ceratocystis*** is similar to ***Stereum*** in its invasion of cortex but is retained here because of the importance of the Dutch elm disease, that it causes. Thus ***Fusarium*** and ***Verticillium*** are true vascular pathogens.

The genus ***Fusarium*** (Fig. 23.1) is a very successful 'soil inhabitant' and once established, persists for several years, rendering the soil unfit for profitable crop production. Taxonomically, ***Fusarium*** is a difficult genus. The classification of Wollenweber and Reinking (1935) and Snyder and Hansen (1940) are in use. Snyder and Hansen have grouped all vascular wilt fusaria into one species, viz. ***F oxysporum*** comprising several ***formae specialis***, which, in the absence of any morphological differences, are named on the basis of their specialized hosts.

Thus, ***F. oxysporum*** f sp ***vasinfectum*** infects cotton;
" " ***lycopersici*** " tomato;
" " ***cubense*** " banana;
" " ***lini*** " flax.

Many of these ***formae*** further comprise of races which are restricted to particular cultivar of the host ***F oxysporum*** f sp ***pisi*** has four such races. Delimitation of races on the basis of their pathogenicity does not find support from Subramanian (1971) specially because of their facultative saprophytic nature. Several fusaria have a wider host range than actually known and, then, pathogenicity is a highly variable character, modified by so many and often little understood factors. Species of ***Fusarium*** are more prevalent in warmer and tropical countries and in soils with acidic pH.

The genus ***Verticillium***, unlike ***Fusarium***, is found in cooler, temperate parts of the world and is favoured by an alkaline soil reaction. It is a 'soil invader', with little saprophytic ability. The fungus causes wilts in wide range of plants like cotton, tomato potato, egg plant, lucerne, hop, mint, apricot, peach, strawberry, raspberry, flax, and several weeds.

Verticillium albo-atrum and *V. dahliae* are the two important species which are easily identifiable by their resting bodies viz, "resting mycelium" in *V. albo-atrum* (Fig.23.2) and "microsclerotia" in *V. dahliae* (Fig. 23.3). The two less pathogenic species are *V. nigriscens* and *V. tricorpus*. The various pathogenic species and their distinguishing features are given in Table 23.1.

TABLE 23.1. SHOWING DIAGNOSTIC FEATURES OF PATHOGENIC SPECIES OF *Verticillium*

Distinguishing features	*V. albo-atrum*	*V. dahliae*	*V. nigriscens*	*V. nubilum*	*V. tricorpus*
Resting bodies	Resting mycelium	Micro-sclerotia	Chlamydospores small (7-10μ)	Chalmydospores bigger (8.5-15.5μ)	Resting mycelium, Microscelerotia and Chalmydospores. (reported from UK only)
Growth in culture at 30°C	Nil	Fairly good	Fairly good	Nil	Fair
Reaction of host	Very rapid	Rapid	Fairly rapid	Slow	Slow

Most of the American plant pathologists refer all the pathogenic verticillia as *V. albo-atrum* (as "resting mycelial" or "sclerotial" forms). However, British workers (Isaac, 194?) prefer to designate them as distinct species.

Symptoms

The most common wilt symptoms are epinasty of leaves, petioles and branches; vein clearing and chlorosis of leaves followed by browning and defoliation; vascular browning; formation of gums and gels, tyloses; stunting of plants and development of adventitious roots. Considerable variation in symptoms is encountered and one or more of the above symptoms may be present. The most pronounced symptom is 'epinasty' (Fig. 23.4). The petioles bend downwards, forming an obtuse angle with the main stem. It may, however, be pointed out that this most important symptom, does not occur in *Verticillium*-wilt of hop and egg plant. Leaves only become necrotic and brittle.

The symptoms (Vein clearing, yellowing or wilting of leaves) develop in acropetal order. In herbaceous plants, with elongated axis, symptoms first appear on the lower, oldest leaves and develop progressively upwards. In woody perennials, the outermost leaves show first symptoms which progress inwards. Sometimes, symptoms do not develop uniformly and only one side of the plant is affected. This may happen even in individual leaves and chlorosis may be on one side only, the other half remaining normal even upto the time of leaf fall. This happens in verticillial wilt of hop. Such unilateral development of symptoms is not known in physiological wilting. Infection of roots only on one side and the consequent confinement of the pathogen on that side only (due to restricted lateral movement). may be the reason of such symptoms. Interveinal necrosis and chlorosis are mineral deficiency symptoms and also the result of chlorophyll degradation.

Browning of vascular elements, is another pronounced symptom, which may be confined only to lower reaches of the stem or may extend through the whole length of the stem including petioles and veins of the leaves. The discolouration is due to deposition of melanin pigments in the wails and lumina of the vessels. The melanin pigments are formed by oxidation and polymerization of phenols of the host.

Gels and gums are formed by breakdown products of the pectin of the middle lamella due to macerative action of pectic enzymes. Tyloses are formed by the extensions of xylem parenchyma, which also liberate gums in the vessels. Excessive tylosing imparts resistance against the disease and also causes death of the plant. The speed of formation of tyloses is an important factor in determining resistance.

Stunting of plants, reported in almost all verticillial wilts, either precedes or follows chlorosis. Shorter internodes cause stunting. Adventitious roots are often formed in infected plants. Perennial plants often show recovery from disease in the next season. This is due to death or "burial" of the pathogen in the wood of the previous year by excessive tylosis and gumosis. The various wilt symptoms and their probable causes are presented in Table 23 2.

TABLE 23.2. WILT SYMPTOMS AND THEIR CAUSES

Symptoms	*Cause(s)*
External symptoms	
Wilting	Impediment of water flow due to occlusion of vessels by polysaccharides, slime, gels, gums, tyloses and wefts of hyphae and bacterial masses.
Yellowing	Breakdown of chlorophyll due to deficiency of minerals and water plus action of host and pathogen metabolites.
Leaf epinasty	Ethylene and increased level of IAA.
Marginal necrosis of leaves	Decreased supply of water or probable action of toxins.
Adventitious roots	Increase in IAA.
Stunting	?
Internal symptoms	
Vascular discolouration	Melanoid pigments, formed by oxidation and polymerization of host phenols.
Tyloses	Increase in IAA level
Gums	" "
Vessel collapse	" "
Dissolution of pectic substances in middle lamella	Pectic enzymes
Gels and gums	Breakdown products of pectin and primary cell wall.
Degradation of cellulose	Cellulolytic enzymes.

ESTABLISHMENT OF INFECTION

The 'host-parasite relationship' starts when the pathogen invades and starts drawing nutrition from the host. This is called establishment of infection (Gäumann, 1950) and is governed by factors outside the host. Wilt pathogens are mostly soil-borne. Mycostasis (fungistasis), known universally in surface soils, which inhibits the germination of resting bodies (chlamydospores, resting mycelium or microsclerotia), is overcome by plant root exudates of the rhizosphere. This stimulation of germination brings conditions favourable for infection. However, it can not explain the specificity of parasitism, and the key, possibly, lies in the inhibitors, released from resistant plants, rather than selective nutrient. Some good evidences (Timonin 1941, Buxton 1957, 1962) are available in this direction. Timonin found that flax varieties, resistant to fusarial

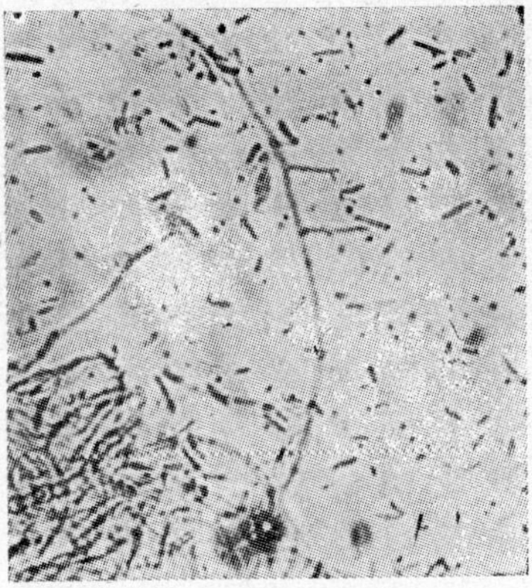

Fig. 23.1. Shows macro (elongated) and microcondia (small spherical) of *Fusarium moniliformis*.

(A)

Fig. 23.2. (*A*) *Verticillium albo-atrum*, showing resting mycelium, verticillate conidiophores and hyaline conidia.

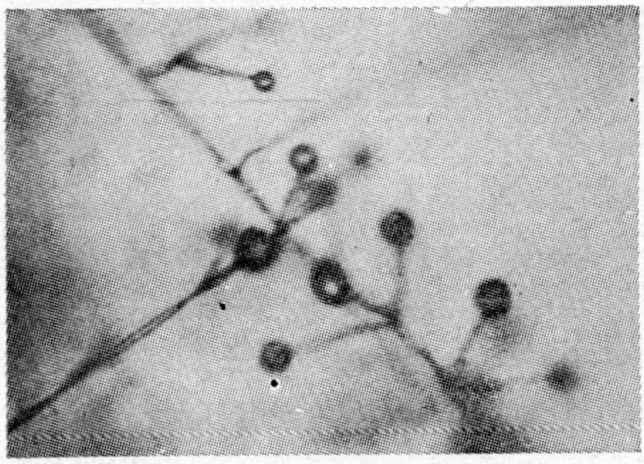

(B)

Fig. 23.2. (*B*) *Verticillium albo-atrum*, showing clusters of conidia at the tips of phialides.

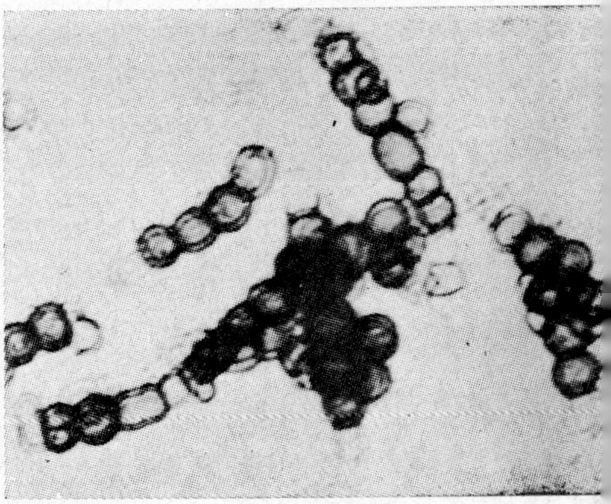

Fig. 23.3. Microsclerotia of *Verticillium dahliae*.

Fig. 23.4. An egg-plant showing typical wilt symptoms.

wilt, excreted HCN into the rhizosphere which depressed the growth of the fungus and this made the host resistant. Buxton reported exudation of inhibitory substances by varieties of pea and banana, resistant to fusarial wilt. An inhibitor of *Verticillium albo atrum* was found in the seed coat of cotton, (Garber and Houston, 1959). These reports point out that biological depression or retardation, rather than 'biological stimulation' of germination and growth, may hold the key to the secret of host-pathogen specificity. Thus, the root exudates of the rhizosphere possibly play a decisive role during the pre-penetration stage of the disease. Nothing is known about the prepenetration factors in branch invaders like the Dutch elm disease fungus. Penetration of roots may be direct or through wounds, caused by emergence of lateral roots or more importantly by nematodes. The 'predisposition', brought about by wounds caused by plant parasitic nematodes, goes a long way to determine the host's susceptibility. Plant parasitic nematodes, in association with fungal parasites, establish "plant disease complexes". In these "complexes" the pathogenic strains of the nematode predispose the otherwise resistant host. Sidhu and Webster (1974) reported that in the disease complex of tomato, caused by the two pathogens (*Fusarium oxysporum* f. sp. *lycopersici* and the root-knot nematode, *Meloidogyne incognita*) the nematode breaks the plant's resistance to the fungus. The branch invaders are typical wound parasites. *Xanthomonas vasculorum* invades sugarcane through wounds in the leaves. During and after actual penetration, resistance mechanisms of the host play decisive role. In compatible host-pathogen combinations, the pathogen advances in the root cortex with ease. The growth may be inter or intracellular or both. Eventually, the pathogen colonises the vasculas tract. Growth in cortex comprises the real parasitic phase of the pathogen, as the life in the dead vessels is not strictly parasitic. However, bacterial wilt pathogens (not wilt fungi) escape from the xylem into the surrounding tissues (xylem parenchyma, pith and cortex) and form bacterial pockets therein. Root injury, though known in certain tree wilts, is negligible. The pathogens show adaptations to the environment of the xylem vessels. They grow and reproduce in flowing dilute nutrient solution under conditions of oxygen deficiency.

MECHANISM OF WILTING

It is agreed that water shortage is the cause of wilting but how this is brought about, is the point of debate and controversy. Excellent reviews on this aspect have been contributed by Dimond (1955, 1970), Sadasivan (1961), Saraswathi Devi (1964) and Talboys (1970).

Does the plant express wilting because: (*i*) its roots cannot absorb enough water, (*ii*) because the stem cannot transport the absorbed water to the leaves with the required speed or (*iii*) because the leaves cannot retain their share of water and lose them excessively? It has been demonstrated conclusively that the infected plants absorb adequate amount of water and thus impaired absorption is not the cause. However, the other two possibilities have been implicated and constitute the well known "plugging" and "toxin" theories of wilts.

THE PLUGGING THEORY (=Occlusion theory)

This explains the water shortage and the accompanying wilting to be the result of retarded rate of water movement in the xylem vessels. The vascular tissues get choked due to occlusion of several types. Support for occlusion theory is based on the following facts:

(*i*) reduced rate of water flow in vessels, (*ii*) presence of obstructions in the vessels, and (*iii*) reduced rate of transpiration in infected plants.

Reduced Rate of Water Flow

Without exception, resistance to water flow is increased markedly in welting plants. Melhus *et al.* (1942) showed 88% and 75% reduction in water transport in *Fusarium*-infected cabbage and alfaalfa respectively. Similar were the observations of Harris (1940) for bacterial wilts of cucumber and corn as well as those of Beckman *et al.* (1953) for Oak wilt and Dimond and Waggoner (1953b) for fusarial wilts of tomato. The reason for this retarded flow is the plugging of vessels.

Presence of Obstructions in the Vessels

In most of the wilt diseases vascular occlusions are prominent. The occlusions may be due to:

(*i*) physical presence of the pathogen,
(*ii*) metabolites produced by the pathogen (slime, and other polysaccharides),
(*iii*) products of host tissue degradation by the enzymic action of the pathogen (gels, gums),
(*iv*) pathogen-induced host responses (tyloses, gums).

Each of these can either cause or contribute to the impediment of water flow.

Physical Presence of the Pathogen

The presence of wefts of hyphae and slimy bacterial masses in the vessels was ascribed as the cause of impeded water flow by Stewart (1897), Smith (1911) and Grieve (1941). However, the adequacy of such blockage, causing all the symptoms of wilting plant, is ruled out by Hutchinson (1913), Ludwig (1952) and Waggoner and Dimond (1954). However, in recent years its significance has been reemphasized by Matta and Dimond (1963), who found good correlation between mycelial blockage and symptom development in fusarial wilt of tomato. The significance of bacterial masses is already conceded. Thus, physical blockage by presence of pathogens merits consideration as a contributory factor to reduced water flow.

Metabolic Products as the Cause of Impeded Water Flow

Polysaccharides. Presence of polysaccharides and other large molecules, such as glycopeptides, has been demonstrated in the xylem vessels. Wilting occurs if such substances are introduced in the vascular tissues. These substances act in two ways. High molecular weight substances block the pit membranes and cause blockage of water movement from roots to stems Wilting of stems and petioles occurs by such blockage. On the other hand smaller molecules get transported to the leaves where they block the small terminal ends of the vascular elements, especially at the leaf periphery. This causes the leaflet or leaf edge wilting. Thus, the activity of the polysaccharides is related to their molecular size rather than to any functional groups. Hodgson *et al.* (1949) examined the wilt inducing capacity of several natural and synthetic polysaccharides on tomato plants and concluded that: (*i*) the toxicity increased with an increase in the molecular weight, and (*ii*) the toxicity was not specific and could be caused by a variety of substances. Polysaccharides have been implicated with pathogenesis in Dutch elm disease, oak wilt, maple verticillial wilt and bacterial wilts caused by *Pseudomonas solanacearum.* Husain and Kelman (1958b) consider the increased viscosity, caused by the polysaccharides, rather than physical blockage, as the mechanism of action in tomato wilt. This is, however, disputed by Dimond (1970).

Husain and Kelman (1958a) working with tomato wilt found that the virulence of *P. solanacearum* was conditioned by slime production. Strongly virulent forms produced copious amounts of slime, which increased the viscosity of the transpiration stream. This resulted in retarded water flow. Slime derived from infected plants, as well as from bacterial cultures, caused wilt symptoms even after autoclaving. This confirms slime as the cause of wilting. Slime production has been linked to pathogenesis in many other bacterial wilt diseases such as wilt of bean caused by *Xanthomonas phaseoli* (Corey and Starr, 1957), wilt of alfa-alfa and wilt of potato caused by *Corynebacterium sepedonicum* (Spencer and Gorin, 1961).

Products of Host Tissue Degradation

Wilt inducing bacteria and fungi produce pectic and cellulolytic enzymes. The products of their action form vascular plugs. Rufianic acid, which inhibits both polygalacturonase and pectinmethyl esterase, when applied to infected plants, reduced severity of symptoms in fusarial wilt of tomato (Grossmann, 1962).

Gels, gums. The pectic enzymes attack and macerate the middle lamella at the pits where they are exposed. This results in the liberation of pectic acid and other products of pectic degradation. These products combine with metals to form gels. The pectic nature of the gels is indicated by their being stained with ruthenium red. Gums are formed by the reaction of these gels with melanoid pigments formed in the browning reaction.

Browning of vessels. Maceration affects the xylem parenchyma cells and the phenols contained in these cells as harmless glucosides (phenols bound to glucose), are freed. The liberated phenols are oxidised by the phenol-oxidase enzymes of the host/pathogen, and subsequent polymerization of the oxidation products results in the formation of brown melanoid pigments. Some phenols are also withdrawn from the lignin synthetic pathway. The brown pigments get deposited in the walls and lumina of the vessels and cause browning of the vascular tissue of wilted plants. Mace (1963) reported the occurrence of free dopamine, a phenol, in the xylem parenchyma of banana roots. Infection with *Fusarium* results in browning of these cells by oxidation and polymerization of dopamine. The pigment later leaks into the vessels. Dopamine is known to inhibit IAA oxidase and probably during its oxidation contributes to additional IAA synthesis.

Thus, the pectic enzymes start a chain of events leading to formation of gels, gums and brown pigments, all contributing to vascular plugging. The role of cellulolytic enzymes is less studied. Several wilt pathogens produce cellulases inductively, *in vitro* and some *in vivo* but their role is yet to be elucidated.

Nevertheless, the production of pectic enzymes is not a *sine qua non* for wilting. Pectic and cellulolytic enzymes, though produced by Dutch elm fungus *in vitro*, has no role in pathogenesis (Husain and Dimond, 1958). McDonnell (1958) found a pathogenic strain of *Fusarium*, causing wilt of tomato, to be a non-producer of enzymes. Kamal and Wood (1955) reported that pectinase-containing filtrate of *V. albo-atrum* caused wilt symptoms in cotton even after autoclaving. All these reports rule out any involvement of pectic enzymes in these wilts.

Vessel collapse. The hyperplastic growth of xylem parenchyma cells exerts pressure on the vessels, which get crushed. This has been seen in fusarial and verticillial wilt of tomato and verticillial wilt of hop. Increased IAA level in diseased tissues is related to the hyperplastic growth.

Tyloses and Gums. 'Tylosis' and 'Gummosis' ensue as a general host response aimed at restricting invasion by the pathogen. Infection sets in motion metabolic processes in xylem parenchyma that lead to tylose and gum formation. Gums formed in these cells mainly comprise glucouronic acid, hexose and pentose sugars and are different from the gums formed by enzymic maceration, described earlier. These gums after formation are released into the vessels. Tyloses are extensions of pit membranes, which extend as balloons into the xylem vessels and may block them completely. Tylosis is induced by IAA and related hormones. (Beckman, 1966; Pegg and Selman, 1959). Since tylosis and gummosis go together, a similar induction mechanism may be responsible for gummosis also. The formation of tyloses and gums is determined by the pore size. If pits are larger, tyloses are formed; if small, gum is produced.

Reduced Transpiration

Infected plants show a reduced rate of transpiration. This is in contrast to other diseases (e.g. mildews, rusts), where increased transpiration is a common feature. In *Fusarium* wilt of tomato, plants were observed to have closed stomata during day time (Dimond and Waggoner, 1953), a feature common in drought-hardened plants. Reduced rate of transpiration, obviously, results from decreased water supply to the leaves, which in turn, is caused by resistance to flow offered by occlusions in the vessels. Beckman *et al.* (1962) with the help of highly sophisticated instrumentation, monitored transpiration and CO_2 absorption from a limited area of banana leaf infected with *P. solanacearum* and found that the transpiration rate started declining much early and reached the lowest ebb by the time the symptoms appeared. Apparent photosynthesis, as measured by CO_2 uptake, too was limited but was restored to normal by increasing ambient humidity. This indicated that physiological damage had not been done, as would happen had a toxin been involved. The observations suggest that vascular occlusion, rather than toxins, caused the symptoms.

TOXIN THEORY

In contrast to occlusion theory that explains water shortage due to obstruction to flow of water, Gäumann (1951) advanced the toxin theory, which abscribes wilting to toxins. The toxin theory envisages that low molecular weight substances (toxins), produced by wilt pathogens in roots, are translocated to the leaves where they adversely affect the semipermeable membranes. The leaves lose power of water retention; excessive loss of water occurs due to breakdown of the osmoregulation. The heavy loss of ions, mainly potassium, causes an 'ionic imbalance'. There is every evidence to show that the ionic derangements and water loss are brought about by low molecular weight toxins. Toxic activity of xylem fluid of wilt infected plants have been shown by Thatcher (1942) and Gottlieb (1944). Lycomarasmin and fusaric acid were detected *in vitro* by Gäumann and his coworkers but their involvement was not taken seriously until detected *in vivo*. Fusaric acid was *in vivo* detected in fusarial wilts of cotton (Lakshminarayanan and Subramanian, 1955), tomato (Kluepfel, 1957), banana (Page, 1959) and watermelon (Nishimura, 1962). But these toxins do not fall in category of pathotoxins which replace the pathogen completely and produce all the symptoms of the disease syndrome.

The viral wilt of tobasco pepper, caused by tobacco etch virus, has been shown (Ghabrial and Pirone, 1966) to be due to excessive loss of electrolytes from the roots, 24-26 hours before the wilt symptoms appear.

Wilt Toxins*

Lycomarasmine

Its action is non-specific and the symptoms do not exactly compare with that of natural infection. Dimond and Waggoner (1953a) discount the role of this toxin because it is a product of lysis rather than of actively growing mycelium. It is not detected *in vivo*.

Fusaric acid is reported *in vivo* in banana, cotton, tomato and watermelon infected by *Fusarium*. It also produces only atypical symptoms.

Ethylene

It is produced in cultures of several wilt-causing pathogens and has been detected in wilted plants also. Dimond and Waggoner (1953b) and Buddenhager and Kelman (1964) assign epinasty to ethylene. The physiological action of ethylene is closely linked to IAA and other auxins, which are produced by several wilt pathogens.

The fact that symptoms produced by toxin treatment do not correspond well with natural symptoms, does not exclude their involvement in pathogenesis. The atypical symptoms could be due to the dissimilar conditions obtaining during natural infection and under experiments with toxins. The absence of water stress and uniform supply of toxins at higher concentrations would make much difference. The different behaviour of toxins should be seen in light of the findings of Keyworth (1963) in his grafting experiments.

Keyworth (1963) grafted resistant and susceptible scions of tomato on infected susceptible root stocks. When the susceptible scion was grafted over susceptible root stock, the pathogen invaded the scion and typical symptoms of tomato wilt appeared. However, in case when resistant scion was grafted over the infected root stock, the pathogen failed to colonise the resistant scion and the symptoms that appeared were exactly similar to those produced by toxin treatment (i.e. browning of vessels, leaf-yellowing and necrosis but no epinasty). Keyworth (1964) concluded that

*For details see chapter on Toxins.

fungal metabolites acting in advance in resistant scions were toxins formed in susceptible root stocks. He, however, did not identify the toxins.

The grafting experiments provide evidence of *in vivo* toxin-production. Susceptibility to fungal invasion resulted in normal symptoms. Atypical symptoms were due to susceptibility to pathogen's metabolites in the absence of colonisation. The two situations can be designated as pathogenic wilting and toxigenic wilting respectively.

The involvement of toxin is not supported by the transpirational study of the infected plants. The reduced rate of transpiration cannot be explained by toxin action. Against this, the physical blockage theory is supported by the fact that occluded stems when cut and put in water regain turgor and wilting is undone. This is not possible under toxaemia.

Polysaccharides have been regarded by some as toxins acting, of course, *via* physical processes, discussed earlier. It is at this point that the two mechanisms of wilting converge.

Sadasivan (1971) points out "It would be interesting to obtain induced mutant strains of the pathogen that would produce the toxin fusaric acid and not the pectic enzymes and *vice versa*, strains that could produce the enzymes but not the toxin."

On the basis of evidences at hand, the reduction in water flow, brought about by occlusion of the vessels rather than toxins, seems to be the major cause of pathological wilting of plants.

DUTCH ELM DISEASE

Pathogen. *Ceratocystis ulmi* (Buisman) Mereau.

This is a wilt disease of historical importance, which also exemplifies that pathogens, when introduced to new areas, become more destructive than in their natural habitat. The disease was reported in 1919 from Holland and soon spread to Central Europe. By 1935, one third elm trees in Britain had to be felled because of the wilt. In USA, the disease appeared in 1930 and within 8 years, 25,000 trees were killed. Even now the disease is feared in USA and other places.

Symptoms

In trees, wilting appears in one or two twigs. The leaves turn yellow to brown, shrivel, and fall. The branches curve down, showing 'dieback' of twigs. Internal symptoms are browning of the outer sap wood, ray and parenchyma cells and occlusion of vessels with tyloses and gums. The trees often overcome the disease in succeeding seasons by burial of the pathogen in older annual rings. This is called natural recovery.

Disease Cycle

The fungus over winters on elm logs in tunnels made by beetles for laying eggs (Fig. 23.5). The fungus grows saprophytically and forms *Graphium* type conidia, which are carried by beetles to elm trees on whose sap they feed. The insect vectors are .two species of *Scolytus*. In xylem vessels, the fungus produces small conidial 'buds', which move up to the petioles and veins in the leaves, and cause extensive invasion of the vessels. In this way, multiple infections result and susceptible trees die

Mechanism of Wilting

Obstruction of vessels by gums and tyloses and the consequent impairment of water flow causes water shortage and wilting in leaves. Husain and Dimond (1958) concluded that pectic enzymes are not involved in this disease.

Control Measures

(1) Use of resistant varieties.
(2) Elimination of insect vectors indirectly, by burning diseased logs, or directly by spraying insecticides.
(3) Spread through roots can be blocked by digging trenches between trees.
(4) By check on movement of diseased elm logs.

Wilt of Pigeon Pea (*Cajanas cajan*, Blackgram, Arhar)

Pathogen. *Fusarium oxysporum* f. sp. *udum* (Butler) Snyder and Hansen.

This is the most destructive disease of pigeon pea and does considerable damage throughout India.

Symptoms

Yellowing of leaves is the first symptom and appears on 2 to 3 months old plants. The leaves later turn brown and fall. Internal browning is traceable up to the top of the shoot. Occasionally, wilting is confined to one side (partial wilting), and in such cases, browning of vascular tissues too appears on the same side. Mild infections result in stunting.

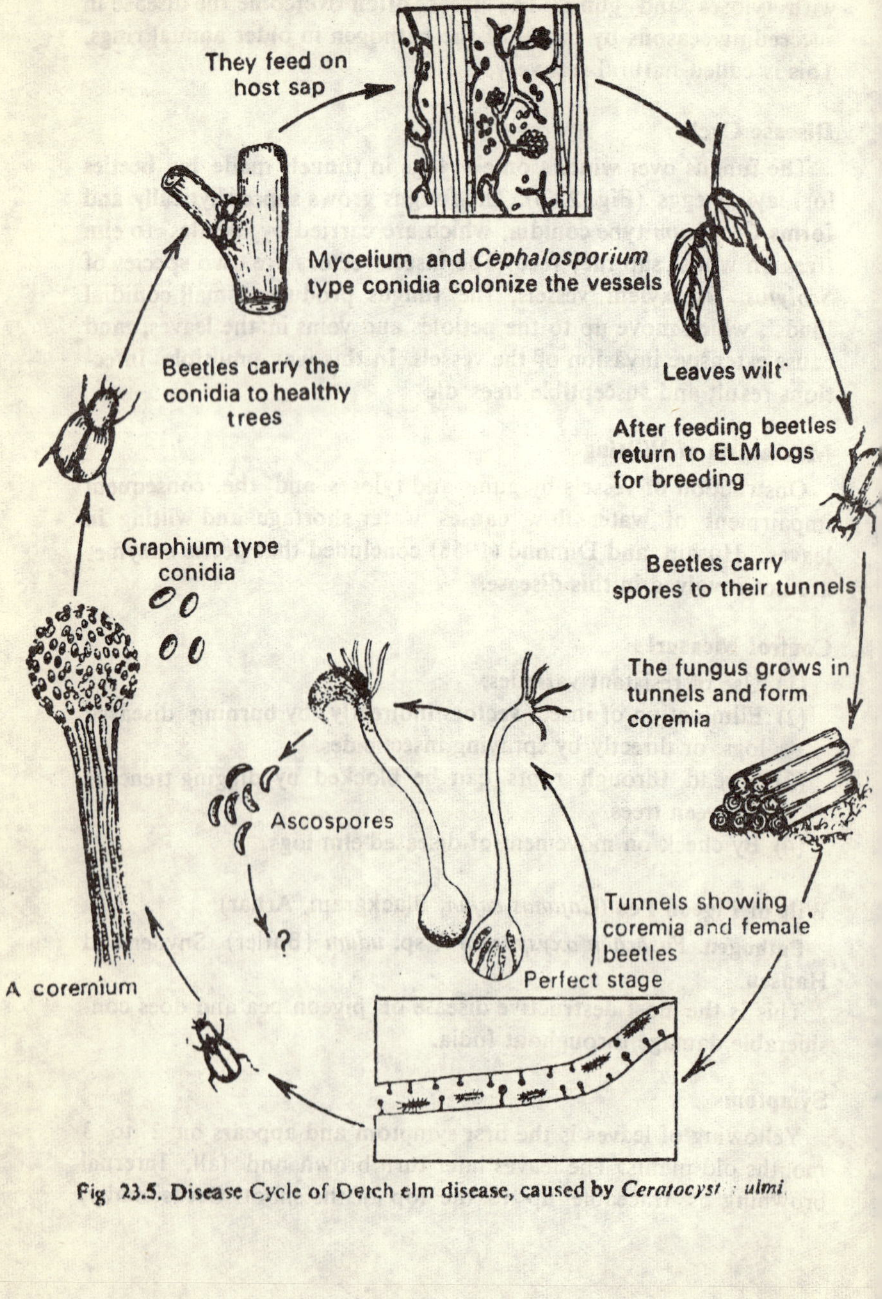

Fig 23.5. Disease Cycle of Detch elm disease, caused by *Ceratocyst : ulmi*.

Disease Cycle

Infection occurs through roots. The fungus grows saprophytically in soil and on availability of the host, embarks on parasitic mode of life. The seeds are not reported to carry the propagules. After death of the host, the fungus moves back into the soil.

Control

Use of resistant varieties; NP 15 and 38 are less susceptible.

WILT OF COTTON

Fusarial Wilt

Pathogen. *Fusarium oxysporum* f. sp. *vasinfectum* (Atk.) Snyder and Hansen.

The fusarial wilt of cotton occurs in several states of India especially Maharashtra, Madhya Pradesh, Rajasthan and Tamil Nadu. The disease is more prevalent in black, acidic soils; alkaline soil-reaction inhibits the fungus.

Symptoms

Plants of any age may be infected and show symptoms. In seedling infections, cotyledonary leaves show yellowing, turn brown, shrivel and fall. Other leaves behave similarly. Older plants show stunting followed by yellowing and wilting of leaves. Internal browning extends throughout the plant. Losses result from poor stands, stunting, small bolls and the poor quality of lint.

Disease Cycle

In addition to usual infection of roots by propagules, present in soil, cotton seeds carry the fungus internally and introduce the disease in distant areas. Infection of roots is aggravated by nematodes, which make openings in the roots for easy penetration. The fungus is known to persist in soil in absence of the host for many years (25 years) and is to be really feared.

Control

(1) By eliminating the nematodes through fumigants, the disease can be limited.

(2) Cultivation in alkaline soil and use of resistant cultivars.

(3) Use of potassium fertilizers in excess controls the disease.

Verticillium Wilt of Cotton

Pathogen. *Verticillium dahliae* Kleb.

This wilt has been reported recently from Coimbatore (Natarajan *et al*, 1968) and Rajasthan (Gour and Dube, 1974). Srinivasan (1971) reported serious losses (70-90 per cent) in seed cotton. The disease needs serious attention in India.

Symptoms

The symptoms include marginal flagging, irregular yellow patches, browning of the veins, cupping, wilting, drying up and falling of leaves.

Disease Cycle

The fungus survives on host debris and infects hosts on 'root contact'. The pathogen infects 'egg plant' also. The fungus, like any other *Verticillium* wilt pathogen, has poor saprophytic ability.

Control

Experiments by Natarajan and Ramakrishnan (1971) showed that percentage and intensity of disease was reduced by deep ploughing, burning trash, drenching wettable Ceresan and spraying potassium chloride. Search for resistant varieties is on. Cotton plants, resistant to *Verticillium*, show increased production of "Gossypol", a phenol. Its role as a 'phytoalexin' has been suggested by Bell (1969).

TOMATO

Tomato wilts are caused by *Fusarium*, *Verticillium* and *Pseudomonas*. The host-parasite relationship in all these wilts has been much studied.

Verticillium Wilt

Pathogen. *Verticillium albo-atrum*.

All the five pathogenic species of *Verticillium* attack tomato plants but only *V albo-atrum* and *V. dahliae* cause serious losses (Isaac, 1962).

Symptoms

Small patches appear on the lower surface of leaves. The leaf margins roll upward along the mid rib, exposing the lower surface (cupping), leaves later become flaccid and fall. Fruits remain small and ripe prematurely, but 'fruit drop' is rare. Other usual wilting symptoms present.

Disease Cycle

As in cotton wilt.

REFERENCES

BECKMAN, C.H., 1966, "Cell irritability and localization of vascular infections in plants", *Phytopathology*, **56**: 821.

BECKMAN, C.H., KUNTZ, J.E., RIKER, A.J., and BERBEE, J.C., 1953, "Host responses associated with the development of Oak wilt," *Phytopathology*, **43**: 448.

BECKMAN, C.H., BRUN, W.A. and BUDDENHAGEN, I.W., 1962, "Water relations in banana plants infected with *Pseudomonas solanacearum*," *Phytopathology*, **52**: 1144.

BELL, A.A., 1969, "Phytoalexin production and *Verticillium* wilt resistance in cotton," *Phytopathology*, **59**: 1119.

BERGESON, G.B., 1972, "Concepts of nematode-fungus associations in plant disease complexes: A Review," *Exptl. Parasit.*, **32**: 301-314.

BUDDENHAGEN, I. and KELMAN, A., 1964, "Biological and physiological aspects of bacterial wilt caused by *Pseudomonas solanacearum*," *Annu. Rev. Phytopath.*, **2**: 203-230.

BUXTON, E.W., 1957 "Differential rhizosphere effects of three pea cultivars on three physiologic races of *Fusarium oxysporum* f *pisi*." *Trans, Brit. Mycol. Soci.*, **40**: 305-317.

———, 1962, "Root exudates from banana and their relationship to strains of the *Fusarium* causing Panama wilt," *Ann. appl. Biol.*, **50**: 269-282.

COREY, R.R. and STARK, M.P., 1957, "Colony types of *Xanthomonas phaseoli*." *J. Bacteriol*, **74**: 137.

DIMOND, A.E., 1955, "Pathogenesis in wilt diseases," *Annu. Rev. Plant. Physiol*, **6**: 329.

———, 1970, "Biophysics and biochemistry of the vascular wilt syndrome. *Annu Rev. Phytopath.*, **8**: 301-322.

———, and WAGGONER, P.E., 1953 (a) "The physiology of lycomarasmine production by *Fusarium oxysporum* f. sp. *lycopersici*," *Phytopathology* **43**: 195.

———, 1953, "The water economy of *Fusarium* infected tomato plants." **43**: 619-623.

———, 1953 (b), "The cause of epinastic symptoms in *Fusarium* wilt of tomatoes," *Phytopathology*, **43**: 663-669.

GARBER R.H. and HOUSTON, B.R., 1959, "An inhibitor of *Verticillium alboatrum* in cotton seed," *Phytopathology*, **49**: 449-450.

GÄUMANN, E., 1950, *Principles of plant infection*, Crosby Lockwood and Sons Ltd., London.

———, 1951, "Some problems in pathological wilting of plants," *Advanc. Enzymol.* 11: 401-437.

GHABRIAL, S.A. and PIRONE, T.P., 1964, "Physiological changes which preceded virus induced wilt of Tobasco pepper," *Phytopathology*, 54: 893 (Abstr.)

GOTTLIEB, D., 1944, "The mechanism of wilting caused by *Fusarium bulbigennum* var. *lycopersici*," *Phytopathology*, 34: 41.

GOUR, H.N. and Dube, H.C., 1974, "Isolation of *Verticillium dahliae*," *Curr. Sci*, 43: 195.

GRIEVE, B.J., 1941, "Studies in the physiology of host parasite relations: I. The effect of *Bacterium solanacearum* on the water relations of plants," *Proc. Roy. Soc.*, Victoria 53: 268-299.

GROSSMANN, F., 1962, "Untersuchungen iiber die Hemmung pektolytischer Enzyme Von *Fusarium oxysporum* f. *lycopersici* III. Wirkung einiger Hemmstoffe *in vivo Phytopathology, Z.*, 45: 139-159.

HARRIS, H.A., 1940, "Comparative wilt induction by *Erwinia tracheiphila* and *Phytomonas stewarti*," *Phytopathology*, 30: 625.

HODGSON, R., PETERSON, W.H. and RIKER, A.J., 1949, "The toxicity of polysaccharides and other large molecules to tomato cuttings," *Phytopathology*, 39: 47-62.

HUSAIN, A. and DIMOND, A.E., 1958, "The function of extracellular enzymes of Dutch elm disease pathogen," *Proc. Nat. Acad. Sci.* U.S. 44: 549-601.

———, and KELMAN, A., 1958 (a), "Relation of slime production to mechanism of wilting and pathogenicity of *Pseudomonas solanacearum*," *Phytopathology*, 48: 155-165.

HUSAIN and KELMAN., 1958 (b), "The role of pectic and cellulolytic enzymes in pathogenesis by *Pseudomonas solanacearum*," *Phytopathology*, 48: 377-386.

HUTCHINSON, C.M., 1913, "Rangpur tobacco wilt," *Mem. Dept. Agr. India Bacteriol. Ser.*, 1,67.

ISAAC, I., 1949, "A comparative study of pathogenic isolates of *Verticillium*," *Trans. Brit. Mycol Soc.* 32: 137-157.

———, 1967, "Speciation in *Verticillium*," *Annu. Rev. Phytopath.*, 5: 201-222.

———, and GRIFFITHS, D.A., 1962, "*Verticillium* wilt of tomatoes," *XVIth International Horticultural Congress*, 333-342.

KAMAL, M. and WOOD, R K.S., 1955, "Role of pectic enzymes in the *Verticillium* wilt of cotton," *Nature*, 175: 264-275.

KEYWORTH, W.J., 1963, "The relation of monogenic resistant and susceptible tomato to inoculation with *Fusarium oxysporum* f. *lycopersici* into stems or through Bonny Best root stocks," *Ann. Appl Biol*, 52: 257-270.

———, 1964, "Hypersensitivity of monogenic resistant tomato scions to toxins produced in Bony Best root stocks invaded by *Fusarium oxysporum* f. *lycopersici*," *Ann. Appl. Biol.*, 54: 99-105.

KLUEPFEL, D., 1957, "Uber die Biosynthese und die Umwandlungen der Fusarinsaure in Tomatienpenpflanzen," *Phytopath, Z.* 29: 349-379.

LAKSHMINARAYANAN, K. and SUBRAMANIAN, D., 1955, "Is fusaric acid a vivotoxin?" *Nature*, Lond., 176: 697-698.

LUDWIG, R.A, 1952, "Studies on the physiology of hadromycotic wilting in the tomato plant." *Mac. Coll Tech. Bull.*, 20: 1-40.

MACE, M.E., 1963, "Histochemical localization of phenols in healthy and diseased banana roots," *Physiol. Plant.*, 16: 1915-1925.

MATTA, A. and DIMOND, A.E., 1963, "Symptoms of *Fusarium* wilt in relation to quantity of fungus and enzyme activity in Tomato stems," *Phytopathology*, 53: 574-578.

MCDONNELL. K, 1958, "Absence of pectolytic enzymes in a pathogenic strain of *Fusarium oxysporum* f. *lycopersici*," *Nature*, 182: 1025.

MELHUS, I E., MUNCIE, J H and HO, W.T.H., 1924, "Measuring water flow interference in certain gall and vascular diseases," *Phytopathology*, 14: 580.

NATARAJAN, M.K., SIVAPRAKASAM. K. and RAMAKRISHNAN, K., 1968, "Record of *Verticillium* wilt of cotton in Madras state," *Madras Agric. Jr.* 55: 455.

———, and RAMAKRISHNAN, K., 1971, "Control of *Verticillium* wilt of cotton by soil treatments," International Symposium on Pathological Wilting of Plants, Madras, (Abstract).

NISHIMURA, S., 1962, "Relation between pathogenicity of *Fusarium oxysporum* f. sp. *niveum* to water melon plant and the production of toxic metabolites *in vitro*," *Trans. Tottori Soc. Agr. Sci.* 14: 4-7.

PAGE, O.T., 1959, "Fusaric acid in banana plants infected with *Fusarium oxysporum* f. *cubense*," *Phytopathology*, 49: 230.

PEGG, G.F. and SELMAN, I.E., 1959, "Analysis of the growth response of young tomato plants to infection by *Verticillium albo-atrum*. II-Production of growth substances," *Ann. Appl. Biol.*, 47: 222-231.

POWELL, N.T., 1971, "Interactions between nematodes and fungi in disease complexes," *Annu. Rev. Phytopathol.*, 9: 253-274.

SADASIVAN, T.S., 1961, "Physiology of wilt disease," *Annu. Rev. Plant Physiol*, 12: 449-468.

———, 1971, "Wilt as we see it," International Symposium on Pathological Wilting of Plants," (Madras). (Abstracts).

SARASWATHI DEVI, L.S., 1964, "Host-parasite relations in plant wilt" *Indian Phytopath. Bull.* 2: 35-50.

SIDHU, G. and WEBSTER, M., 1974, "Genetics of resistance in the tomato root-knot nematode-wilt fungus complex." *Jr. of Heredity*, 65: 153-156.

SMITH, E.F., 1911, *Bacteria in relation to plant disease.* Carnegie Institution of Washington, Washington.

SNYDER, W.C. and HANSEN, H.N., 1940, "The species concept in *Fusarium*," *Amer. J. Bot.*, 27: 64-67.

SPENCER, J.F.T. and GORIN, P.A.J. 1961, "The occurrence in the host plant of physiologically active gums produced by *Corynebacterium insidiosum* and *Corynebacterium sepedonicum*," *Can. J. Microbiol.*, 7: 88-188.

SRINIVASAN, K.V., 1971, "Some observations on resistance," International Symposium on Pathological Wilting of Plants, (Abstracts.).

STEWART, F.C., 1897, "A bacterial disease of sweet corn," *N. Y. State Agri. Expt. Stn.* (Geneva, N. Y.) Bull. 130

SUBRAMANIAN, C V., 1971, *Hyphomycetes*, Indian Council of Agricultural Research, New Delhi.

TALBOYS, P.W., 1970, "Water deficits in vascular disease," in *Water Deficits and Plant Growth*, Kozlowski (Ed.) Vol II. A.P.

THATCHER, F.S., 1942, "Further studies of osmotic and permeability relations in parasitism," *Cand. J. Res.*, **20**: 283.

TIMONIN, M.I, 1941, "The interaction of higher plants and soil microorganisms. III Effect of by-products of plant growth on activity of fungi and actinomycetes," *Soil. Sci.*, **52**: 395-413.

WAGGONER, P.E. and DIMOND, A.E., 1954, "Reduction in water flow by mycelium in vessels," *Am. J. Bot.*, **41**: 637-640.

WOLLENWEBER, H.W. and REINKING, O.A., 1935, *Die Fusarien*, 355 pp. Paul Parey, Berlin.

24

LEAF SPOTS AND LEAF BLIGHTS

LEAF spots and leaf blights reduce the photosynthetic area of the plants and thus impair the food manufacturing capacity. Such diseases also involve production of toxins which cause injury to the host cells. Leaf spot diseases outnumber all other diseases. A close look at any plant will reveal that it expresses leaf spot symptoms at one stage or the other. Environmental conditions and age of the plant determine the frequency of the spots. The symptoms comprise a varied pattern of expression. Infection may be apical, marginal or scattered. Lesions may be of various shapes and shades. They are usually self-limiting and necrotic areas which present a discrete pattern. Association of imperfect fungi and bacteria is more common with these diseases. Usually such diseases result in destruction of the cellular contents and collapse of the cell walls. Changes in the protoplast are brought about by the proteolytic enzymes. In general, the primary effect of the pathogen results in degradation of protoplast constituents. Many of the leaf-spot causing fungi like *Alternaria*, *Helminthosporium*, *Pestalotia* and *Phyllosticta* are capable of degrading pectic and cellulosic material *in vitro* and this may possibly be an indirect evidence regarding functioning of these enzyme systems in the affected host tissues. Pectolytic enzymes are also known to be formed by certain bacterial leaf spot pathogens like *Pseudomonas tabaci* and *P. angulatum*. The necrogenic capabilities of the leaf spot pathogens have naturally led to the conclusion that they form highly potent specific toxins. Unfortunately, such conclusions are not supported by experimental evidences. Severe leaf spot infections sometimes result in total defoliation. Akai (1951) showed the formation of abscission cells around the necrotic areas in peach leaf spots induced by *Xanthomonas pruni*. Infection by *Cladosporium carpophilum* (Samuel, 1927) leads to perforation of spots in the leaves of *Prunus amygdalus*.

Leaf spots and leaf blights considerably resemble each other in the expression of symptoms as well as mechanism of pathogenesis. Many leaf blight-causing organisms are, however, capable of killing the host partially or totally not only by direct destruction of the tissue but also by systemic dispersal of the toxic substances far beyond the original region of infection.

Leaf spot diseases of crop plants have not received much attention. Bilgrami (1963) and Tandon (1967) have given a comprehensive survey of the leaf spot diseases of ornamental plants and fruit trees respectively. Some important leaf spots and leaf blights have been included in this chapter.

BLAST DISEASE OF RICE

Pathogen. *Piricularia oryzae* Cav.

This is the most important disease of rice plants and occurs all over the world. It is common throughout India.

Fig. 24.1. Symptoms of Blast disease of Rice.

Spindle shaped lesions appear on leaves (Fig. 24.1). The spots are ash coloured in the middle with brown margin. Culms and glumes also show symptoms. The neck of the panicle shows black

R. N. Tandon

necrotic areas, 'neck blast', and bend down due to weight of the ears. Infected grains are shrivelled.

The pathogen causes localized infections The conidiophores emerge through the stomata or broken epidermis and form conidia at the tip sympodially. Conidia are pyriform, usually two-septate and have a small hilum at the base.

The conidia are wind-borne and may also be carried through seeds. It is desirable to use resistant varieties. The pathogen survives on several collateral hosts e.g., *Panicum repens, Digitaria marginata, Brachiaria mutica, Leersia hexandra* and *Dinebra retroflexa.* These grasses may be the source of primary inoculum. At least 28 races of the pathogen are known. In Japan, the disease is being controlled by use of systemic fungicide Blasticidin, an antibiotic.

The physiology of the host parasite relation has been studied extensively in India, at Madras (see Sadasivan *et al*, 1965) and abroad especially in Japan. Suryanarayanan (1958) demonstrated that a low night temperature of about 20°C is essential for infection. Several toxins are involved in the production of the disease syndrome.

BROWN LEAF SPOT OF RICE (*Helminthosporium* disease)

Pathogen. *H. oryzae* Breda de Hann [=*Cochliobolus miyabeanus* (Ito et Kuribay) Dickinson.]

This is also a very serious disease of rice and is known from all the paddy-growing regions of the world. In India, it was first recorded in 1922. It is presently known from all the southern and eastern states of India.

Symptoms

The most characteristic symptom of the disease is the development of dark-brown, isolated, oval to 'eye' shaped lesions on the upper surface of the leaves. Infection may also occur on coleoptile, leaf sheaths and glumes. In early stages, a yellowish halo usually surrounds the spots. In severe cases, spots coalesce and leaves become brittle and dry. Infection of inflorescence may result in failure of grain development. The diseased grains express dark brown spotting on the glumes and brown discoloration of the kernels inside.

Considerable work has been done on the physiology of *Helminthosporium oryzae*. The fungus is known to produce proteolytic enzymes both *in vivo* and *in vitro*. It produces a toxin cochliobolin which is toxic to young seedlings of rice and adversely affects the respiration of plants.

Disease Cycle

The disease is both seed and soil-borne. Some collateral hosts like *Leersia hexandra* and *Echinochloa colona* are also known to help in disease perpetuation. Nutritional deficiencies like those of nitrogen and potash, predispose rice to infection. Rains, cloudy weather, higher day temperature (28 to 30°C), during the later part of the growing season (September-October) are conducive to rapid disease development.

Control

Primary infection can be reduced by treating the seeds with organo-mercurial fungicides. Hot water treatment (at 55°C for 10 mts.) is also quite effective in reducing seed-borne infection. Secondary spread, which is air-borne, is best checked by spraying 5:5:50. Bordeaux mixture or Dithane Z-78, twice or thrice. Resistant varieties like T-141 (Orissa), C.0.20 (Madras) and T.498-2A (Bihar) are reported to be quite promising in India.

BACTERIAL BLIGHT OF RICE

Pathogen. *Xanthomonas oryzae* (Uyeda and Ishiyama) Dowson.

Bacterial blight of rice is known from Japan since the beginning of this century. In India, it was first recorded by Srinivasan *et al.* (1959) from Maharashtra. The disease is also known from the Philippines, Thailand, Indonesia, China, Malaysia, Cambodia, Taiwan, Burma, Vietnam, Bangladesh, USSR, Mexico and Australia.

The disease has caused extensive damage in India during the last decade. High yielding dwarf varieties like Taichung are very susceptible.

Symptoms

The symptoms of the disease are distinguishable in three distinct phases. During the initial phase, straw coloured to yellow stripes with wavy margins appear on the leaves (Fig. 24.2). This is followed by drying and twisting of the leaf tips. Blighting may extend

Fig. 24.2. Leaf blight of rice caused by *Xanthomonas oryzae*.

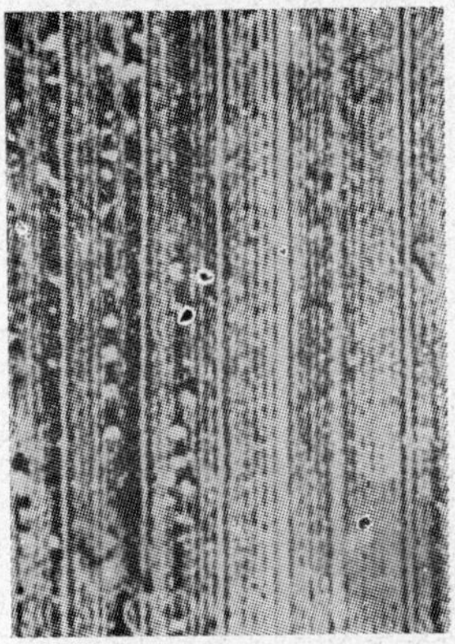

Fig. 24.3. Leaf streak of rice caused by *X. transluscens*.

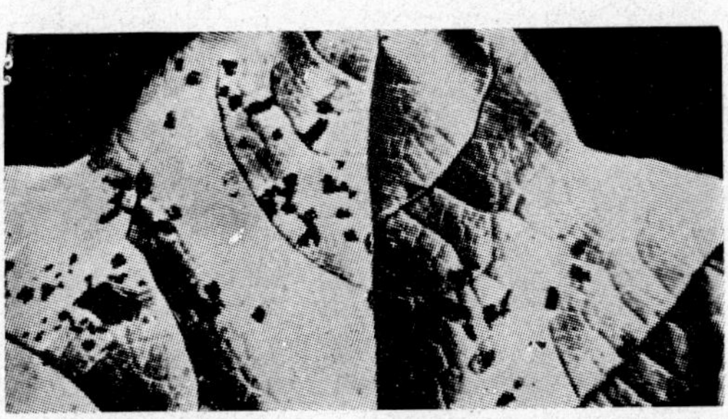

Fig. 24.4. A portion of cotton leaf showing angular leaf spots caused by *X. malvacearum*.

Fig. 24.5. Lesions on stems.

Fig. **24.6.** Citrus Canker, lesions on leaves (*Courtesy* S. Porwal).

Fig. 24.7. Citrus Canker, lesions on leaves (*Courtesy* S. Porwal)

Fig. 24.9. Leaf spots of Tikka disease of groundnut.

Fig. 24.10. (*a*) Blister blight of tea leaves.

to leaf sheaths, culms and glumes. Bacterial masses ooze and dry on surface of the lesions. The second phase of the disease is most destructive. It is referred as Kresek (wilt) phase. It results in complete rolling and drooping of leaves, which turn yellow or grey. In severe cases, the entire plant is kiiled. The third phase is known as the pale-yellow leaf phase and is recorded from the Philippines (Goto, 1964) only. The leaves turn yellowish brown, wither and die.

Disease Cycle

The disease cycle is not clearly worked out in India. It is believed that in double and triple cropped areas, the pathogen survives in infected stubbles and straw, besides the living host. Perpetuation of disease in north and Central India is supposed to be perpetuated through infected seeds of the previous season (Srivastava and Rao, 1963). In Japan the weed hosts like *Leersia oryzoides* constitute the chief source of primary inoculum. The pathogen perpetuates in the rhizosphere of these weeds.

The pathogen generally invades the plants through wounds, hydathodes or stomata and progresses intravascularly. The bacterial exudates, produced on diseased leaves, constitute a rich source of secondary inoculum and its dispersal is carried out by rain splashes, insects and irrigation water etc.

Control

Spray of certain chemicals and antibiotics have been used in India to control the disease. Srivastava and Rao (1963) reported that 95 per cent eradication of infection in seeds could be accomplished by soaking the seeds for 12 hours in 0.025 per cent solution of Agrimycin plus 0.05 per cent wettable Ceresan and then transferring the seeds to hot water at 52°-54°C for 30 minutes. Application of high doses of nitrogen and water logging should be avoided. Several chemicals and antibiotics like Sankel, Celdion, Phenazine, Cellomate and Streptomycin etc. have been extensively used in Japan to control the disease but they are not of much use in Indian conditions.

LEAF STREAK OF RICE

Pathogen. *Xanthomonas transluscens* f. sp. *oryzicola* (Fang.) Brand.

Leaf streak of rice is another important bacterial disease which was first described from the Philippines in 1918 by Reinking as bacterial leaf stripe. Srivastava *et al.* (1967) recorded this disease from different rice-growing parts of Eastern and Central India. Except for a single report from Andhra Pradesh (Shekhawat and Srivastava, 1971), the disease is not known from South India.

Symptoms

It is essentially a foliage disease and under favourable conditions may be as severe as bacterial blight of rice. At the initial stage, small water-soaked areas of pinhead size appear on interveinal parts of the leaf. These enlarge gradually and their size may range upto 10 cm. These areas are generally vein-limited and assume a yellow or brown colour. After about 15 days of infection, characteristic yellow or amber coloured prominent bacterial beads get deposited in linear rows all along the streak (Fig. 24-3). Sometimes, several streaks may coalesce and form blotchy lesions resulting in leaf blighting.

Disease Cycle

The pathogen is capable of surviving in infected seeds (Shekhawat and Srivastava, 1972). In mature seeds, the bacterium hibernates under the glumes. Glumes catch infection at the time of seed germination. The first leaf carries the bacteria to the aerial parts from where secondary spread occurs through injured surface and stomata. Atmospheric humidity is a decisive factor for infection. Continuous high humidity (RH 90 per cent or above) or dew during morning hours for 2-3 days are necessary for rapid infection. A temperature between 26-30 C is more favourable for the spread of the disease. Young leaves are more susceptible. Wind, rain splashes and irrigation waters facilitate secondary infection.

ANGULAR LEAF SPOT OF COTTON

Pathogen. *Xanthomonas malvacearum.* (Smith) Dowson.

Angular leaf spot of cotton is one of the most serious diseases of cotton in India. The disease was first observed in Alabama, USA in 1891. Now it is known from all cotton growing regions of the world. The bacterium infects different plant parts and on that basis the disease is distinguished in several phases like angular leaf spot (attacking the leaf), blackarm (attacking the stem), boll rot (attacking the bolls) and seedling blight (attacking the seedlings).

Symptoms

The leaves are the first target of attack and the earliest symptoms comprise the appearance of water-soaked areas on lower leaf surface. The spots subsequently enlarge, turn brown and form angular areas surrounded by veinlets (Fig. 24.4) and become visible on the upper surface. Often, the disease spreads along the veins, hence called vein blight or black vein. Through the veins the infection extends upto stem and forms dark grey lesions. The affected stems show cracks (Fig. 24.5) and get easily broken by wind (black arm phase); subsequently the infection is initiated on the bolls where small rounded water-soaked spots appear. The lesions are invariably sunken and bolls fall prematurely.

Disease Cycle

The source of primary inoculum are the seeds which may carry the pathogen both externally as well as internally. Plant debris may also serve as an important source of inoculum. Secondary infection is carried through rain splashes and dew. There are several collateral hosts like ***Thurbaria thespisioides***, ***Eriodendron anfructuosum***, ***Jatropha curcas*** and ***Lochnera pusilla***, but their role in disease cycle is not definite.

Control

The infection is essentially seed-borne. Following measures are suggested for the control of the disease:

(1) The delinted seeds should be treated with organomercurial compounds @ 2g/kg seed before sowing.

(2) A hot water treatment to the seeds for 10 mts at 56°C is beneficial.

(3) Proper sanitation and regular destruction of diseased plants and alternate hosts is essential.

(4) Crop rotation, late sowing and addition of potash to the soil minimize the incidence of disease.

(5) Some resistant varieties like HG-9, BJA-592, P-14, T-12, 101-102 B and Reba-B-50 have been recommended for use.

CITRUS CANKER

Pathogen. *Xanthomonas citri* (Hasse) Dowson.

The disease was first recorded in Japan by Abe in 1904. Now it is known practically from every part of the world.

Symptoms

Young leaves, twigs and fruits of citrus are readily attacked. On leaves (Fig. 24.6) small water soaked spots appear at the initial phase. The affected regions become dark green and the surface gets raised. With the advancement of disease, the spots turn greyish and rupture exposing a light brown spongy mass with a crater like depression in the centre. The lesions subsequently turn brown, scaby and are surrounded by dark brown margins and yellow haloes. Lesions on leaves are generally produced on highly susceptible varieties like sweet oranges. Heavily infected twigs (Fig. 24.7) start drying from the tip and severe defoliation occurs. Symptoms on fruits are generally similar to those produced on leaves, except for the absence of yellow haloes.

Disease Cycle

The inoculum is provided by the diseased twigs or leaves. The pathogen enters through natural openings and wounds caused by various agencies. The bacteria are also capable of surviving in soil for 4 to 5 years. Rao and Hingorani (1963) are, however, of the view that soil does not perpetuate the bacterium and does not serve as a potential source of inoculum.

Control

Removal and destruction of the infected twigs and leaves and spraying of Bordeaux mixture or Bordeaux oil emulsion are the most effective method of control. Use of Perenox, Ultra sulphur, Colloidal sulphur and Coppesan has also been suggested. Some antibiotics like Streptomycin sulphate, Phytomycin and Agrimycin 100, have been tried but they are not yet popular.

LATE BLIGHT OF POTATO (*Solanum tuberosum* L.)

Pathogen. *Phytophthora infestans* (Mont.) De Bary.

This disease is of wide occurrence. It has been recorded practically on all the varieties of potato. Varieties of tomato (*Lycopersicon esculentum* L.) are also susceptible. In India it occurs in epidemic form only on the hills at elevation of 6000 ft or above. The disease has been responsible for the famous Irish famine in 1845-46 which resulted in death and migration of millions of people in that country. Due to cold storage facilities of potato tubers, the disease is now of wide occurrence in the plains of India as the fungus survives at the cold temperature. The disease has been reported from several states including Assam, West Bengal, Bihar and Uttar Pradesh.

Symptoms

The symptoms of the disease appear at about the flowering time. During the initial stages, purplish brown spots appear on the foliage. The infected areas subsequently increase in dimensions. During moist weather these areas extend rapidly and there is profuse hyphal and sporangial development. In case of severe infections entire crown of leaves may perish within 2-3 days. After the destruction of the foliage, the infection advances on the underground parts specially the tubers. The tubers, which are immediately below the soil surface, are the first to take the infection. Tubers may also catch infection at the time of harvesting or during storage. Early infection is also responsible for reducing the size and number of tubers. Primary infection to tubers is caused by zoospores or sporangia. The rot may be either dry or wet, depending upon the availability of moisture. Dry rot of tubers is usually expressed by formation of brownish areas which extend to about ½ an inch below the surface. Sometimes the decay of the tubers is very rapid. In moist weather the surface of the infected tubers turn light creamy due to formation of sporangia.

Disease Cycle

The disease is propagated by sporangia which may germinate either directly or indirectly (by zoospore formation). 90 per cent RH is essential for sporangial germination. Sporangial production is accomplished in a saturated atmosphere at a wide range of temperature (20°C to 26°C); 21°C is, however, optimum. It has been observed that infection declines significantly at temperatures above 20°C, because of fall in germination percentage and the slow growth of germ tube. Epiphytotics occur whenever environmental conditions favour the production and germination of sporangia and zoospores. A cool weather, combined with abundant moisture, is ideal for bulk production of inoculum. In India, it is generally the moisture (and not the low temperature, which is readily available during later part of winter) that becomes limiting factor.

Various theories have been put forward regarding the disease cycle and availability of inoculum. However, to visualize that the fungus persists in soil is out of question because temperature during summers becomes very high. The only source of inoculum is through infected seed tubers. The increasing practice of storing the seeds in cold storage has mainly contributed in perpetuation of the disease in plains of India. After formation of sporangia or zoospores, the disease spreads to large number of plants. Infection may take place through any part of leaf or stem. In tubers infection takes place through 'eyes', lenticels and wounds.

Control

(1) Certified and disease-free tubers should be used for cultivation.

(2) Care should be taken not to allow the debris of the crop to remain in the fields.

(3) Harvesting should be carried out in perfectly dry weather. Diseased plants should be destroyed and burnt before harvesting.

(4) Regular foliar spray after a gap of ten to fifteen days should be carried out. Spraying with suitable fungicides should usually be started much before the appearance of the disease. First spray is usually carried out when the plants are about a month old. Bordeaux mixture (4:4:50) is most popular fungicide. It is, however, being replaced by dithiocarbamates like Dithane D-14, Dithane Z-78 and Dithane M-22 etc. Use of Perenox, Blitox-50 and Fytolan is also recommended.

Attempts have been made to breed resistant varieties but the programme has not been a success due to existence of several physiologic races of the fungus. Three distinct races (i.e. 0, 1 and 4) are known from India.

COLOCASIA BLIGHT

Pathogen. *Phytophthora colocasiae* Racib.

This is a very common disease in all the Arvi (*Colocasia antiquorum* Schott.) growing regions of India.

Symptoms

The disease appears in rainy season when dark brown, circular or oval specks appear on the leaves. These spots enlarge, and adjoining lesions coalesce involving a large portion of the leaf. Sometimes, drops of light yellow liquid are secreted from the affected areas. The affected areas finally get dried, turn brittle and form shot holes.

Under highly humid conditions, the disease spreads rapidly and entire leaf may collapse.

Disease Cycle and Control

It is believed that the disease perpetuates from season to season through infected corms or through viable zoospores available in the soil. Spraying with Bordeaux mixture (2:2:50 or 4:4:50) gives complete control of the disease. Spraying should be started in late July and should be carried out at regular fortnightly intervals till the end of September. Blitox-50 and Dithane Z-78 have also been found to be effective against the disease. Healthy and disease free corms should be used for cultivation. Certain varieties have been claimed to be resistant (Paharia and Mathur, 1964) on regional basis.

LEAF SPOT OF TURMERIC (*Curcuma longa* L.)

Pathogen. *Taphrina maculans* Butl.

The leaf spot of turmeric (Fig. 24.8) is common throughout India. It is often reported to occur as epiphytotics in Orissa and Andhra Pradesh. Some other species of *Curcuma* and *Zingiber* are also attacked by the same fungus.

Symptoms

The leaves manifest reddish-brown spots on both surfaces of the leaves. These leaves subsequently become yellow. Sometimes smaller spots coalesce to form bigger spots. The spots are usually small, 1-2 mm in diameter. Attacked plants are not killed, but due to excessive spotting and destruction of foliage the yield is adversely affected.

Disease Cycle and Control

The mode of transmission of the disease is not clearly understood. It is quite likely that the dried leaves, having spots with numerous asci, remain as debris in fields and serve as primary source of inoculum in the next season. Field sanitation is essential. Destruction and burning of diseased leaves should be regularly carried out. Spraying with Bordeaux mixture (4:4:50), Fytolan, Blitox-50 and Dithane Z-78 has been found to be useful.

Upadhya and Pavgi (1967) reported that "China" and "Jaweli" varieties of *Curcuma longa* and *C. amada* Roxb were immune to *Taphrina maculans*.

EARLY BLIGHT OF POTATO

Patnogen. *Alternaria solani* (Ell. and Mart.) Jones and Grout.

'Early blight' is a common and widespread disease of potato in Indian plains. The disease appears on about three weeks-old plants. The disease also occurs on other solanaceous crops like tomato, brinjal and chillies.

Early blight of potato

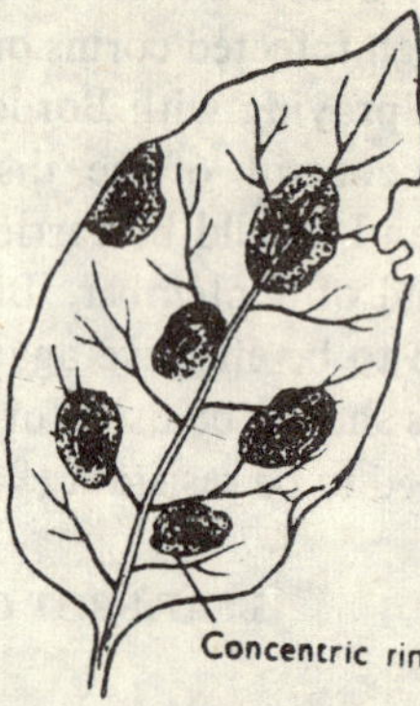

Fig. 24.9.

Symptoms

Scattered pale brown spots appear on the leaflets. The spots subsequently get covered with greyish fungal growth. The tissues turn necrotic, and concentric ridges are formed in these areas (Fig. 24.9). This is the most characteristic symptom of early blight. In humid weather, affected areas coalesce, big rotting patches appear which get shrivelled. In dry weather, the leaves show curling. In severe cases spots may extend to petioles and stems. Photosynthetic area of plants is considerably reduced due to infection and this leads to decrease in size and number of potato tubers.

Disease Cycle and Control

Conidia and mycelium remain viable in soil and plant debris and serve as the source of primary inoculum for the next season. Secondary spread is accomplished by conidia, produced in new lesions. Field sanitation and rotation of crops minimize infections. Regular and timely spray with fungicides effectively controls the disease Blitox-5 (0.25%), Difolatan, Daconil and Captan are some of the commercial fungicides which have been found to be effective. Dithane Z-78 @ 2 lbs gallon is, however, considered to be most effective.

LEAF SPOT OF GROUNDNUTS

Pathogen. 1. *Cercospora Personata* (Berk et Curt.) Ell. and Ever.
2. *C. arachidicola Hori.*

Serious leaf spotting of groundnuts is recorded from all parts of the country. The disease is popularly known as *Tikka* disease. *C. personata*(Fig. 24.10) is comparatively more virulent.

Symptoms

In initial stages it is difficult to distinguish the spots produced by the two species. In severe cases the stems may also be affected. Spots produced by *C. personata* are usually circular in shape, dark brown in colour and they are upto 8 mm in diameter. Sometimes these spots are surrounded by a yellow halo at maturity. Conidia are confined to the lower surface and development of the conidiophores is in concentric rings. The spots produced by *C. arachidicola* are circular to irregular. They are much bigger in dimension and have a tendency to coalesce. The spots are surrounded by a yellow halo and later turn reddish-brown to black n colour. The conidial production is generally confined to the upper surface of the leaf.

Disease Cycle

Yearly perpetuation of the disease is accomplished by conidia which lie in the soil either in the fruit-shells or in the debris of the plants. Seed treatment minimizes the percentage of infection to a great extent.

Control

Bordeaux mixture (4:4:50), Dithane Z-78 (0.2%), Dithane M-45 and Fycol-8E are some of the effective fungicides. Spray schedules should be carried out at regular intervals and both the upper and lower surfaces of the leaves should be covered. Dusting with copper-sulphur (1:1) dust has also been recommended.

Mineral deficiency of certain elements like magnesium contributes to the spread of the disease. It is, therefore, recommended that mineral deficiency of the soil should be made up by soil amendments. Crop rotation and deep ploughing also help in reducing the availability of the inoculum. Some early-maturing exotic varieties are known to show considerable resistance to the disease. Susceptibility and resistance are reported to be determined by increased ascorbic acid and riboflavin contents of the leaves, respectively.

BLISTER BLIGHT OF TEA

Pathogen. *Exobasidium vexans* Massee.

Blister blight is the most destructive disease of tea in India, Sri Lanka, and Indonesia.

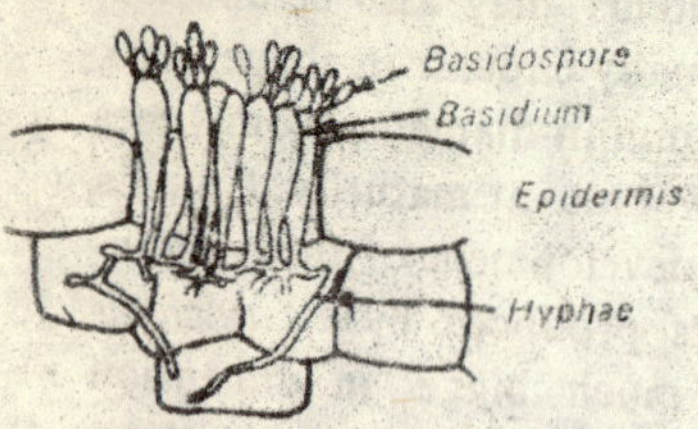

Fig. 24.11 Blister blight of tea basidiospores on the lower surface of leaf.

Symptoms

Pale green, pale yellow or pinkish translucent spots, which are sunken on the upper surface and convex on the lower (hence the name blisters), appear on young leaves. The blisters also occur on tender stems and cause their breakage at the point of infection. Mature leaves are not affected. More than 50 discrete blisters may occur on a single leaf. Sometimes the blisters coalesce also and form a big patch. The lower raised surface of the blister bear the basidiospores (Fig. 24.11). Infected leaves curl and get deformed.

Disease Cycle

The pathogen is an obligate parasite and, since the crop is grown throughout the year, the inoculum exists endemically in the plantation. The disease becomes severe during rainy seasons. Blister blight is a multiple cycle disease and the fungus completes the life cycle in 10-28 days. Basidiospores are wind and insect-transmitted but infection is by wind-borne spores only. The basidiospores germinate and penetrate the leaf epidermis by forming appresorium and infection thread. Blisters start appearing after 3 days of penetration. Sporulation on the lower raised leaf surface starts after 10-12 days of infection. Spores are discharged upto 8 days. According to one study (cf. Venkataram, 1975) one million spores are produced in 24 hours. A relative humidity of 80 per cent is required for sporulation and infection.

Control

Preventive control is achieved by spraying copper fungicides. Cuprous oxide and copper oxychloride effectively control the disease at usage rate of 0.21 Kg. metallic copper per hectare. Two organic fungicides viz., Daconil and Difolatan (trade names) give excellent control.

Eradicant control. Weekly sprays of nickel chloride effectively check the disease after infection. In addition, nickel spray does not stimulate mite population which accompanies the continuous use of copper sprays. Maximum benefit is achieved by using mixture of nickel chloride and copper fungicides.

Systemic Fungicides. Tridemorph (Calixin) and Pyracarbolid (Sicarol) are proving very successful in controlling the disease by preventing sporulation and thus aborting the lesions.

Genetic control. Unfortunately no resistant variety is available. Several races of the pathogen are known.

REFERENCES

ABE, K., 1904, "Citrus canker in Japan. M-37,V: 162-165" (Translated by T. Tanaka, *Phytopathology* 8: 443-444, 1918.

AKAI, S., 1951, "On the anatomy of necrotic lesions on the leaves and stems of the plants," *Mem. Coll. Agr. Kyoto Univ.*, 61: 1-30.

APPA RAO, A., "Survival of *Piricularia oryzae* in soil," *Indian Phytopathology.*, 18: 321.

BILGRAMI, K.S., 1963, "Leaf spot diseases of some ornamental plants," *Proc. Natl. Acad. Sci.* India., 33: 429-452.

FANG, C.T., REN, H.C., CHEN, T.Y., CHU, Y.K., FANN, H.C. and WU, S.C., 1957, "A comparison of rice bacterial leaf blight organism with bacterial leaf streak organism of rice and *Leersia hexandra* Swartz," *Acta Phytopath, Sinica.*, 3: 99-124.

GOTO, M., 1964, "Nomenclature of the bacteria causing bacterial leaf streak and bacterial leaf stripe of rice," *Rep. Fac. Agri. Shizuoka Univ.*, 14: 3-10.

PAHARIA, K.D. and MATHUR, P.N., 1964, "Screening of colocasia varieties for resistance to colocasia blight (*Phytophthora colocasiae* Racib)," *Sci and Cul.*, 30: 44.

RAO, Y.P. and HINGORANI, M.K., 1963, "Survival of *Xanthomonas citri* (Hasse) Dowson. in leaves and soil," *Indian Phytopath.*, 26: 362-364.

REINKING, O.A., 1918, "Philippine economic plant disease. *Oryza sativa* L. Rice bacterial leaf stripe," *Philippine J. Sci. Sec.*, A, 13: 225-226.

SADASIVAN, T.S., SURYANARAYANAN, S. and RAMAKRISHNAN, L., 1965, "Influence of temparature on rice blast disease," in *The Rice Blast Disease*, 163-171, Johns Hopkins Press, Baltimore, Maryland.

SAMUEL, G., 1927, "On the shot-hole disease caused by *Cladosporium carpophilum* and on the shot hole effect," *Ann. Botany.*, (London) 41: 375-404.

SHEKHAWAT, G.S. and SRIVASTAVA, D.N., 1971, "Control of bacterial leaf streak of rice," *Indian. J. Agric. Sci.*, 41: 1098-1101.

———, 1972, "Mode of seed infection and transmission of bacterial leaf streak of rice," *Ann. Phytopath. Soc. Japan.*, 38: (1).

SRIVASTAVA, D.N. and RAO, Y.P., 1963, "Epidemic of bacterial blight disease of rice in North India," *Indian. Phytopath.*, 16: 393-394.

———, DURGAPAL J.C. and JINDAL, J.K., 1967, "Bacterial leaf streak of rice in India," *Plant Dis. Reptr.*, **51**: 928-929.

SRINIVASAN, M.C., THIRUMALACHAR, M.J. and PATEL, M.K., 1959, "Bacterial blight of rice," *Curr. Sci.*, **28**: 469-470.

SURYANARAYANAN, S., 1958, "Role of nitrogen in host susceptibility to *Pyricularta oryzae*," *Curr. Sci.*, **27**: 447-448.

TANDON, R.N., 1967, "Post Harvest diseases of tropical and sub-tropical fruits," *Tech. Report.*, P.L. 480 Project, Fg-In-133 Univ. of Allahabad.

UPADHYA, R. and PAVGI, M.S., 1967, "Varietal resistance in Turmeric to leaf spot disease," *Indian. Phytopath.* **20**: 29-31.

VENKATARAM, C.S , 1975 "Blister blight of tea," in *Advances in Mycology and Plant Pathology*. Raychaudhary *et al.* (eds.) Published by Prof. R.N. Tandon's 70th Birthday Celebration Committee, New Delhi.

25

MOSAIC, LEAF CURL AND OTHER ABNORMALITIES
(VIRUS AND MYCOPLASMA DISEASES)

SYMPTOMS like variegation of leaves, curling, flower-breaking or stunting have been known since centuries but it was only in the beginning of the present century that such symptoms were attributed to viruses. Virus diseases have been intensively investigated during the last four decades. Precise number of diseases with such abnormal symptoms is not known. It is, however, speculated that the number of plant diseases which can be attributed to viruses exceed 500. Some diseases which were previously ascribed to viruses have been now attributed to a comparatively new group of organisms known as *Mycoplasma* or *Bedsonia*. It is usually the yellows type of diseases, like Mulberry dwarf, Aster yellows and Potato witches' broom etc., which have been referred to *Mycoplasma*. Viruses and mycoplasma are similar so far as size and filterability are concerned. The main difference, however, being that viruses contain either RNA or DNA, never both and can replicate only in a living cell.

From the point of view of parasitism viruses are true obligates in nature as they are incapable of growing on non-living media. Like other plant pathogens the viruses also cause morphological, histological as well as metabolic changes in their hosts. The morphological changes which are expressed in form of symptoms are mainly of the following types.

MORPHOLOGICAL CHANGES

(1) Mosaic or mottling of various types. Such symptoms may be accompanied by varying amounts of blistering and distortion of the infected leaf. Sometimes (specially on the leaves of monocot plants) these symptoms may get modified to definite streaks or stripes. In

extreme cases the leaf may get completely chlorotic and the disease is then referred to as "Yellows".

(2) Rolling, curling, epinasty, dwarfing and thickening or a combination of all these symptoms is also common.

(3) Outgrowths (enations) are also a common expression of several viral diseases. Formation of witches' broom and tumours are seen in many infections.

(4) Vein-thickening, vein-banding or vein-clearing are also frequently observed.

(5) Leaves are sometimes reduced "little leaf" and lamina may get greatly suppressed. Occasionally, leaves become stiff and brittle. Galls or pouch like structures may be produced on the leaves.

(6) Sometimes, flowers are affected and they may become multicoloured. Virescence (greening) of petals, (phyllody) is also common. Dwarfing of flowers and enlargement of sepals is also occasionally seen.

(7) Fruiting may be reduced, fruits may be undersized and get aborted. Their surface may become warty, exhibit cracks and blisters.

(8) Elongation, stunting or rosetting of entire plants may also occur.

(9) Some viruses cause internal and external necrosis of tissues.

(10) There is loss of germinability of seeds and other organs of propagation.

Symptoms listed above are far from comprehensive. In many cases the symptoms may be masked and it may be difficult to ascertain infection on the basis of host morphology.

HISTOLOGICAL AND CYTOLOGICAL CHANGES

(1) Chlorotic tissues of the mottled leaves have generally thinner walls, shorter palisade and smaller chloroplasts.

(2) Necrosis may lead to breakdown of cells of pith and cortex affecting ultimately the xylem.

(3) Overgrowth is usually the result of extra-cambial activity producing extra-vascular tissue, which is generally abnormal in nature.

(4) Palisade tissue may be replaced by spongy parenchyma or *vice versa.*

(5) Certain viruses cause formation of tyloses in the xylem.

(6) Several inclusion bodies like X-bodies, crystalline inclusions and intranuclear inclusions are developed in many virus-infected plants.

(7) In certain extreme cases lobing and distortion of nucleolus, breaking of spindles and scattering of chromosomes, resulting in formation of giant nuclei, is also known. The sterility of plants infected by certain viruses could perhaps be associated with such nuclear abnormalities.

The morphological and histological changes, which develop in virus infected plants, are the result of virus-host interaction. Esau (1967) has discussed at length the anatomical changes introduced in virus infected plants.

PHYSIOLOGY OF VIRUS DISEASES

Diener (1963) considers that physiological derangements most commonly associated with virus infections are:

(*i*) decreased photosynthetic activity,
(*ii*) increased rate of respiration,
(*iii*) accumulation of soluble nitrogen compounds,
(*iv*) increased polyphenol oxidase activity and accumulation of oxidized polyphenol products, and
(*v*) decreased activity of growth-regulating substances.

Decreased Photosynthetic Activity

Colour changes of leaves is characteristic of virus infections. This could be due to destruction or inhibition of chlorophyll synthesis. Even the existing chloroplasts when isolated and studied showed a decrease in photosynthetic phosphorylation and in the Hill Reaction (Zaitlin and Jagendorf, 1960).

Respiration

There are great variations and contradictions in the rates of respiration of virus infected plants, even for plants of uniform size, age and appearance. In tobacco mosaic disease, Owen (1956) obtained higher, lower and identical rates of respiration as compared to the healthy controls. Immediately after inoculation, the respiration of virus-infected leaves increased in winter, while during summer a decrease was recorded. Variations in the results depend upon: (*a*) the time of inoculation, (*b*) physiological state of the plants, (*c*) the environmental conditions during growth, and (*d*) the leaves. Obviously, statistical tests of significance must be done before deriving inferences from the results.

Accumulation of Soluble Nitrogen Compounds

Diseased leaves contain greater amount of soluble nitrogen, particularly amides, than their healthy counterparts. Total nitrogen and protein nitrogen are more in healthy than in diseased leaves.

Increased Polyphenol Oxidase Activity

Hypersensitive defense reaction is caused by accumulation of toxic substances produced by oxidation of polyphenols, by polyphenoloxidase activation. The increased activity of these enzymes, which occurs invariably due to virus infection, is more pronounced in host-virus combinations resulting in local lesions than in those resulting in systemic spread of viruses. In local lesions, a fall in chlorogenic acid content is accompanied by accumulation of quinones. The importance of polyphenol oxidases in localization of viruses and the development of local lesion symptom is proved by the fact that when ascorbic acid is exogenously supplied, the lesions are limited but not the virus multiplication. Whether localization is the result of destruction of virus translocation by quinone, induced death of host tissues or by inactivation of viruses is not known.

Decreased Activity of Growth Regulating Substances

The virus-infected plants, invariably, show a decrease in growth regulating substances or an increase in the activity of growth inhibiting substances (antiauxin, oxidase enzymes etc). Decreased auxin level has been reported in tomato leaves infected with spotted-wilt-virus, potato leaves and tubers infected with several viruses, including leaf-roll virus and potato viruses X Y or A. The decrease is due to destruction or retarded synthesis of auxins. Scopoletin, which stimulates auxin action when applied to plants by inhibiting IAA oxidase enzyme, has been reported to accumulate in certain virus-infected plants (Andreae and Andreae, 1953). But scopoletin also causes photooxidation of IAA in light, and therefore, it can act both ways—inhibit or promote IAA destruction. Several workers have shown reversal of stunting of virus-infected plants by application of gibberellic acid (Maramorosch, 1957 and Chessin, 1958). Thus, stunting of growth can be assigned to virus-induced reduction in concentration of gibberellins.

Translocation of Viruses in Plants

The movement of viruses is of two kinds (*i*) slow cell-to-cell movement via plasmodesms, and (*ii*) rapid movement via phloem.

Esau (1967), by electronmicrophotographs showed virus particles of beet yellows virus in the act of passing through the plasmodesms. However, Kassanis, Tinsley and Quak (1958) observed movement, though slow, (1 mm in one week) in mosaic callus devoid of plasmodesms, in tissue culture experiments. This suggests the possibility of other route also. The rate of movement is dependent on several factors viz., the rate of protoplasmic streaming, the rate of virus synthesis and the number of plasmodesms.

In phloem, the viruses are shown to move through the sieve tubes. Viruses have been shown to move rapidly in the direction of food utilization and storage and slowly in the opposite direction. Bennett, 1960 (cf. Smith, 1968), examined such movement of three viruses of crown sugarbeets. He also observed that mosaic virus moved most rapidly; yellows virus was intermediate while curly top virus was slowest.

Viruses confined to parenchyma, obviously, are handicapped and hence form local lesions. Then, there are viruses which are restricted to either phloem or the xylem vessels. Those that inhabit the phloem e.g., curly top sugarbeet virus, tobacco mosaic virus, raspberry leaf-curl virus, are unable to pass through xylem bridges separating phloem portions. If such viruses are injected into xylem vessels, they fail to escape from there. Similarly, some viruses (alfalfa dwarf virus) remain confined to xylem vessels and are transmitted by leaf-hoppers that feed in the xylem.

In what form viruses are translocated

Whether the viruses are translocated as naked ribonucleicacid or as complete particles is debated, although the electronmicrophotographs by Esau (1967) clearly demonstrated the movement as particles for beet yellows virus. Siegel *et al* (1962) have been able to obtain two incomplete forms' of tobacco mosaic virus: one fails to synthesize the protein coat while the other fails to form the RNA. The movement of these may be helpful in solving the problem.

MOSAIC OF SUGARCANE

This is the most important viral disease of **sugarcane. The disease** is reported from different parts of the world. It was first recorded in Java in 1892. In India it is known since 1921. There are contradictory opinions about this disease. Chona (1944) observed that even 100 per cent affliction of a crop with mosaic did not cause more than 10 per cent loss in yield and that juice-quality remained unimpaired. Kumar *et al* (1963) have listed mosaic as one of the most serious diseases of sugarcane in India. Shah (1972), however, does not include it in major diseases.

Symptoms of sugarcane mosaic appear after about six weeks of planting. They are more pronounced during pre-monsoon and monsoon seasons. With advancing age of the plant, the symptoms get obliterated. Pale oval or elongated patches of different dimensions appear on the leaves (Fig. 25.1). Several strains of this virus have been distinguished on the basis of minor variations in symptoms. Abbot and Tippett (1966) have recognized seven strains viz. A, B, C, D, E, F, and H, primarily on basis of their reaction on different hosts. Strain G was considered as a variant of B. More than a dozen alternate hosts of this virus have been listed by Hutchinson (1972). In addition to maize, six alternate hosts of sugarcane mosaic virus are known from India which serve as perennial source of inoculum and also of mutants (Bhargava 1975).

Vectors play a very important part in the spread of this disease. Pemberton and Charpentier (1969) have listed seven species of aphids which serve as vectors of SCMV. Bhargava *et al* (1971) have given a comprehensive list of vectors of sugarcane common in India. The efficiency of aphid species, as transmitting agents, depends on several factors. *Rhopalosiphum maidis* and *Melanaphis sacchari* are comparatively more efficient vectors.

Several methods of control, including use of resistant varieties, have been tried from time to time. Koike and Tippett (1972) have recommended roguing with applications of herbicides like Erbon or Glytoc. Roguing of weeds, which act as alternate hosts in sugarcane fields, is also recommended. Leu (1972) obtained sugarcane plants, free from mosaic virus by meristem cultivation of 3-4 mm long growing points. Shah (1972) reported that treatment of setts with 1000 ppm of 2-thiouracil was effective against mosaic virus. Heat therapy against mosaic has been tried in India but without much success (Singh, 1971).

VEIN MOSAIC OF BHINDI

The disease was first recorded by Uppal *et al.* (1940) from Maharashtra. It is now known from every part of the country. Vein clearing, followed by veinal chlorosis, is the distinguishing character of the disease. The veins and veinlets get thickened and appear as a yellowish network on the entire leaf. Sometimes, the entire leaf becomes chlorotic. Fruits turn yellow and get reduced in size.

The virus is transmitted by white fly (*Bemisia tabaci*). Jha and Mishra (1955) noted the presence of bhindi leaf hopper (*Empoasca devastans*) in abundance from July to September but there is no conclusive evidence to show that transmission is carried out by them. Many weed hosts like *Croton sparsiflora* and *Malvastrum tricuspidatum*, provide a ready source of inoculum. Protection of crops from white flies and other insects by regular spraying of insecticides may be helpful in reducing the incidence of disease but which has so far remained uncontrolled.

TOBACCO MOSAIC

Tobacco mosaic is caused by tobacco mosaic virus (*Nicotiana* virus I). Affected plants show chlorotic and dark green raised patches presenting a mosaic pattern. Yellowish blotches may also, sometimes, appear. The virus particles perennating in plant debris and tobacco products are the source of primary inoculum. Processed manufactured tobacco, used in cigars and cigarettes, remains infectious. Different agriculturalloperations and mechanical contacts are responsible for secondary infections. The virus has a longevity of more than 25 years and is highly infectious even in extremely diluted state.

Cultivation in infested soils should be avoided.

MILD MOSAIC OF TOMATO

Nicotiana virus I is responsible for mosaic disease of tomato. Foliage of affected plants (Fig. 25.2) exhibit mottling with raised dark green areas. This symptom may, however, get masked at low temperatures and low light intensity. Stunting of plant and distortion of leaves are prominent symptoms. Besides TMV, other viruses like cucumber mosaic virus, potato mild or latent mosaic virus and potato vein-banding virus, are also responsible for mild mosaic symptoms on tomato plants.

MILD MOSAIC OF POTATO

Inter-veinal mottling of leaves is the most characteristic symptom (Fig. 25.3). The virus has large number of strains and has been designated by various names like potato mottle virus, potato latent virus, potato virus X, and *Solanum* virus I. The mottling symptom of the disease is generally masked and slight dwarfing of the plants or deformation of the foliage normally go unnoticed. Potato virus X may retain its viability for one year or so. The virus remains infective even in extreme dilutions. It also attacks a number of other solanaceous crops like *Petunia*, *Datura*, tobacco and tomato. This is a sap-transmitted virus, which, under field conditions, spreads by contact.

MOSAIC OF CUCURBITS

This is a very common disease of cucurbits. Pumpkin (*Cucurbita maxima*), bottle gourd (*Lagenaria vulgaris*) and Cucumber (*Cucumis sativus*) are some of the important hosts. Mottling of leaves, with dark green, light-green or yellowish-green patches, are the chief symptoms of the disease. Fruiting is reduced; fruits may show warty growths. In cucumber, the fruits produce various types of malformations. The disease is both sap and insect transmitted. *Myzus persicae* and *M. ornatus* are the important aphid vectors.

Mosaic diseases, affecting a number of cereals, have been studied in India. Some of the better known are: mosaic of maize (Chona and Seth, 1960); barley mosaic (Dhanraj and Raychaudhuri, 1969); mosaic streak of wheat (Raychaudhuri and Ganguly, 1968) and streak disease of Bajra (Seth *et al*, 1971). A number of leguminous crops like broad bean (Azad *et al*, 1961), cowpea (Chenulu *et al*, 1968); mung (Nariani, 1960); soyabean (Nariani and Pingaley 1960); and pea (Sreenivasan and Nariani, 1966) also suffer with mosaic diseases.

LEAF ROLL OF POTATO

This is a very important disease of potato causing extensive losses. The virus has been designated as ***Potato virus* I**, ***Solanum virus*** or *leaf-roll virus*. Inward rolling of leaves is the most characteristic

Fig. 25.1. Mosaic of sugarcane, patches on leaf.

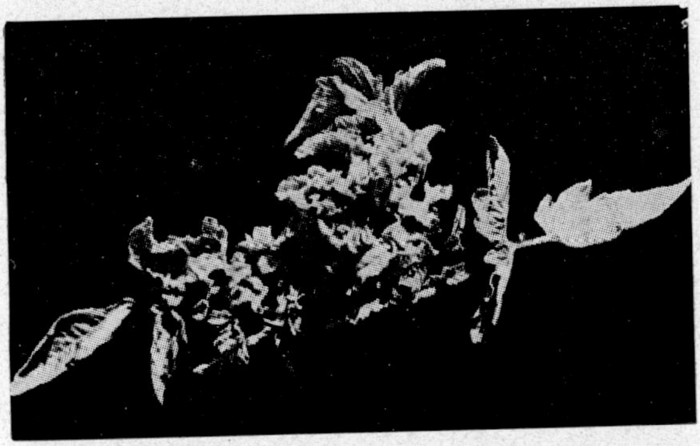

Fig. 25.2. Mild mosaic of tomato; mottling of leaves.

Fig. 25.3. Mild mosaic of potato, intervenial mottling of leaves.

Fig. 25.4. Leaf curl of tomato; note curling and drooping of leaves.

Fig. 25.5. 'Bunchy top' of banana; infected plants stunted in growth. (*Courtesy* P. P. Mehta)

symptom of the disease. Rolling usually starts from the lower leaflets. The rolled leaflets are usually more stiff and rigid. The leaves may also become thick and leathery. The affected plants have less tuber production and show necrosis in severe cases. Diseased seed tubers are the chief source of transmission of the virus. *Myzus persicae*, *M. pseudosolani* and *M. circumflexum* are the important vectors.

RUGOSE MOSAIC OF POTATO

The disease appears in scattered form. It is characterized by dwarfing of plants and mottling, crinkling and puckering of the leaves. Sometimes, the margins of leaflets may express downward rolling. The leaves become rugose, rough and hairy. In severe cases, plants die prematurely. Two viruses *viz.*, potato virus-X and potato virus-Y, are jointly responsible for the disease.

The virus is transmitted through infected tubers. *Myzus persicae* is known to transmit virus-Y.

LEAF CURL OF PAPAYA

This is a common disease of papaya and is reported from different parts of the country. The disease is caused by *Tobacco* virus-16 or *Nicotiana* virus-10. Curling, crinkling, distortion and reduction in leaf size are the characteristic symptoms of the disease; leaves become rugose and brittle, and their curling is usually inwards and downwards. Veins may get thickened. Sometimes entire petiole may exhibit inward rolling and twisting. According to Nariani (1956), the virus is easily transmitted by grafting or juice inoculation. Transmission in nature is, however, brought about by white fly, *Bemisia tabaci*. The virus has a wide host range; infects tobacco, tomato, zinnia and holyhock. Uprooting and destruction of the affected plants is the only remedy.

LEAF CURL OF TOMATO

The disease, which is widespread in India during winters, is also caused by *Tobacco* virus-16 or *Nicotiana* virus-10. Drooping and curling of leaves, due to inward rolling of the leaflets along the mid-rib, is quite common (Fig. 25.4). Plant growth is stunted and leaves become stunted and crisp. Reduction in fruit size and their premature ripening is common. In extreme cases, the affected plants die due to killing of the roots. *Bemisia tabaci* is the chief transmitting agent under natural conditions.

BUNCHY TOP OF BANANA

Bunchy top of Banana is known from different parts of Eastern and Southern India. The disease may appear in plants of different ages. The affected plants usually remain stunted (Fig. 25.5) and the leaves develop into a dense rosette like structure at the apex. During the initial phase, green streaks appear on the midrib, petiole and underside of the lamina. Sometimes, symptoms with whitish streaks are visible even in the emerging leaves. Marginal chlorosis and curling of the leaves is an associated feature of the disease. The virus responsible for the disease has been designated as Banana virus I. Infection is systemic and the virus renders the phloem non-functional. In severe infections, the entire root system may decay completely.

Infection is initiated by diseased suckers. Secondary infection is carried out by aphid (*Pentalonia nigronervosa*) during the growth of the plant. The aphid has to feed on the diseased plants for about 17 hours to retain their infectivity for few days. There is seen heavy concentration of the aphids at the base of the pseudostems.

Practically all the seedless varieties of banana are highly susceptible to the disease. Diseased plants should be uprooted and burnt. Planting material should be properly checked before use and propagation.

LITTLE LEAF OF BRINJAL

This disease was first reported in India by Thomas and Krishna Swami in 1939. Diseased plants show extreme reduction in size of the leaves. Axillary buds grow with short internodes, giving a bushy and stunted appearance. In case of heavy infections, flowering and fruiting is completely inhibited. The disease is now known to be caused by *mycoplasma*. The pathogen remains confined to the phloem and possibly interferes with the transport of metabolites.

RICE YELLOW DWARF

This disease was first observed in Japan in 1910. In India it is known to occur in many parts of Orissa, Bihar, West Bengal, Andhra Pradesh and Delhi (Raychaudhuri *et al*, 1967). It is characterized by general chlorosis with pronounced stunting and profuse tillering. Chlorotic leaves are uniformly pale green to pale yellow. Infected plants do not bear flowers and die prematurely. The disease is caused by *mycoplasma*.

SANDAL SPIKE

This is also a *mycoplasma*-incited disease. It was first reported by McCarthy (cf. Barber, 1903) from Mysore state in 1839. The disease is fast spreading in South Mysore and is threatening the whole sandal industry.

Reduction of leaf size and shortening of the internodes, gives spike-like look to the branches. The affected plants fail to form flowers and fruits; diseased plants die in 2-3 years. The disease is suppressed by tetracycline antibiotics and benlate (a systemic fungicide).

REFERENCES

ABBOT, E.V. and TIPPETT, R.L., 1966, "Strains of sugarcane mosaic virus," *U.S.Deptt. Agric. Tech. Bull.* 1340: 25 pp.

ANDREAE, W.A and ANDREAE, S.R., 1953, "Studies on indoleacetic acid metabolism II The effect of methyl ambelliferone maleic hydrazide and 2, 4-D on indole acetic acid oxidation." *Can. Jr. Bot*, 31: 426-37.

AZAD, R.N. NAGAICH, B.B. and SEHGAL, O.P., 1961, "Broad bean mosaic virus in India," *Indian Phytopathol.* 14: 169-173.

BHARGAVA, K.S., JOSHI, R.D., and RIZVI, S.M.A., 1971, "Some observations on the insect transmission of sugarcane mosaic virus in India," *Sugarcane. Path. Newsletter* 6: 20-21.

BHARGAVA, K.S., 1975, "Sugarcane mosaic—retrospect and prospects" Presidential address to Indian Phytopathological Society, 1975.

CHENULU, V.V., SACHIDANANDA, J. and MEHTA, S. C., 1968, "Studies on mosaic disease of cowpea from India" *Phytopathol.* Z. 63: 381-87.

CHESSIN, M., 1958, "Growth substances and stunting in virus infected plants." *Proc. 3rd conf.* Potato Virus Diseases, 80-84.

CHONA, B.L., 1944, "Sugarcane mosaic and its control". *Indian Farming.* 4: 178-141.

CHONA, B.L. and SETH, M.L., 1960, "A mosaic disease of maize (*Zea mays*) in India". *Indian J. Agric. Sci.*, 30: 25-32.

DHANRAJ, K.S. and RAYCHAUDHURI, S.P., 1969, "A note on barley mosaic in India." *Plant. Dis. Reptr.*, 53: 967-968.

DIENER, T.O., 1963, "Physiology of virus-infected plants," *Annu. Rev. Phytopath* 1: 197-218.

ESAU, K., CRONSHAW, J. and HOEFERT, L.L., 1967, "Relation of beet yellows virus to the phloem and the movement in the sieve tube." *J. Cell. Biol.* 32: 71-81

HUTCHINSON, P.B, 1972, "Alternate hosts for diseases of sugarcane. Sugarcane *Path.*" Newsletter 8: 36-38.

JHA, A. and MISHRA, J.N., 1955, "Yellow-vein-mosaic of bhindi". *Proc. Bihar Acad. Agric. Sci.* 4: 124-130.

KASSANIS, B., TINSLEY, T.W., and QUAK, F, 1958, "The inoculation of tobacco callus tissue with tobacco mosaic virus," *Ann. Appl. Biol.* 46: 11-19.

KOIKE, H. and TIPPETT, R.L., 1972, "Rouging mosaic diseased ratoon seed plots with herbicides in louisiana". Sugarcane *Path.* Newsletter. 8: 20

KUMAR, L.S.S., AGARWAL, A.C., ARAKERI, H.R., KAMATH, M.G., and DOUAHNE, R., 1963, *Agriculture in India*, vol II. London 243 pp.

LEU, L.S., 1972, "Freeing sugarcane from mosaic virus by apical meristem culture and tissue culture". *Rep. Taiwan. Sugar. Exp. Stat. No.* 57: 57-63.

MARAMOROSCH, K., 1957, "Reversal of virus caused stunting in plants by Giberellic acid" *Science*, 126, 651-652.

NARIANI, T.K., 1956, "Leaf curl of Papaya" *Indian Phytopathol.* 9: 151-155.

———,1960, "Yellow mosaic of Mung (*phaseolus aureus* L.)" *Indian Phytopathol.* 13: 24-29.

———and PINGALEY, K.V., 1960, "A mosaic disease of soyabean". *Indian Phytopathol.* 13: 130-136.

OWEN, P.C., 1956 "The effect of infection with tobacco mosaic virus on the respiration of tobacco leaves of varying ages in the period between inoculation and systemic infection". *Ann. Appl. Biol.* 44, 227-232.

PEMBERTON, C.E. and CHARPENTIER, L.J., 1969, "Pests of sugarcane." Elsevier, Amsterdam, 417 pp.

RAYCHAUDHURI, S.P. and GANGULY, B., 1968, "A mosaic streak of wheat." *Phytopathol. Z.* 62: 61-65.

SETH, M.L., RAYCHAUDHURI, S.P. and SINGH, D.V., 1977, "A streak disease of Bajra (*Pennisetum typhoides*) (Burun. Stap and Hubb) in India." *Curr. Sci.* 40: 272.

SHAH, S.S., 1972, "Sugarcane Agriculture". *Indian Farming*, 22: 38-43.

SIEGEL, A., ZAITLIN, M. and SEGHAL, O.P., 1962, "The isolation of defective tobacco mosaic virus strains". *Proc. Nat Acad. Sci. US.* 48, 1845-51.

SINGH, K., 1971, "Virus disease of sugarcane and seed programme," *Advances in Agriculture* (Kanpur) 1: 69-87.

SMITH, K. M., 1968, *Plant Viruses*, Methuen & Co. Ltd., London.

SREENIVASAN, T.N. and NARIANI, T.K., 1966, "Studies on a mosaic disease of pea (*Pisum sativum* L.)" *Indian Phytopathol*, 19: 189-193.

UPPAL, B.N. *et al*, 1940, "Yellow mosaic of Bhindi," *Curr. Sci.*, 9: 227-228.

ZAITLIN, M. and JAGENDORF, A.T, 1960, "Photosynthetic phosphorylation and Hill reaction activities of chloroplasts isolated from plants infected with tobbcco mosaic virus", *Virology*, 12: 477-86

GLOSSARY

Abscission: The shedding of leaves or fruits.

Anthracnose: Lesions on stems, leaves and fruits caused by acervulus fungi like *Colletotrichum* and *Gloeosporium.*

Antibiosis: Antagonism (toxicity) between two organisms, particularly microorganisms in soil or between one organism and a metabolic product of another organism, harmful to one of them.

Antibody: Specific protein produced in the blood of mature animals against any macromolecule as a defense mechanism (called antigen) produced by pathogen (or chemical), which is foreign particle to the host.

Antigen: A substance capable of inducing antibody formation. Most macromolecules, proteins, nucleic acids and even polysaccharides act as antigens.

Appresorium (*Pl.* Appresoria): An enlarged, usually disc shaped, structure formed at the tip of hypha of some parasitic fungi from which entry is made into the host.

Autotrophic: An organism capable of utilizing energy from light or from oxidation of inorganic compounds for metabolic synthesis.

Autoecious: See index.

Benign (Virus): A mild or avirulent type, not producing visible symptoms.

Biological control: Control of a disease by the activity of organisms.

Biotype: A subdivision of races of fungi.

Blight: A non-restricted, necrotic symptom on leaves which is characterized by sudden and serious leaf damage.

Blotch: A necrotic spot having fibrillose margins of visible mycelial strands.

Callus: Parenchymatous tissue formed by cambial activity for protection after wounding.

Canker: A necrotic symptom of woody parts. Spreading wounds (usually on bark) surrounded by a raised tumor-like margin.

Chemotherapy: Treatment of disease by chemicals which act internally without damaging the host seriously.

Clone: A group of plants derived from an individual by vegetative propagation.

Chlorosis (*Pl.* Chloreses)*:* Symptom of chlorophyh deficiency.

Control: The prevention of, or reduction in the development of disease.

Cross protection: See index.

Damping off: Disease of germinating seeds (pre-emergence damping off) and seedlings (post-emergence damping off) characterized by their toppling due to rotting of stem tissues at soil level.

Dieback: Symptom of disease in which apical shoots die first followed by lower shoots.

Disease: A malfunctioning of physiological processes due to continuous irritation from the presence or absence of some agent.

Disinfect: To kill a pathogen within plant tissues.

Disinfest: To kill a pathogen or organisms on the surface of plants or on/within non-living substances.

Epidemic: Widespread and destructive development of a disease in a large human population. For plants, the equivalent word is epiphytotic; and epizootic for animal diseases.

Epidemiology: The study of epidemics.

Epinasty: Downward curvature of leaf.

Epiphyte: A non-parasitic plant attached to another plant for mechanical support, *e.g.* orchids, lichens, mosses, etc.

Epiphytology: The study of epiphytotics; analogous to "epidemiology".

Epiphytotic: Widespread and destructive development of a plant disease in large populations.

Eradication: See index.

Escape: See index.

Etiology: The study of the causal agent of disease and its relation to the susceptible plant.

Exclusion: See index.

Forma specialis (Pl. *formae specialis*, abb. *f. sp.*)*:* Special form; and group of biotypes of species.

Fructification: Spore production in fungi or the structure in which spores are formed.

Fumigant: A chemical that is used in the vapour phase to kill microorganisms, insects etc.

Fungicide: A chemical that kills or inhibits fungi.

Fungistat: A chemical that reversibly inhibits the growth of a fungus as long as it is in contact with fungus.

Gall: A tumor; formed by hyperplastic growth of unorganized cells.

Heteroecious: See index.

Heterokaryosis: The state of a fungus having genetically different nuclei in its cells.

Heterothallism: See index.

Heterotopy: Hyperplastic symptom in which an organ grows at an unusual place (*e.g.*, the develonment of ears in the tissues of corn plants).

Heterotrophic: Organism deriving food from other living or dead (parasitic or saprophytic) organisms; are unable to synthesize organic materials from inorganic sources (*Cf.* autotrophic).

Host: A living organism providing sustenance to a parasite.

Host-indexing: Determination whether a given plant is carrier of a pathogen.

Hypersensitive plant: A plant that shows extreme resistance against a pathogen by quick death of cells around the point of attack, thus preventing spread.

Hyperparasite: An organism that parasitizes another parasite.

Hyperplasia: Outgrowth due to abnormal increase in number of cells.

Hypertrophy: Outgrowth due to abnormal increase in size of cells.

Immune: Free from disease by virtue of inherent properties of the plant.

Infect: To establish a pathogenic relationship with the host plant (*Cf.* Disinfect).

Infest: To introduce a pathogen into the environment of the host (*Cf.* Disinfest).

Infection court: The site of initiation of infection.

Infectious: Refers to a pathogen that can transmit from one host to another by an external agent.

Injury: Transient irritation by some causal agent. Different from disease.

Inoculum: Portion of pathogen capable of dissemination and initiation of disease.

Lesion: A localized area of injured or diseased tissue (*e.g.*, a spot or scab) or functional change (biochemical lesion).

Micro-environment: Any localized environment.

Mosaic: Irregular dark and light areas in leaves; a type of virus disease.

Necrosis (*Pl.* necroses): Death or disintegration of cells.

Parasite: An organism that derives its food from a different species of plant.

Pathogen: Causal agent of a disease.

Pathogenesis: The chain of events, from origination to development of disease.

Pathogenic: Causing or capable of causing disease.

Pathogenicity: The capabilities of the pathogen to cause disease.

Phyllody: Transformation of petals into leafy structures.

Phytoalexin: See index.

Protectant: A substance, usually a fungicide that prevents infection.

Physiological race: One or more biotypes or clones of a pathogenic species producing similar reactions in selected varieties of host plants.

Pustule: A pimple like eruption of host tissue formed by fruiting body such as uredium or telium of a rust.

Resistance: The inherent capacity of a plant to prevent, restrict or withstand disease.

Rhizosphere: The zone around roots of a plant under its influence.

Rogue: To remove unwanted plants from a planting of the same species.

Rugose: Rough, crinkled leaves *e.g.* rugose mosaic.

Russetting: Brownish, rough areas on fruits or tubers due to for mation of abundant cork-cells.

Scab: Rough, crusty lesions formed by excessive cork-cells.

Sign: Visible structures of pathogens appearing on or in diseased tissue.

Symptom: A visible expression of disease in the host plant.

Syndrome: The pattern of symptoms and signs of a disease.

Suscept (=Host): A plant susceptible to disease.

Symptomatology: Study of symptoms of diseases.

Systemic: 1. A disease in which the pathogen is spread throughout the plant body.

2. A pesticide which spreads throughout the plant body.

Tolerance: The ability of a host to survive and give satisfactory yield when other varieties of the same species suffer great loss.

Toxin: See index.

Variety: Subdivision of a species that differs from the type species in certain characteristics.

Vector: An agent that causes dissemination of a pathogen.

Vein-banding: A symptom of virus infected leaves a which tissue close to veins are darker-green than laminar tissues.

Vein-clearing: A symptom of virus infected leaves in which vein tissues are lighter-green, clearer than the laminar tissues.

Viriscence: A symptom in which normally non-green tissues develop green pigmentation.

Virulence: Relative ability to cause disease; a measure of pathogenicity.

Viruliferous: An insect vector containing virus and capable of inoculating it into a host.

Virus: See index.

Witches-broom: A symptom of disease due to clustering of shoots.

Yellowing: Green tissues turning yellow.

SUBJECT INDEX

AUTHOR INDEX